Basic Palaeontology

Basic Palaeontology

MICHAEL J. BENTON

University of Bristol, United Kingdom

DAVID A.T. HARPER

University College Galway, Ireland

 LONGMAN

Addison Wesley Longman
Addison Wesley Longman Limited,
Edinburgh Gate, Harlow,
Essex CM20 2JE, England
and associated companies throughout the world

First published 1997

British Library Cataloguing in Publication Data
A catalogue entry for this title is available from the British Library

ISBN 0-582-22857-3

Library of Congress Cataloging-in-Publication Data
A catalog entry for this title is available from the Library of Congress

Typeset in 9/11pt Times
Produced by Longman Asia Limited, Hong Kong

Contents

Preface

The history of life is documented by fossils. Recent palaeontological research has shown the fascinating information that can be found by studying ancient organisms. To look into the history of life on Earth is to see a conveyor belt passing by which is loaded with the most astonishing organisms, many of them more remarkable than the wildest dreams (or nightmares) of a science fiction writer. Indeed, palaeontology reveals a seemingly endless catalogue of alternative universes, landscapes and seascapes that look superficially familiar, but which contain plants that do not look quite right, animals that are very different from anything now living.

The last 30 years has seen an explosion of applied palaeontological research, that is, the attempt to use fossil evidence to study wider questions, such as rates of evolution, mass extinctions, high-precision dating of sedimentary sequences, the palaeobiology of dinosaurs and Cambrian arthropods, the structure of Carboniferous coal-swamp plant communities, ancient molecules, the search for oil and gas, the origin of humans, and many more. Palaeontologists have benefited enormously from the growing interdisciplinary nature of their science, with major contributions from geologists, chemists, evolutionary biologists, physiologists, and even geophysicists and astronomers. Many areas of study have also been helped by an increasingly quantitative approach.

There are many palaeontology texts that describe the major fossil groups, or that give a guided tour of the history of life. In this book we have tried to give students a flavour of the excitement of modern palaeontology. In addition, we have tried to present all aspects of palaeontology; not just invertebrate fossils or dinosaurs, but fossil plants, trace fossils, macroevolution, palaeobiogeography, biostratigraphy and microfossils. Where possible, we have tried to show how palaeontologists are currently tackling controversial questions, and to highlight what is known, and what is not known. This, we hope, will show that palaeontology is an active and dynamic research field.

The book is intended for first- and second-year geologists and biologists who are taking courses in palaeontology or palaeobiology. It should also be of interest to keen amateurs and others interested in current scientific evidence about the origin of life, the history of life, mass extinctions, human evolution, and related topics.

Michael J. Benton
David A.T. Harper
Bristol and Galway

Acknowledgements

We thank Dick Aldridge (Leicester), Stefan Bengtson (Stockholm), Peter Bowler (Belfast), Derek Briggs (Bristol), Pat Brenchley (Liverpool), Richard Bromley (Copenhagen), David Bruton (Oslo), Chris Cleal (Cardiff), Peter Crimes (Liverpool), Steve Donovan (Kingston, Jamaica), Euan Clarkson (Edinburgh), Dianne Edwards (Cardiff), Tony Ekdale (Salt Lake City), Elsie Marie Friis (Stockholm), Ken Higgs (Cork), Gareth Jones (Conodate, Dublin), Paul Kenrick (Stockholm), Andrew Knoll (Cambridge, Mass.), Martin Lockley (Denver), John Murray (Southampton), Michael O'Connell (Galway), Alan Owen (Glasgow), John Peel (Uppsala), Robert Riding (Cardiff), J. William Schopf (Los Angeles), Colin Scrutton (Durham), George Sevastopulo (Trinity College, Dublin), Peter Sheldon (Milton Keynes), David Siveter (Leicester), Peter Skelton (Milton Keynes), John Thackray (Natural History Museum, London), Barry Thomas (Cardiff), Henry Williams (Newfoundland), Mike Williams (Galway), Rachel Wood (Cambridge), Tony Wright (Belfast), and Jeremy Young (Natural History Museum, London) for reading individual chapters and for giving many valuable comments. Colin Scrutton (Durham) read the entire manuscript, and he made many useful comments about the content and presentation, and in particular about shortening the text. These kind referees have saved us from many embarrassing blunders, but needless to say, we are doubtless guilty of others. We thank our wives, Mary and Maureen, for their help and forbearance, and Jo Wright for the Index.

Illustrations are based on a great range of previously published works, cited here. Uncited figures are original. We are grateful to the following for permission to reproduce copyright material:

Figs. 1.3 and 1.4, courtesy of E. Buffetaut; Fig. 1.5, based on data in Benton, M. J. and Storrs, W. G., 1994, *Geology*, 22, 111–114; Fig. 1.6, courtesy of R. F. Vaughan; Fig. 1.7, painting by Pam Baldaro; Figs 1.11 and 1.12(a), based on Allison, P. A., 1988, *Paleobiology*, 14, 331–344; Fig. 12(b), courtesy of P. Wilby; Fig. 1.12(c)–(e), courtesy of D. E. G. Briggs; Fig. 1.13, based on Seilacher, A., 1985, *Phil. Trans. R. Soc. Lond., Ser. B*, 311, 5–23; Fig. 2.1(b), based on Sheppard, T., 1917, *Proc. Yorks. Geol. Soc.*, 19; Fig. 2.2, based on Holland, C. H., 1986, *J. Geol. Soc. Lond.*, 143, 3–21; Fig. 2.3, replotted from various sources; Fig. 2.4, replotted from various sources; Fig. 2.5, based on Temple, J. T., 1988, *J. Geol. Soc. Lond.*, 145, 875–879; Fig. 2.8, replotted from various sources; Fig. 2.9, based on data in Staff, G. M., Powell, E. N., Stanton, R. J. Jr. and Cummins, H., 1986, *Bull. Geol. Soc. Amer.*, 97, 428–443; Fig. 2.10, replotted from data in Whittington, H. B., 1980, *Proc. Geol. Ass.*, 91,

127–148; Fig. 2.11, modified and redrawn from various sources, Fig. 2.12, redrawn from various sources; Fig. 2.13, based on Copper, P., 1988, *Palaios*, 3, 136–152; Fig. 2.14(a), modified from Benton, M. J., 1990, *Vertebrate palaeontology*, Chapman and Hall, London; Fig. 2.14(b), modified from Hollingworth, N. and Pettigrew, T., 1988, Zechstein reef fossils and their palaeoecology. *Palaeontological Association Field Guide to Fossils*, 3; Fig. 2.15(a), replotted from data in Brett, C. E., Boucot, A. J. and Jones, B., 1993, *Lethaia*, 26, 25–40; Fig. 2.15(b), replotted from data in Pickerill, R. P. and Brenchley, P. J., 1991, *Geoscience Canada*, 18, 119–138; Figs. 2.17 (a)–(f), based on McKerrow, W. S., 1978, *Ecology of fossils*. Duckworth Press, London; Fig. 2.18, based on Smith, P., 1990, *Geoscience Canada*, 15, 261–279; Fig. 2.19, based on Benton, M. J., 1990, *Vertebrate palaeontology*, Chapman and Hall, London: Fig. 2.20, replotted from data in Harper, D. A. T., 1992, *Terra Nova*, 4, 204–209. Fig. 2.21, based on Smith, P., 1990, *Geoscience Canada*, 15, 261–279; Fig. 2.22, replotted from information in Jones, G. Ll., 1992, *Terra Nova*, 4, 238–244; Fig. 3.6, based on data in Cano, R. J., Poinar, H. N., Pieniazek, N. J., Acra, A., and Poinar, G. O., Jr., 1993, *Nature*, 536–538; Fig. 3.9, based on Williamson, P. G., 1981, *Nature*, 293, 437–443; Fig. 3.10, based on Sheldon, P. R., 1987, *Nature*, 330, 561–563; Fig. 3.11, based on Vrba, E. S., 1984, p. 115–142, in Ho, M.-W. and Saunders, P. T. (eds), *Beyond Neo-Darwinism: an introduction to the new evolutionary paradigm*, Academic, London; Fig. 3.12, courtesy of W. J. Kennedy and P. Skelton; Figs. 3.13 and 3.14, courtesy of M. Manabe; Fig. 3.16, based on McNamara, K. J., 1976, *J. Paleontol*, 57, 461–473; Fig. 3.17, based on Benton, M. J. and Kirkpatrick, R., 1989, *Palaeontology*, 32, 335–353; Fig. 4.3, courtesy of M. E. Tucker; Fig. 4.4, courtesy of J. W. Schopf; Fig. 4.5, drawn from photographs in Barghoorn, E. S. and Tyler, S. A., 1965, *Science*, 147, 563–577; Fig. 4.6, based on Lake, J. A., 1990, *Proc. Natn. Acad. Sci., USA*, 87, 763–766; Wainright, P. O., Hinkle, G., Sogin, M. L., and Stickel, S. K., 1993, *Science*, 260, 340–342; Woese, C. R., 1987, *Microbiol. Rev.*, 51, 221–271; Fig. 4.7, based on Schopf, J. W. and Blacic, 1971, *J. Paleontol.*, 45, 925–960; Fig. 4.8, courtesy of A. H. Knoll; Fig. 5.1, modified and redrawn from a variety of sources, Fig. 5.2, based on data in Williamson, D. I., 1992, *Larvae and evolution. Toward a new zoology*. Chapman and Hall, London; Fig. 5.3, modified and redrawn from various sources; Fig. 5.4, based on Seilacher, A., 1989, *Lethaia*, 22, 229–239; Fig. 5.6, drawn from photographs in Matthews, S. C. and Missarzhevsky, V. V., 1975, *J. Geol. Soc. Lond.*, 131, 289–304; Fig. 5.7, replotted from data in Conway Morris, S., 1993, *Nature*, 361, 219–225; Fig. 5.8, based on Dzik, J. and Krumbiegel, G., 1989, *Lethaia*, 22, 169–181; Fig. 5.9, based on Rigby, J. K. and Scrutton, C. T., 1985, Sponges, chaetetids and stromatoporoids, in Murray, J. W. (ed.) *Atlas of invertebrate macrofossils*, pp. 3–10, Longman, Harlow: Fig. 5.10, modified and redrawn from various sources; Fig. 5.11, drawn from information in the *Treatise on Invertebrate Paleontology*, part E, Geol. Soc. Amer. and Univ. Kansas Press; Fig. 5.12, redrawn from *Treatise on Invertebrate Paleontology*, part E, Geol. Soc. Amer. and Univ. Kansas Press; Fig. 5.13, replotted from data in Wood, R., 1991, Problematic reef-building sponges. In Simonetta, A. M. and Conway Morris, S. (eds) *The early evolution of Metazoa and the significance of problematic taxa*, pp. 113–124, Cambridge University Press; Fig. 5.14, based on Rigby, J. K and Scrutton, C. T., 1985, Sponges, chaetetids and stromatoporoids, in Murray, J.W. (ed.) *Atlas of invertebrate macrofossils*, pp. 3–10. Longman, Harlow; Fig. 5.15, based on Kershaw, S., 1984, *Palaeontology*, 27, 113–130; Fig. 5.16(a), based on

Savarese, M., 1992, *Paleobiology* 18, 464–480; Fig. 5.16(b), based on Wood, R., Zhuravlev, A. Yu., Debrenne, F., 1992, *Palaios*, 7, 131–156; Fig. 5.17, courtesy of R. Wood; Fig. 5.18, based on Wood, R., Zhuravlev, A. Yu., Chimed Tseren, Anaaz, 1993, *Sedimentology*, 40, 829–858; Fig. 5.19, based on Savarese, M., 1992, *Paleobiology* 18, 464–480; Fig. 5.20, based on Wood, R., Zhuravlev, A. Yu., Debrenne, F., 1992, *Palaios*, 7, 131–156; Fig. 6.1, based on various sources; Fig. 6.2, based on various sources; Fig. 6.3, modified and redrawn from various sources; Fig. 6.4, drawn from various sources; Fig. 6.5, based on the *Treatise on Invertebrate Paleontology*, part F, Geol. Soc. Amer. and Univ. Kansas Press; Fig. 6.6, redrawn from a variety of sources; Fig. 6.7, replotted from data in Coates, A. G. and Oliver, W. A. Jr., 1973, Coloniality in zooantharian corals, in Boardman, R. S., Cheetham, A. H. and Oliver, W. A. Jr. (eds) *Animal colonies – development and function through time*, pp. 3–27. Dowden, Hutchinson and Ross, Stroudsberg, Pennsylvania; Fig. 6.8, replotted from data in Scrutton, C. T., 1993, *Cour. Forsch.-Inst. Senckenberg*, 164, 273–281; Fig. 6.9, based on *Treatise on Invertebrate Paleontology*, part F, Geol. Soc. Amer. and Univ. Kansas Press; Fig. 6.11, modified and redrawn from Neuman, B. E. E., 1988, *Lethaia*, 21, 97–114; Fig. 6.12, courtesy of C. T. Scrutton: Fig. 6.13, modified from various sources; Fig. 6.14, based on Scrutton, C. T., 1989, *Lethaia*, 23, 61–75. Fig. 6.15, based on various sources; Fig. 6.17, courtesy of C. T. Scrutton; Fig. 6.18, based on Copper, P., 1988, *Palaios*, 3, 136–152; Fig. 6.19, courtesy of C T. Scrutton; Fig. 6.20, replotted from data in Mitchell, M. and Scrutton, C. T., 1991, Excursion A2, The Lower Carboniferous coral faunas of England. *6th International Symposium on fossil Cnidaria including Archaeocyatha and Porifera*, Münster, Germany; Fig. 6.21, courtesy of C. T. Scrutton; Fig. 6.22, replotted from data in Scrutton, C. T. and Clarkson, E. N. K., 1990, *Palaeontology*, 34, 179–194; Fig. 6.23(a), replotted from Rowell, A. J., 1981, in Lophophorates notes for a short course. *University of Tennessee, Department of Geological Sciences Studies in Geology*, 5; Fig. 6.23(b), replotted from Carlson, S., 1991, in MacKinnon, D. I., Lee, D. L. and Campbell, J. D. (eds) *Brachiopods through time*, A. A. Balkema, Rotterdam; Fig. 6.23(c), replotted from Popov, L. E., Bassett, M. G., Holmer, L. E. and Laurie, J. 1993, *Lethaia*, 26, 1–5; Fig. 6.24, modified and redrawn from a variety of sources; Fig. 6.25, based on Schumann, D., 1991, in MacKinnon, D. I., Lee, D. L. and Campbell, J. D. (eds) *Brachiopods through time*, A. A. Balkema, Rotterdam; Fig. 6.26, courtesy of the late R.E. Grant (f, i, j, l); Fig. 6.27, redrawn from photographs in Jaanusson, V., 1971, *Smithson, Contr. Paleobiol.* 3, 33–46; Fig. 6.28, based on Williams, A., 1968, *Lethaia*, 1, 268–287; Fig. 6.29(a), based on Bassett, M. G., 1984, *Spec. Pap. Palaeontol.*, 32, 237–263; Fig. 6.29 (b), based on Grant, R. E., 1981, in Lophophorates notes for a short course, *University of Tennessee, Department of Geological Sciences Studies in Geology*, 5; Fig. 6.29 (c), based on Surlyk, F., 1972, *Kong. Danske Skr. Videns. Sels. Biol. Skr.* 19; Fig. 6.30, based on Rudwick, M. J. S., 1964, *Palaeontology*, 7, 135–171; Fig. 6.31, based on Harper, D. A. T., Ryan, P. D. and Whalley, J. S., 1994, *Manual for PALSTAT package*, Chapman and Hall, London; Fig. 6.32, redrawn from data in Ager, D.V., 1965, *Palaeogeogr. Palaeoclimatol. Palaeoecol.*, 1, 143–173; Fig. 6.33, redrawn from Ager, D. V., 1961, *Q. J. Geol. Soc. Lond.*, 117, 1–10; Fig. 6.34, based on various sources; Fig. 6.35, courtesy of P. Wyse Jackson; Fig. 7.1, based on Sprinkle, J., 1980, Echinoderms notes for a short course, *University of Tennessee, Department of Geological Sciences Studies in Geology*, 3; Fig. 7.2, based on *Treatise on Invertebrate Paleontology*, part S, Geol. Soc. Amer. and Univ. Kansas Press;

Fig. 7.3, based on various sources; Fig. 7.4, courtesy of S. K. Donovan; Fig. 7.5, based on *Treatise on Invertebrate Paleontology*, part S, Geol. Soc. Amer. and Univ. Kansas Press; Fig. 7.6, based on Smith. A.B and Murray, J.W., 1985, Echinodermata, in Murray, J.W. (ed.) *Atlas of invertebrate macrofossils*, pp. 153–190. Longman, Harlow; Fig. 7.7, based on various sources; Fig. 7.8, modified from *Treatise on Invertebrate Paleontology*, part S, Geol. Soc. Amer. and Univ. Kansas Press; Fig. 7.9, based on Smith, A., 1984, *Echinoid palaeobiology*, Chapman and Hall, London; Fig. 7.10, replotted from data in Smith, A., 1984, *Echinoid palaeobiology*, Chapman and Hall, London; Fig. 7.11(a), based on Kier, P.M., 1982, *Palaeontology*, 25, 1–10; Fig. 7.11(b), based on Kier, P.M., 1972, *Smithson. Contr. Paleobiol.*, 13; Fig. 7.12, based on Rose, E. P. F. and Cross, N. E., 1994, *Geology Today*, 9, 179–186; Fig. 7.13, courtesy of S.K. Donovan (a–d, f); Fig. 7.14, based on *Treatise on Invertebrate Paleontology*, part U, Geol. Soc. Amer. and Univ. Kansas Press; Fig. 7.15, based on Jefferies, R.P.S. and Daley, 1996, in Harper, D. A. T. and Owen, A. W. (eds) *Fossils of the Upper Ordovician*. Palaeontological Association, Field Guide to Fossils, 7; Fig. 7.16, based on *Treatise on Invertebrate Paleontology*, part U, Geol. Soc. Amer. and Univ. Kansas Press; Fig. 7.17, redrawn and modified from a variety of sources; Fig. 7.18, courtesy of S. H. Williams; Fig. 7.19, courtesy of D. E. B. Bates; Fig. 7.20, courtesy of D. E. B. Bates; Fig. 7.21, based on Underwood, C., 1993, *Lethaia*, 26, 189–202; Fig. 7.22, modified and drawn from a variety of sources; Fig. 7.23, modified and drawn from a variety of sources; Fig. 7.24, based on Clarkson, E. N. K., 1993, *Invertebrate palaeontology and evolution*, Chapman and Hall, London; Fig. 8.1, based on various sources; Fig. 8.2, based on Harper, D. A. T., Owen, A. W. and Doyle, E. N., 1996, in Harper, D. A. T. and Owen, A. W. (eds) *Fossils of the Upper Ordovician*. Palaeontological Association, Field Guide to Fossils, 7; Fig. 8.4, redrawn from Barrande, J., 1852, *Système Silurien du Centre de la Bohème. Ière Partie: Recherches Paléontologiques*, Vol. I planches, Crustacés: Trilobites. Prague and Paris; Fig. 8.5, courtesy of E. N. K. Clarkson; Fig. 8.6, based on *Treatise on Invertebrate Paleontology*, part O, Geol. Soc. Amer. and Univ. Kansas Press; Fig. 8.7, modified from various sources; Fig. 8.8(a), based on Fortey, R. A., 1975, *Fossils and Strata* 4, 331–352; Fig. 8.8(b), redrawn from data in Price, D., 1979, *Geol. J.*, 16, 201–216; Fig. 8.8(c), redrawn from data in Thomas, A. T., 1979, *Spec. Publ. Geol. Soc. Lond.*, 8, 447–451: Fig. 8.9, courtesy of A. W. Owen; Fig. 8.10, based on Clarkson, E. N. K., 1993, *Invertebrate palaeontology and evolution*, Chapman and Hall, London; Fig. 8.11, courtesy of E. N. K. Clarkson; Fig. 8.12, redrawn from McKinney, F. K., 1991, *Exercises in invertebrate paleontology*, Blackwell Scientific Publications, Boston; Fig. 8.13, based on Clarkson, E.N.K., 1993, *Invertebrate palaeontology and evolution*, Chapman and Hall, London; Fig. 8.14, replotted and modified from data in Peel, J. S, 1991, *Grøns. Geol. Unders.*, 161; Fig. 8.15, based on Peel, J. S., 1991, *Grøns. Geol. Unders.*, 161; Fig. 8.16, replotted from Raup, D. M. 1966, *J. Paleont.*, 40, 1178–1190 and other sources; Fig. 8.17, based on *Treatise on Invertebrate Paleontology*, part N, Geol. Soc. Amer. and Univ. Kansas Press; Fig. 8.18, based on various sources; Fig. 8.21, based on Skelton, P. W., 1985, *Spec. Pap. Palaeont.* 33, 159–173; Fig. 8.24, based on various sources; Fig. 8.25, courtesy of J. S. Peel; Fig. 8.26, based on Peel, J. S., Skelton, P. W. and House, M. R., 1985, Mollusca, in Murray, J. W. (ed.) *Atlas of invertebrate macrofossils*, pp. 79–152. Longman, Harlow; Fig. 8.27, based on Peel, J. S., Skelton, P. W. and House, M. R., 1985, Mollusca, in Murray, J. W. (ed.) *Atlas of invertebrate macrofossils*, pp. 79–152. Longman,

Harlow. Fig. 8.28, based on Peel, J. S., Skelton, P. W. and House, M. R., 1985, Mollusca, in Murray, J.W. (ed.) *Atlas of invertebrate macrofossils*, pp. 79–152. Longman, Harlow; Fig. 8.31, based on Peel, J. S., Skelton, P. W. and House, M. R., 1985, Mollusca, in Murray, J.W. (ed.) *Atlas of invertebrate macrofossils*, pp. 79–152, Longman, Harlow; Fig. 8.32(a), based on Truman, A.E., 1940, *Q. J. Geol. Soc. Lond*, 96, 339–383; Fig. 8.32(b), based on Batt, R., 1993, *Lethaia*, 26, 49–63; Fig. 8.33, based on Peel, J. S., Skelton, P. W. and House, M. R., 1985, Mollusca, in Murray, J.W. (ed.) *Atlas of invertebrate macrofossils*, pp. 79–152. Longman, Harlow; Fig. 8.34, based on Doyle, P. and MacDonald, D. I. M., 1993, *Lethaia*, 26, 65–80; Fig. 9.2(a),(b), based on Gagnier, P.-Y., 1993, *Ann. Paléontol.*, 79, 19–51; Figs 9.2(c), (d), 9.3, 9.6(a), based on Moy-Thomas, J. A. and Miles, R. S., 1971, *Palaeozoic fishes*, 2nd edn, Chapman & Hall, London; Fig. 9.4, courtesy of N. H. Trewin; Fig. 9.6(b), after Woodward, A. S., 1891–1901, *Catalogue of the fossil fishes in the British Museum*, British Museum (Natural History), London; Fig. 9.6(c), modified from Grande, L., 1988, *J. vertebr. Paleontol.*, 8, 117–130; Fig. 9.7(b), (c), modified from Coates, M. I. and Clack, J. A., 1990, *Nature*, 347, 66–69; Fig. 9.7(d), modified from Jarvik, E., 1955, *Sci. Monthly*, 80, 141–154; Fig. 9.8, courtesy of J. A. Clack and S. Bendix-Almgreen; Fig. 9.9(b), based on Gregory, W. K., 1929, *Our face from fish to man*, Putnam, New York; Fig. 9.9(c), based on White, T. E., 1939, *Bull. Mus. comp. Zool.*, 85, 325–409; Fig. 9.10(a), (b), modified from Carroll, R. L., 1969, *J. Paleontol.*, 43, 151–170; Fig. 9.13(a), modified from Carroll, R. L. and Lindsay, W., 1985, *Can. J. Earth Sci.*, 22, 1571–1587; Fig. 9.13(b), modified from Gaffney, E. S. and Meeker, L. J., 1983, *J. vertebr. Paleontol.*, 3, 25–28; Fig. 9.14, courtesy of R. R. Reisz; B. Rubidge; Fig. 9.15(a), after Jenkins, F. A., Jr., 1971, *Bull. Peabody Mus. nat. Hist.*, 36, 1–216; Fig. 9.15(b), after Jenkins, F. A., Jr. and Parrington, F. R., 1976, *Phil. Trans. R. Soc., Ser. B*, 173, 387–431; Fig. 9.16(b), courtesy of D. M. Unwin; Fig. 9.16(c), courtesy of D. R. Grange; Fig. 9.17(a), after Fraser, N. C. and Walkden, G. M., 1984, *Palaeontology*, 27, 575–595; Fig. 9.17(b), after Estes, R., 1983, *Handb. Paläoherpetol.*, 10A, 1–249; Fig. 9.18(a), courtesy of D. B. Weishampel; Fig. 9.19(a), modified from Ostrom, J. H., 1969, *Bull. Peabody Mus. nat. Hist.*, 30, 1–165; Fig. 9.19(b), modified from Newman, B. H., 1970, *Biol. J. Linn. Soc.*, 2, 119–123; Fig. 9.20(a), after Gilmore, C. W., 1914, *Bull. US Natn. Mus.*, 89, 1–1443; Fig. 9.20(b), after Carpenter, K., 1982, *Can. J. Earth Sci.*, 19, 689–697; Fig. 9.20(c), after Brown, B., 1917, *Bull. Am. Mus. nat. Hist.*, 37, 281–306; Fig. 9.22, courtesy of R. Wild; Fig. 9.23, modified from Yalden, D., 1984, *Zool. J. Linn. Soc.*, 82, 17–188; Fig. 9.24(a), after Riggs, E. S., 1934, *Trans. Am. phil. Soc.*, 24, 1–32; Figs 9.24(b), 9.26, 9.29(d), after Flower, W. H. and Lydekker, R., 1891, *An introduction to the study of mammals, living and extinct*, Black, London; Fig. 9.25, based on information in Novacek, M. J., Wyss, A. R., and McKenna, M. R., 1988, pp. 31–71, in Benton, M. J. (ed.) *The phylogeny and classification of the tetrapods, vol. 2*, Clarendon, Oxford; Fig. 9.27(a),(b), after Matthew, W. D., 1909, *Mem. Am. Mus. nat. Hist.*, 9, 291–567; Fig. 9.27(c), modified from Mitchell, E. D., 1975, *Rapp. Proc.–verb. Réun. Cons. intern. Explor. Mer*, 169, 12–26; Fig. 9.27(d), after Jepsen, G. L., 1970, *Biology of bats, V1. 1*, Academic, New York; Fig. 9.27(e), after Wood, A. E., 1962, *Trans. Am. phil. Soc.*, 52, 1–261; Fig. 9.28(a), courtesy of J. L. Franzen; Fig. 9.28(b),(c), based on Flower, W. H. and Lydekker, R., 1891, *An introduction to the study of mammals, living and extinct*, Evans, London; Fig. 9.29(a), modified from Thewissen, J. G. M., Hussain, S. T., and Arif, M., 1994, *Science*, 263,

210–212; Fig. 9.29(b), after Kellogg, R. M., 1936, *Publ. Carnegie Instn., Washington*, 482, 1–366; Fig. 9.29(c), after Andrews, C. W., 1906, *A descriptive catalogue of the Tertiary Vertebrata of the Fayûm, Part 1*, British Museum (Natural History), London; Fig. 9.30(a), modified from Tattersall, I., 1970, *Man's ancestors*, John Murray, London; Fig. 9.30(b), modified from Lewin, R., 1989, *Human evolution*, Blackwells, Oxford; Figs 10.1(a), 10.5(a)–(d), courtesy of D. E. Edwards; Fig. 10.1(b)–(d), courtesy of C. J. Cleal; Fig. 10.1(e), courtesy of R. Riding; Fig. 10.2(a),(b), based on Kidston, R. and Lang, W. H., 1917–1921, *Trans. R. Soc. Edinb.*, 51, 761–784; 52, 603–627, 643–680, 831–854, 855–902; Fig. 10.2(c), redrawn from photographs in Stewart, W. N. and Rothwell, G. W., 1993, *Paleobotany and the evolution of plants*, 2nd edn, Cambridge University Press, Cambridge; Fig. 10.3, redrawn from Andrews, H. N., 1960, *Palaeobotanist*, 7, 85–89; Figs 10.5(e), 10.9, redrawn from Thomas, B. A. and Spicer, R. A., 1987, *The evolution and palaeobiology of land plants*, Croom Helm, London, by permission of Chapman & Hall; Fig. 10.10, redrawn from Morgan, J., 1959, *Illinois Biol. Monogr.*, 27, 1–108; Fig. 10.12, based on Gould, R. E. and Delevoryas, T., 1977, *Alcheringa*, 1, 387–399; Fig. 10.13, modified from Thomas, B. A. and Cleal, C. J., 1993, *The Coal Measures forests*, National Museum of Wales, Cardiff; Fig. 10.14(c), redrawn from Delevoryas, T. and Hope, R. C., 1971, *Postilla*, 150, 1–21; Fig. 10.14(d), redrawn from Delevoryas, T., 1971, *Proc. N. Am. Paleontol. Conv.*, 1, 1660–1674; Fig. 10.16, based on information in Friis, E. M., Chaloner, W. G. and Crane, P. R., 1987, *The origins of angiosperms and their biological consequences*, Cambridge University Press, Cambridge; Fig. 10.17, courtesy of P. R. Crane; Fig. 10.18, based on data in Crane, P. R., 1989, pp. 153–187, in Allen, K. C. and Briggs, D. E. G. (eds) *Evolution and the fossil record*, Belhaven, London; Fig. 10.19, based on information in Upchurch, G. R., Jr. and Wolfe, J. A., 1987, in Friis, E. M., Chaloner, W. G. and Crane, P. R. (eds) *The origins of angiosperms and their biological consequences*, Cambridge University Press, Cambridge; Fig. 11.1, replotted from data in Bignot. G., 1985. *Elements of micropalaeontology*, Graham and Trotman, London: Figs 11.2 and 11.3, modified and redrawn from various sources: Figs 11.4 and 11.5, modified and redrawn from various sources: Fig. 11.6, courtesy of J. Smith, Figs 11.7 and 11.8, modified and redrawn from various sources; Fig. 11.9, courtesy of J. Young; Figs 11.10–11.12, modified and redrawn from various sources, Fig. 11.13, courtesy of J. W. Murray (b, d–e, g, h, j, k) and E. N. K. Clarkson (a,c,f,i,l); Fig. 11.14, based on Brasier, M.D., 1980, *Microfossils*, Chapman and Hall, London; Fig.11.15, replotted from data in Bignot, G., 1985, *Elements of micropalaeontology*, Graham and Trotman, London; Figs 11. 16 and 11.17, modified and redrawn from various sources: Fig. 11.18, based on Brasier, M.D., 1980, *Microfossils*, Chapman and Hall, London; Fig. 11.19, courtesy of David J. Siveter; Fig. 11.20, based on Olempska, E., 1983. *Lethaia*, 22, 159–168; Fig. 11.21, based on Brasier, M. D., 1980. *Microfossils*, Chapman and Hall, London; Figs 11.22–11.24, courtesy of R. J. Aldridge; Fig. 11.24, replotted from data in Aldridge, R. J., Jeppsson, L. and Dorning, K. J., 1993, *J. Geol. Soc. Lond.*, 150, 501–513; Fig. 11.25, replotted from data in Davis, J.C., 1973. *Statistics and data analysis in geology*, John Wiley and Sons, New York; Fig.11.26, courtesy of S. Turner (a, b, Natural History Museum; c, e, f, Australian Geological Survey Organization); Fig.11.27, modified and redrawn from various sources; Figs 11.28 and 11.29, courtesy of K. Higgs; Fig.11.30, modified and redrawn from Molloy, K. and O'Connell, M., 1988, *Archaeol. Ireland*, 2, 67–70, Fig. 11.31, replotted from data in Bignot, G., 1985 *Elements of micropalaeontology*, Graham and Trotman, London; Figs 12.1–12.4, 12.7, based on Ekdale *et al.*, 1984, *Ichnology: the use of trace fossils in sedimentology and stratigraphy*, Soc. Econ. Paleontol. Mineral., Tulsa, Oklahoma; Fig. 12.5, courtesy of T. P. Crimes; Fig. 12.6, courtesy of M. G. Lockley; Fig. 12.8, modified from Frey, R. W., Pemberton, S. G., and Saunders, T. D. A., 1990, *J. Paleontol.*, 64, 155–158, and other sources; Fig. 12.9, based on Ekdale *et al.*, 1984, *Ichnology: the use of trace fossils in sedimentology and stratigraphy*, Soc. Econ. Paleontol. Mineral., Tulsa, Oklahoma., Frey, R. W., Pemberton, S. G., and Saunders, T. D. A., 1990, *J. Paleontol.*, 64, 155–158, and other sources; Fig. 12.10, based on Ekdale, A. A. and Bromley, R. G., 1991, *Palaios*, 6, 232–249, and other sources; Fig. 12.11, based on information in Crimes, T. P., 1987, *Geol. Mag.*, 124, 97–119, Landing, E., 1993, *Geology*, 22, 179–182, and other sources; Fig. 12.12, based on information in Frey, R. W. and Seilacher, A. 1980, *Lethaia*, 13, 183–207, McCann, T., 1990, *Lethaia*, 23, 243–255, and other sources; Fig. 12.13, courtesy of T. P. Crimes; Fig. 12.14, based on Pemberton, S. G. and Frey, R. W., 1984, *Can. Soc. Petrol. Geol., Mem.*, 9, 281–304; Fig. 12.15, based on information in Lockley, M. G., Houck, K. J., and Prince, N. K., 1986, *Geol. Soc. Am., Bull.*, 97, 1163–1176. Fig. 13.1(a), modified from Valentine, J. W., 1969, *Palaeontology*, 12, 684–709; Fig. 13.1(b), modified from Raup, D. M., 1972, *Science*, 177, 1065–1071; Figs 13.2(a) and 13.3, modified from Sepkoski, J. J., Jr., 1984, *Paleobiology*, 10, 246–267; Fig. 13.2(b), modified from Benton, M. J., 1985, *Nature*, 316, 811–814; Fig. 13.2(c), modified from Niklas, K. J., Tiffney, B. H., and Knoll. A. H., 1983, *Nature*, 303, 614–616; Fig. 13.4, based on Macfadden, B. J., 1992, *Fossil horses: systematics, paleobiology, and evolution*, Cambridge University Press, Cambridge, and other sources; Fig. 13.5, based on Gould, S. J. and Calloway, C. B., 1980, *Paleobiology*, 6, 383–396, and other sources; Fig. 13.6, based on data in Benton, M. J., 1983, *Q. Rev. Biol.*, 58, 29–55; Fig. 13.8, based on Keller, G. and Barrera, E., 1990, *Geol. Soc. Am. Spec. Pap.*, 247, 563–575; Fig. 13.9, based on data in Raup, D. M. and Sepkoski, J. J., Jr., 1984, *Proc. Natn. Acad. Sci. USA.*, 81, 801–805; Fig. 13.10, based on information in Orth, C.J., Gillmore, J. S., Knight, J. D., Pillmore, C. L., Tschudy, R. H., and Fassett, J. E., 1981, *Science*, 214, 1341–1343; Fig. 13.12, courtesy of P. Claeys; Fig. 13.13, modified from Gingerich, P. D., 1984, *Univ. Tennessee Stud. Geol.*, 8, 1–16; Fig. 13.14, based on data in Benton, M. J., 1995, *Science*, 268, 52–58.

We welcome comments and corrections from instructors and from students. E-mail us:

mike.benton@bristol.ac.uk
david.harper@ucg.ie

1 Palaeontology as a science

Key Points

- Classical and medieval views about fossils were often magical.
- Observation in the 16th and 17th centuries showed that fossils were the remains of ancient plants and animals.
- By 1800, most scientists accepted the idea of extinction.
- By 1830, most geologists accepted that the Earth was very old.
- By 1840, the major divisions of the geological column had been established by the use of fossils.
- By 1840, it was seen that fossils showed direction in the history of life.
- Plants and animals with hard tissues are most frequently preserved.
- Soft tissues are usually labile, and decay rapidly, but rapid burial or early mineralization may prevent decay in cases of exceptional preservation.
- Physical and chemical processes may damage hard tissues during transport and in the rocks.

Steps to understanding

Earliest fossil finds

Fossils are very common in certain kinds of rocks, and they are often beautiful objects. It is probable that people picked up fossils long ago, and perhaps wondered why shells of sea creatures are now found high in the mountains, or how a perfectly preserved fish specimen came to lie buried deep within layers of rock. Prehistoric peoples picked up fossils and used them as ornaments, presumably with little understanding of their meaning.

Some early speculations about fossils by the classical authors seem now very sensible to modern observers. Early Greeks such as Xenophanes (576–480 BC) and Herodotus (484–426 BC) recognized that some fossils were marine organisms, and that these provided evidence for earlier positions of the sea. Other classical and medieval authors, however, tried to argue in favour of magical interpretations of fossils (see Box 1.1).

Fossils as fossils

The debate about 'plastic forces' was terminated abruptly by the debacle of Beringer's figured stones, but it had been resolved rather earlier. Leonardo da Vinci (1452–1519), a brilliant scientist and technologist (as well as a great artist), used his observations of modern plants and animals, and of modern rivers and seas, to explain the fossil sea shells found high in the Italian mountains. He interpreted them as the remains of ancient shells, and he argued that the sea had once covered these areas.

Box 1.1 Fossils as magical stones

In Roman and medieval times, fossils were often interpreted as mystical or magical objects. Fossil sharks' teeth were known as 'glossopetrae' (= tongue stones), in reference to their supposed resemblance to tongues, and many people believed they were the petrified tongues of snakes. This interpretation led to the belief that the glossopetrae could be used as protection against snake bite and other poisons. The teeth were worn as amulets to ward off danger, and they were even dipped into drinks in order to neutralize any poison that might have been placed there.

Most fossils were recognized as *looking like* the remains of plants or animals, but they were said to have been produced by a 'plastic force' (*vis plastica*) which operated within the Earth. Numerous authors in the 16th and 17th centuries wrote books presenting this interpretation. For example, the Englishman Robert Plot (1640–1696) argued that ammonites (p. 183) were formed 'by two salts shooting different ways, which by thwarting one another make a helical figure'. These interpretations seem ridiculous now, but there was a serious problem in explaining how such specimens came to lie far from the sea, why they were often different from living animals, and why they were made of unusual minerals.

The idea of 'plastic forces' had been largely overthrown by the 1720s, but some extraordinary events in Wurzburg in Germany at that time must have dealt the final blow. Johann Beringer (1667–1740), Professor at the University, began to describe and illustrate 'fossil' specimens brought to him by paid collectors from the surrounding area. However, it turned out that the collectors had been paid by an academic rival to manufacture 'fossils' by carving the soft limestone into outlines of shells, flowers, butterflies and birds (Figure 1.1). There was even a slab with a pair of mating frogs, and others with astrological symbols and Hebrew letters. Beringer resisted evidence that the specimens were forgeries, and wrote as much in his book, the *Lithographiae Wirceburgensis* (1726), but realized the awful truth soon after publication.

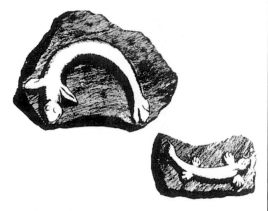

Fig 1.1   Magical stones: two of the remarkable 'fossils' described by Professor Beringer of Wurzburg in 1726. He believed these specimens represented real animals of ancient times that had crystallized into the rocks by the action of sunlight.

Later, Niels Steno (or Stensen) (1638–1686) demonstrated the true nature of glossopetrae simply by dissecting the head of a huge modern shark, and showing that its teeth were identical to the fossils (Figure 1.2). Robert Hooke (1625–1703), a contemporary of Steno's, also gave detailed descriptions of fossils, using a crude microscope to show some fine details of porous wood and to illustrate a microfossil. This simple descriptive work showed that magical explanations of fossils were without foundation.

The idea of extinction

Robert Hooke was one of the first to hint at the idea of extinction, a subject that was hotly debated during the 18th century. The debate fizzed quietly until the 1750s and 1760s when accounts of fossil mastodon remains from North America began to appear. Explorers sent large teeth and bones back to Paris and London for study by the anatomical experts of the day (which was normal practice at the time, since the pursuit of science as a profession was barely beginning in North America). William Hunter noted in 1769 that the 'American *incognitum*' was quite different from modern elephants and from mammoths, and was clearly an extinct animal, and meat-eating at that. 'And if this animal was indeed carnivorous, which I believe cannot be doubted, though we may as philosophers regret it,' he wrote, 'as men we cannot but thank Heaven that its whole generation is probably extinct'.

The fact of extinction was demonstrated most convincingly by Georges Cuvier (1769–1832). He showed that the African and Indian elephants were two species that differed from the mammoth from Siberia and the mastodon from North America (Figure 1.3). Cuvier extended his studies to the rich Eocene mammal deposits of the Paris Basin, describing skeletons of horse-like animals, an opossum, carnivores, birds, and reptiles, all of which differed markedly from living forms. He also wrote accounts of Mesozoic crocodilians, pterosaurs, and the giant mosasaur of Maastricht. Cuvier's work on these extinct species had implications for the ways in which fossils were named and classified (see Box 1.2).

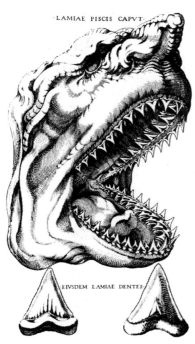

Fig 1.2 Niels Steno's (1667) classic demonstration that fossils represent the remains of ancient animals. He shows the head of a dissected shark together with two fossil teeth, which had previously been called glossopetrae, or tongue stones. The fossils are exactly like the modern shark's teeth.

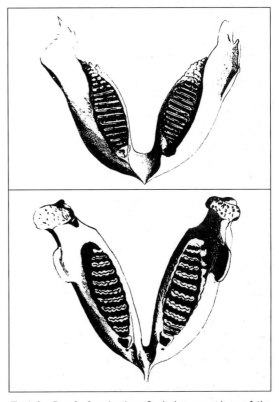

Fig 1.3 Proof of extinction: Cuvier's comparison of the lower jaw of a mammoth (top) and a modern Indian elephant (bottom).

Taxonomy is the study of the morphology and relationships of organisms. Systematics is the broader science of taxonomy and evolutionary processes, while classification refers particularly to the business of naming organisms. When a fossil is described for the first time, the author must name it. Biologists and palaeontologists use a modified version of the principles established by the Swedish naturalist and scientist Carl Gustav Linnaeus (1707–1778). In Linnaean nomenclature a species is given a generic and specific name. These names are latinized and printed in italics, and the author and date of publication follow. If, subsequently, the species is transferred to another genus, the original author and date must then be placed in parentheses. Where several named species turn out to be the same, the subsequent names are synonymized with the first name to have been given to the form.

When a new species is established, a type specimen is designated, and it is housed in a major museum, accessible to future investigators. The new species is defined by a short diagnosis, a few lines emphasizing the distinctive and distinguishing features of the fossil. A fuller description, supported by photographs, drawings, and measurements, is also given, together with information on geographic and stratigraphic distribution.

Fossils, like living animals and plants, are classified in a hierarchical system, where species are included in genera, genera in families, and up through orders, classes and phyla. Techniques of classification are presented in Chapter 3.

Cuvier used detailed comparative anatomical techniques to work out the relationships of fossil skeletons. He believed in the unity of animal structure; that elephants, whether living or fossil, all share certain anatomical features. His public demonstrations became famous: he claimed to be able to identify and reconstruct an animal from just one tooth or bone, and he was usually successful. After 1800, Cuvier had established the reality of extinction.

Geology and palaeontology

The vastness of geological time

In the late 18th century, the idea was forming that the sedimentary rocks, and their contained fossils, documented the history of long spans of time. The Comte de Buffon (1707–1788), wrote about the history of the Earth, from an initially molten state to the present, with several phases in the history of life, and in the formation of modern continents. He accepted the age of Adam as 6000–8000 years ago, a figure calculated from evidence in the Bible. Buffon, however, argued for much longer spans of time before the creation of Adam, making the Earth 75 000 to 3 000 000 years old.

The vast age of the Earth was championed also by James Hutton (1726–1797), who described his geological and geomorphological observations in Scotland. He noted the great cyclical process of mountain uplift, followed by erosion, sediment transport by rivers, deposition in the sea, and then uplift again, and argued that such processes had been going on all through Earth's history. Indeed, Hutton, like many at the time, saw cyclical processes as paramount, and did not appreciate the unidirectional nature of time in the history of the Earth. He wrote, in his *Theory of the Earth* (1795), that his understanding of geological time gave 'no vestige of a beginning, no prospect of an end'.

Most geologists and palaeontologists after 1800 accepted the notion of a vast and unimaginable age for the Earth, so-called deep time, although it was hard to find ways of estimating that age. The discovery of natural radioactivity in the late 19th century gave rise to the technique of radiometric dating in the 20th century, based on measurement of rates of natural radioactive decay to give precise geological ages. By 1900, however, geologists and palaeontologists already had a sophisticated understanding of the divisions of geological time, based on their studies of fossils.

Fossils and time

Field geologists in the early 1800s noticed that certain fossils often occurred together, and that there appeared to be predictable sequences of fossil assemblages. William Smith (1769–1839), a surveyor and canal engineer, realized that he could map geological formations over large areas of southern England, and that he could recognize rock units by their associations of fossil corals and shells. He published his first map and table of strata, showing the broad sequence of rocks in England, in 1815.

Cuvier also published detailed accounts of the sequences of rocks and their contained fossils from the Paris Basin, and he noted major changes in the rock types and fossils. He argued that these came about by local 'revolutions' which wiped out the faunas. William Buckland (1784–1856) extended Cuvier's idea of a sequence of faunas separated by local revolutions, to a full-scale catastrophic model of the history of the Earth and of life. In particular, he collected evidence for the last great revolution, the Biblical Deluge, but his evidence turned out to have been produced by the Pleistocene glaciations of the northern hemisphere.

Field mapping and fossil collecting in England and Germany in particular led to the piecemeal completion of the stratigraphic column. From 1820 to 1840, the broad outlines of the international time-scale were established, and the geological periods and eras given their names. Further subdivision of the periods became possible when palaeontologists carried out more detailed studies of the fossils. It became clear that certain groups of readily identifiable globally-distributed fossils were powerful markers of time.

Fossils and evolution

Progressionism and evolution

Knowledge of the fossil record in the 1820s and 1830s was patchy, and palaeontologists debated whether there was a progression, from simple organisms in the most ancient rocks to more complex forms later. The leading British geologist, Charles Lyell (1797–1875), was an anti-progressionist. He believed that the fossil record showed no evidence for long-term one-way change, but rather for cycles of change. He would not have been surprised to find evidence of human fossils in the Silurian, or for ichthyosaurs to come back at some time in the future if the conditions were right.

Progressionism was linked to the idea of evolution. The first serious considerations of evolution took place in 18th century France, in the work of naturalists such as Buffon and Jean-Baptiste Lamarck (1744–1829). Lamarck explained the phenomenon of progressionism by a large-scale evolutionary model termed the 'Great Chain of Being' or the *Scala naturae*. He believed that all organisms, plants and animals, living and extinct, were linked in time by a unidirectional ladder leading from simplest at the bottom to most complex at the top; indeed, running from rocks to angels. Lamarck argued that the *Scala* was more of a moving escalator than a ladder, that in time present-day apes would rise to become humans, and that present-day humans were destined to move up to the level of angels.

Darwinian evolution

Charles Darwin (1809–1882) developed the theory of evolution by natural selection in the 1830s by abandoning the usual belief that species were fixed and unchanging. Darwin realized that individuals within species showed considerable variation, and that there was not a fixed central 'type' which represented the essence of each species. He also accepted the idea of evolution by common descent, that all species today had evolved from other species in the past. The problem he had to resolve was to explain how the variation within species could be harnessed to produce evolutionary change.

Darwin found the solution in a book published in 1798 by Thomas Malthus, who demonstrated that human populations tend to increase more rapidly than the supplies of food. Hence, only the stronger can survive. Darwin realized that such a principle applied to all animals, that only those individuals which were best fitted to obtain food and to produce healthy young would survive, and that their particular adaptations would be inherited. This was Darwin's theory of evolution by natural selection, the core of modern evolutionary thought.

The theory was published 21 years after Darwin first formulated the idea, in his book *On the Origin of Species* (1859). The delay was a result of Darwin's fear of offending established opinion, and of his desire to bolster his remarkable insight with so many supporting facts that no-one could deny it. Indeed, most scientists accepted the idea of evolution by common descent in 1859, or soon after, but very few accepted (or understood) natural selection. It was only after the beginning of modern genetics early in the 20th century, and its amalgamation with 'natural history' (systematics, ecology, palaeontology) in the 1930s and 1940s, in a movement termed the 'Modern Synthesis' that Darwinian evolution by natural selection became fully accepted.

Palaeontology today

Dinosaurs and fossil humans

Much of 19th century palaeontology was dominated by remarkable new discoveries. Collectors fanned out all over the world, and knowledge of ancient life on Earth increased enormously. The public were keenly interested then, as now, in spectacular new discoveries of dinosaurs. The first isolated dinosaur bones were described from England and Germany in the 1820s and 1830s, and tentative reconstructions were made (Figure 1.4). However, it was only with the discovery of complete skeletons in Europe and North America in the 1870s, that a true picture of these astonishing beasts could be presented. The first specimen of *Archaeopteryx*, the oldest bird, came to light in 1861: here was a true 'missing link', predicted by Darwin only two years before.

Darwin hoped that palaeontology would provide key evidence for evolution; he expected that, as more finds were made, the fossils would line up in long sequences showing the precise pattern of common descent. *Archaeopteryx* was a spectacular start. Rich finds of fossil mammals in the North American Tertiary were further evidence. O. C. Marsh (1831–1899) and E. D. Cope (1840–1897), arch-rivals in the search for new dinosaurs, also found vast numbers of mammals, including numerous horse skeletons, leading from the small, four-toed *Hyracotherium* of 50 million years ago to modern, large, one-toed forms. Their work laid the basis for one of the classic examples of a long-term evolutionary trend (see p. 296).

Human fossils began to come to light around this time: incomplete remains of Neanderthal man in 1856, and fossils of *Homo erectus* in 1895. The revolution in our understanding of human evolution began in 1924, with the announcement of the first specimen of the 'southern ape' *Australopithecus* from Africa, an early human ancestor.

Evidence of earliest life

At the other end of the evolutionary scale, palaeontologists have made extraordinary progress in elucidating the earliest stages in the evolution of life. Cambrian fossils had been known for some time, but the spectacular discovery of the Burgess Shale in Canada in 1909 showed the extraordinary diversity of soft-bodied animals that had otherwise been unknown.

Even older fossils from the Precambrian had been avidly sought for years, but the breakthroughs only happened around 1950. In 1947, the first soft-bodied Ediacaran fossils were found in Australia, and have since

Fig 1.4   The first dinosaur craze in England in the 1850s was fuelled by new discoveries, and dramatic new reconstructions of the ancient inhabitants of that country. This picture, inspired by Sir Richard Owen, is based on his view that dinosaurs were almost mammal-like.

been identified in many parts of the world. Older, simpler, forms of life were recognized after 1960, by the use of electron microscopy, and some aspects of the first 3000 million years of the history of life are now understood.

Macroevolution

Collecting fossils is still a key aspect of modern palaeontology, and remarkable new discoveries are announced every week. However, palaeontologists have made dramatic contributions to the study of large-scale evolution (see Chapter 3), in attempting to understand rates of evolution, the nature of speciation, the timing and extent of mass extinctions, the diversification of life, and other topics where long time-scales are involved.

Studies of macroevolution often push the limits of precision in the fossil record. Perhaps the quality of dating is too poor, and fossils are so rarely preserved that it is foolhardy to attempt to interpret anything about evolution from fossils. It is impossible to say whether palaeontologists know 1%, 50% or 90% of major fossil groups. However, a recent test (Benton and Storrs, 1994) showed that palaeontological knowledge had improved by 5% from 1967 to 1993 (Figure 1.5).

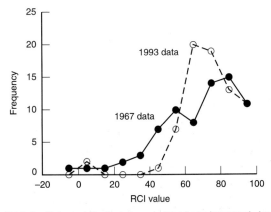

Fig 1.5 Palaeontological knowledge has improved by about 5% in the past 26 years. According to 1993 data, there is 5% less gap, as assessed by a relative completeness index (RCI), implied in the fossil record of tetrapods than in 1967. This figure was obtained by comparing the order of branching points in cladograms with the order of appearance of fossils in the rocks. Will there be a further 5% shift to the right (i.e. towards 100% completeness) by the year 2019?

Palaeontological research

Big science costs money. The newspapers are full of precise details of how many millions of dollars a new space probe costs, or how much has been invested in developing a new drug. Small science, like most of palaeontology, also costs money. Gone are the days when most of the work could be done at home by wealthy people. Most palaeontological research is focused in scientific institutions equipped with computers, scanning electron microscopes, geochemical analytical equipment, and well-stocked libraries, and staffed by laboratory technicians, photographers and artists. However, important work is done by amateurs, who frequently discover new sites and specimens, and many develop expertise in a chosen group of fossils.

Two examples of palaeontological research projects are given, to show how a mixture of luck and hard work is crucial, as well as the co-operation of many people. The spectacular Burgess Shale fauna (Whittington, 1985; Gould, 1989) was found by the geologist Charles Walcott in 1909. The discovery was partly by chance: the story is told of how Walcott and his wife were riding through the Canadian Rockies, and her horse stumbled on a slab of shale bearing beautifully preserved examples of *Marella splendens*, the 'lace crab'. During five subsequent field seasons, Walcott collected over 60 000 specimens, now housed in the National Museum of Natural History, Washington, DC. The extensive researches of Walcott, together with those more recently of Harry Whittington of Cambridge University, England, and his colleagues, have documented a previously unknown assemblage of remarkable soft-bodied animals. The success of the work depended on new technology in the form of high-resolution microscopes, scanning electron microscopes, X-ray photography, and computers to enable 3D reconstructions of flattened fossils. In addition, the work was only possible because of the input of thousands of hours of time in skilled preparation of the delicate fossils, and in the production of detailed drawings and descriptions. In total, a variety of government and private funding sources must have contributed hundreds of thousands of dollars to the continuing work of collecting, describing and interpreting the extraordinary Burgess Shale animals.

A less ambitious palaeontological project has recently been carried out on an assemblage of small bones and teeth associated with dinosaur bones from the Middle Jurassic of Gloucestershire, England. The site was discovered in 1987 by Kevin Gardner, a keen amateur fossil collector. Bones of the large plant-eating dinosaur

Fig 1.6 Excavating large dinosaur bones in a Middle Jurassic site in Gloucestershire, England.

Cetiosaurus (Figure 1.6) were excavated, as well as some 20 tonnes of sediment containing fossils. The material was worked on first at Gloucester City Museum and then at the University of Bristol, and 20 000 teeth and other bones of salamanders, turtles, lizards, crocodiles, pterosaurs and mammals were extracted by sieving (Metcalf *et al.*, 1993).

Geological studies showed that the site was a pond. It lay on a tropical coast, formed from weathered limestone that was covered with thin soils supporting low waterside plants and occasional trees. Pools containing fishes, amphibians and crocodiles lay here and there. The preserved bones include many that had been weathered first in the tropical soils, some that had been washed into the pond, and others that were derived from aquatic creatures living in the water. The nearest modern analogue is the Everglades region of Florida, very different from the modern rolling temperate farmland of Gloucestershire today.

In this study, the huge collections have allowed detailed assessments of the composition of the assemblage (see Box 1.3), and of the modes of damage and fossilization of the bones and teeth. The project has cost £80 000, and it has relied on the labours of many people over a five-year period. The end results include ten papers published in the scientific literature, giving details of the vertebrate life of an otherwise little-known time period, as well as accounts of the modes of preservation of the fossils, and the geology of the site. In addition, the fossils formed the basis for a painted reconstruction (Figure 1.7) which has been used in presentations of the discoveries in books and museums.

Fossil preservation

Fossilization

A key aspect of the Gloucester dinosaur project was the analysis of how the fossils had been buried and preserved; in other words, their taphonomy. Taphonomy is the study of the biological and geological processes that occur between the death of an organism and its final state in the rock. When a plant or an animal dies, it is likely that it will not end up as a fossil (see Chapter 2). For those that do, there is a sequence of stages that normally occur in the transition from a dead body to a fossil (Figure 1.10):

1. decay of the soft tissues of the plant or animal;
2. transport and breakage of hard tissues;
3. burial and modification of the hard tissues.

Fig 1.7 Reconstruction of the scene in the subtropical conditions of Gloucestershire 165 million years ago, showing dinosaurs, pterosaurs, early mammals, lizards, frogs and fishes.

Box 1.3 Palaeobiostatistics

Traditionally, the documentation of fossils in journals and monographs relies heavily on verbal descriptions and illustrations. However, the description, comparison and communication of fossil morphology can be aided by measurements and statistics. Moreover, hypotheses may be designed and tested using inferential statistics. With the wide availability of microcomputers, a large battery of statistical and graphical techniques is now available (e.g. Ryan *et al.* 1994). Two simple examples demonstrate some of the techniques widely used in taxonomic studies, firstly to summarize and communicate precise data, and secondly to test hypotheses.

The smooth terebratulide brachiopod *Dielasma* is common in dolomites and limestones associated with Permian reef deposits in the north of England. Do the samples approximate to living populations, and do they all belong to one or several species? Two measurements (Figure 1.8a) were made on specimens from a single site, and these are plotted as a frequency polygon (Figure 1.8a), to show the population structure. The graph suggests that the population is skewed (has an imbalance) towards smaller size classes, and hence that there is a high rate of juvenile mortality. This is confirmed when the frequency of occurrence of size classes is summed to produce a cumulative frequency polygon (Figure 1.8b). It is possible to test ways in which this population diverges from a normal distribution (i.e. equal areas under the curve to the left and right of the mean value).

It is interesting also to consider growth patterns of *Dielasma*: does the shell grow in a uniform fashion, or does it grow more rapidly in one dimension than the other? The two measurements are compared on logarithmic scales, and the slope of the line equals 1 here. Thus, both features grow at the same rate (Figure 1.8c), but negative and positive rates frequently occur also (see p. 57).

In the Gloucestershire dinosaur study, a sample of 500 specimens was made at random, and the teeth and bones were sorted into taxonomic groups: the results are shown as a pie chart (Figure 1.9a). It is also possible to sort these 500 specimens into other kinds of categories, such as types of bones and teeth, or taphonomic classes (Figure 1.9b, c). A further analysis was made of the theropod (carnivorous dinosaur) teeth, which were very common, to test whether they represented a single population of young and old animals, or whether they came from several species. Tooth lengths and widths were measured, and frequency polygons (Figure 1.9d) suggest the presence of two species.

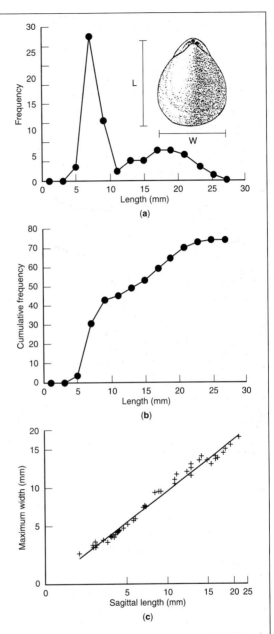

Fig 1.8 Statistical study of the Permian brachiopod *Dielasma*. Two measurements, sagittal length (*L*) and maximum width (*W*) were made on all specimens. The size frequency distributions (a, b) indicate an enormous number of small shells, and far fewer large ones, thus suggesting high juvenile mortality. When the two shape measurements are compared (c), the plot shows a straight line (*y* = 0.819*x* + 0.262); on a previous logarithmic plot, the slope (α) did not differ significantly from unity, so an isometric relationship is assumed, and the raw data have been replotted.

Box 1.3 (cont.)

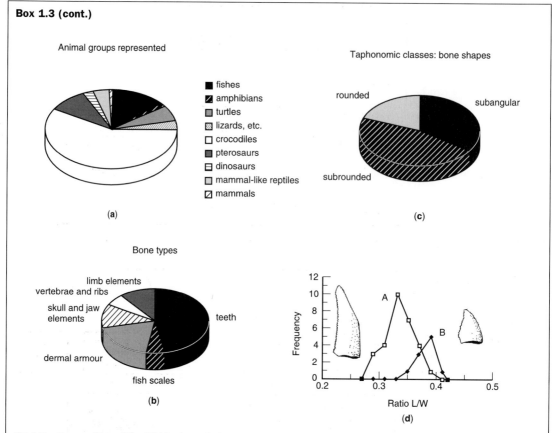

Fig 1.9 Composition of the Middle Jurassic fauna of Gloucestershire, England. The proportions of the major groups of vertebrates in the fauna, shown as a pie chart (a). The sample can be divided into categories also of bone types (b) and taphonomic classes, which depend on the amount of transport (c). Dimensions of theropod dinosaur teeth show two frequency polygons (d) which are statistically significantly different (t-test), and hence indicate two separate forms.

In rare cases, soft parts may be preserved, and these cases of exceptional preservation are crucially important in reconstructing the life of the past.

There are two kinds of fossils: body fossils, which are the partial or complete remains of plants or animals; and trace fossils, which are the remains of the activity of ancient organisms, such as burrows and tracks. In most of the book, 'fossil' is used to mean 'body fossil', which is the usual practice. Trace fossils are treated separately in Chapter 12.

Hard parts and soft parts

Fossils are typically the hard parts – shells, bones, woody tissues – of previously existing plants and animals. In many cases, these skeletons, materials used in

supporting the bodies of the animals and plants when they were alive, are all that is preserved. Skeletons may none the less give useful information about the appearance of many extinct animals since they can show the overall body outline and may hint at the location of muscles, and woody tissues of plants may allow whole tree trunks and leaves to be preserved in some detail. The fossil record is biased in favour of organisms that have hard parts. Soft-bodied organisms may represent up to 60% of the individuals in a marine setting, and these would all be lost under normal conditions of fossilization.

There are a variety of hard materials in plants and animals that contribute to their preservation (Table 1.1). These include inorganic mineralized materials, such as forms of calcium carbonate, silica, phosphates and iron oxides. Calcium carbonate ($CaCO_3$) makes up the shells of foraminifera, some sponges, corals, bryozoans, bra-

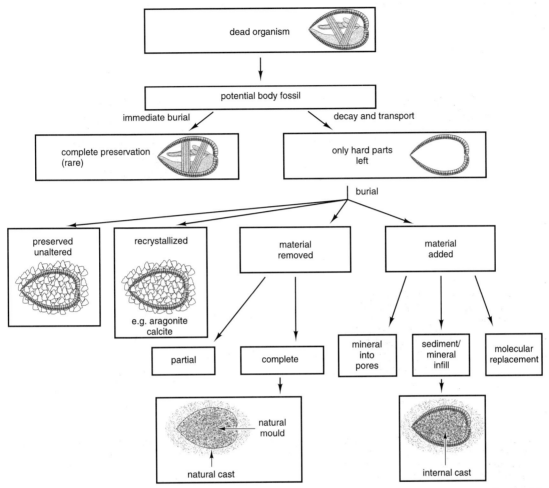

Fig 1.10 How a dead bivalve becomes a fossil. The sequence of stages between the death of the organism, and its preservation in various ways.

chiopods, molluscs, many arthropods, and echinoderms. Silica (SiO_2) forms the skeletons of radiolarians and most sponges, while phosphate, usually in the form of apatite ($CaPO_4$), is typical of vertebrate bone, conodonts, and certain brachiopods and worms. There are also organic hard tissues, such as lignin, cellulose, sporopollenin and others in plants, and chitin, collagen and keratin in animals, which may exist in isolation or in association with mineralized tissues.

Decay

Decay processes typically operate from the moment of death until either the organism disappears completely, or until it is mineralized, although mineralization does not always halt decay. If mineralization occurs early, then a great deal of detail of both hard and soft parts may be preserved (see Box 1.4). If mineralization occurs late, as is usually the case, decay processes will have removed or replaced all soft tissues and may also have affected many of the hard tissues.

Decay processes exist because dead organisms are valuable sources of food for other organisms. When large animals feed on dead plant or animal tissues, the process is termed scavenging, and when microbes, such as fungi or bacteria, transform tissues of the dead organism, the process is termed decay. Well-known examples of scavengers are hyenas and vultures, both of which strip the flesh from large animal carcases. After these large scavengers have had their fill, smaller animals, such as meat-eating beetles, may continue the process of defleshing. In

Table 1.1 Mineralized materials in protists, plants and animals. The commonest occurrences are indicated with XX, and lesser occurrences with X

| | Inorganic | | | | | Organic | | | |
| | Carbonates | | | | | | | | |
	Aragonite	Calcite	Phosphates	Silica	Iron oxides	Chitin	Cellulose	Collagen	Keratin
Prokaryotes	XX	X	X		X		X		
Algae	XX	XX		X		X	XX		
Higher plants		X		X	X		XX		
Protozoa		XX		XX	XX	X	X		
Fungi		X	X		X	XX	XX		
Porifera	X	XX		XX	X			XX	
Cnidaria	XX	XX				X		X	
Bryozoa	XX	XX	X			XX		X	
Brachiopoda		XX	XX			XX		X	
Mollusca	XX	X	X	X	X	X		X	
Annelida	XX	XX	XX		X	X		XX	
Arthropoda		XX	XX	X	X	XX		X	
Echinodermata		XX	X	X				XX	
Chordata		X	XX		X		X	XX	XX

many cases, all flesh is removed in a day or so. Decay is dependent on three factors.

The first factor controlling decay is the supply of oxygen. In aerobic (oxygen-rich) situations, microbes break down the organic carbon of a dead animal or plant by converting carbon and oxygen to carbon dioxide and water, according to the following equation:

$$CH_2O + O_2 \rightarrow CO_2 + H_2O.$$

Microbial decay can also take place in anaerobic conditions, i.e. in the absence of oxygen, and in these cases nitrate, manganese dioxide, iron oxide, or sulphate ions are necessary to allow the decay to occur.

The second set of factors controlling decay – temperature and pH – may be the most important. High temperatures promote rapid decay. Decay proceeds at normal high rates when the pH is neutral, as is the case in most sediments, since this creates ideal conditions for microbial respiration. Decay is slowed down by conditions of unusual pH, such as those found in peat swamps, which are acidic. Fossils preserved in peat or lignite (brown coal) become tanned, like leather, and many of the soft tissues are preserved. Examples are the famous 'bog bodies' of northern Europe, in which skin and internal organs are preserved, and silicified fossils in the lignite of the Geiseltal deposit in Germany (Eocene) which show muscle fibres and skin.

Decay depends, thirdly, on the nature of the organic carbon, which varies from highly labile to highly decay-resistant. Most soft parts of animals are made from volatiles, forms of carbon that have molecular structures that break down readily. Other organic carbons, termed refractories, are much less liable to break down, and these include many plant tissues, such as cellulose.

The normal end-result of scavenging and decay processes is a plant or animal carcass stripped of all soft parts. In rare cases, some of the soft tissues may survive, and these are examples of exceptional preservation (see Box 1.4).

Breakage and transport

The hard parts left after scavenging and decay have taken their toll may simply be buried without further modification, or they may be broken and transported. There are several processes of breakage (Figure 1.14), some physical (disarticulation, fragmentation, abrasion) and some chemical (bioerosion, corrosion and dissolution).

Skeletons that are made from several parts may become disarticulated, i.e. separated into their component parts. For example, the multi-element skeletons of vertebrates may be broken up by scavengers and by wave and current activity on the sea bed (Figure 1.14a). Disarticulation happens only after the scavenging or decay of connective tissues that hold the skeleton together. This may occur within a few hours in the case of crinoids, where the ligaments holding the separate ossicles together decay rapidly. In trilobites and vertebrates, normal aerobic or anaerobic bacterial decay may take weeks or months to remove all connective tissues.

Skeletons may also become fragmented, that is, individual shells, bones or pieces of woody tissue break up into smaller pieces (Figure 1.14b), usually along lines of weakness. Fragmentation may be caused by predators and scavengers such as hyenas which break bones, or such as crabs which use their claws to snip their way into shelled prey. Much fragmentation is caused by physical processes associated with transport: bones and shells may bang into each other and into rocks as they are transported by water or wind. Wave action may cause such extensive fragmentation that everything is reduced to a fine sand.

Shells, bones and wood may be abraded by physical grinding and polishing against each other and against other sedimentary grains. Abrasion removes surface details, and the fragments become rounded (Figure 1.14c). The degree of abrasion is related to the density of the specimen (in general, dense elements survive physical abrasion better than porous ones), the energy of currents and grain size of surrounding sedimentary particles (large grains abrade skeletal elements more rapidly than small grains), and the length of exposure to the processes of abrasion.

In certain circumstances shells, bones and wood may undergo bioerosion, i.e. the removal of skeletal materials by boring organisms such as sponges, algae and bivalves (Figure 1.14d). Minute boring sponges and algae operate even while their hosts are alive, creating networks of fine borings by chemical dissolution of the calcareous shell material. This process continues after death, and some fossil shells are riddled with borings which may remove more than half of the mineral material of any single specimen. Other boring organisms eat their way into logs, and heavily modify the internal structure.

Before and after burial, skeletal materials are commonly corroded and dissolved by chemical action (Figure 1.14e). The minerals within many skeletons are chemically unstable, and they break down after death while the specimen lies on the sediment surface, and also for some time after burial. Carbonates are liable to corrosion and dissolution by weakly acidic waters. The most stable skeletal minerals are silica and phosphate.

Burial and modification

Animal and plant remains are typically buried after a great deal of scavenging, decay, breakage and transport. Sediment is washed or blown over the remains, and the specimen becomes more and more deeply buried. During and after burial, the specimen may undergo physical and chemical change.

The commonest physical change is flattening by the weight of sediment deposited above the buried speci-

Box 1.4 Exceptional preservation

There are many famous examples of exceptional preservation. Certain fossil-bearing formations of different ages, termed Lagerstätten, have produced hundreds of remarkable fossil specimens, and in some cases soft parts are preserved. In the most spectacular cases, soft tissues such as muscle, which is composed of labile forms of organic carbon, may be preserved. Usually, however, only the rather more decay-resistant soft tissues, such as chitin and cellulose, are fossilized. Plant and animal tissues decay in a sequence that depends on their volatile content, and the process of decay can only be halted by mineralization (Figure 1.11). In the process of fossilization, then, it is possible to think of a race between rates of decay and rates of pre-burial mineralization: the point of intersection of those rates determines the quality of preservation of any particular fossil.

Early mineralization of soft tissues may be achieved in pyrite, phosphate or carbonate, depending on three factors: (1) rate of burial, (2) organic content, and (3) salinity (Figure 1.12a). Early diagenetic pyritization (Figure 1.12b) of soft parts is favoured by rapid burial, a low organic content, and the presence of sulphates in the sediment. Early diagenetic phosphatization (Figure 1.12c) requires a low rate of burial and a high organic content. Soft-part preservation in carbonates (Figure 1.12d) is favoured by rapid burial in organic-rich sediments; at low salinity levels, siderite is deposited, and at high salinity levels, carbonate is laid down in the form of calcite. In rare cases, decay and mineralization do not

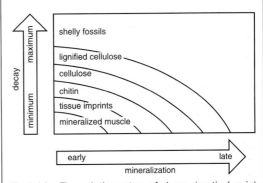

Fig 1.11 The relative rates of decay (vertical axis) and mineralization (horizontal axis) determine the kinds of tissues that may be preserved. At minimum decay rate and with very early mineralization, highly labile muscle tissues may be preserved. When decay has gone to a maximum, and when mineralization occurs late, all that is left are the non-organic tissues such as shells.

Box 1.4 (cont.)

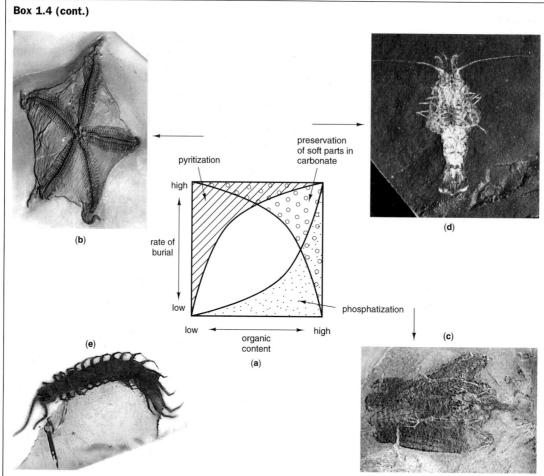

Fig 1.12   The conditions for exceptional preservation. (a) Rate of burial and organic content are key controls on the nature of mineralization of organic matter in fossils. Pyritization (high rate of burial; low organic content) may preserve entirely soft-bodied worms, as in an example of the starfish *Loriolaster* (b) from the early Devonian Hunsrückschiefer of Germany. Phosphatization (low rate of burial; high organic content) may preserve feathers as in (c), an unidentified bird from the Eocene of Germany. Soft parts may be preserved in carbonate (high rate of burial; high organic content), such as the limbs and antennae of a shrimp (d), from the Carboniferous of Scotland. If decay never starts, small animals may be preserved organically and without loss of material, as in a centipede in amber from the early Tertiary of the Baltic region (e).

occur, when the organism is instantly encased and preserved in a medium such as amber (Figure 1.12e) or asphalt.

Mineralization of soft tissues occurs in three ways. Rarely, soft tissues may be replaced in detail, or replicated, by phosphates. Permineralization occurs very early, probably within hours of death, and may preserve highly labile structures such as muscle fibres (Figure 1.12b), as well as more refractory tissues such as cellulose and chitin. The commonest mode of mineralization of soft tissues is by the formation of mineral coats of phosphate, car-

bonate, or pyrite, often by the action of bacteria. The mineral coat preserves an exact replica of the soft tissues which decay away completely. The third mode of soft tissue mineralization is the formation of tissue casts during the early stages of sediment compaction. Examples of tissue casts include siliceous and calcareous nodules which preserve the form of the organism and prevent it from being flattened or dissolved.

The mode of accumulation of fossils also determines the nature of fossil Lagerstätten. Fossil assemblages may be produced by concentration,

Box 1.4 (cont.)

i.e. the gathering together of remains by normal processes of sedimentary transport and sorting to form fossil-packed horizons (see p. 11), or by conservation, the fossilization of plant and animal remains in ways that avoid scavenging, decay, and diagenetic destruction (Figure 1.13). Exceptionally preserved fossil assemblages are produced mainly by processes of conservation. Certain sedimentary regimes, in the sea or in lakes, are stagnant, where sediments are usually anoxic, and are devoid of animals that might scavenge carcasses. In other sit-

uations, termed obrution deposits, sedimentation rates are so rapid that carcasses are buried virtually instantly, and this may occur in rapidly migrating river channels or at delta fronts and other situations where turbidites are deposited. Some unusual conditions of instant preservation are termed conservation traps. These include amber, fossilized resin that oozes through tree bark, and may trap insects, and tar pits and peat beds where plants and animals sink in and their carcasses may be preserved nearly completely.

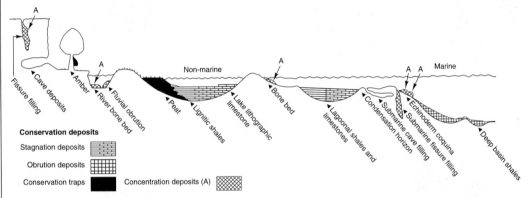

Fig 1.13 An imaginary cross-section showing possible sites of exceptional fossil preservation, most of which are conservation deposits, but a few of which are concentration deposits.

men, and it may occur soon after burial. The flattening forces flatten the specimen in the plane of sedimentary bedding. The nature of flattening depends on the strength of the specimen: the first parts to collapse are those with the thinnest skeleton and largest cavity inside. Greater forces are required to compress more rigid parts of skeletons. Ammonites, for example, have a wide body chamber cavity which would fill up with sand or water after the soft body decayed. This part collapses first (Figure 1.14f) and, because the shell is hard, it fractures. The other chambers are smaller, fully enclosed, and hence mechanically stronger: they collapse later. Plant fossils such as logs are usually roughly circular in cross-section, and they flatten to a more ovoid cross-section after burial. The woody tissues are flexible and they generally do not fracture, but simply distort.

Physical effects such as flattening, and chemical effects that occur after burial, are termed diagenesis. In sedimentological terms, diagenesis may occur very soon after burial (e.g. flattening and some chemical changes) or long after, often thousands or millions of years later, as a result of the passage of chemicals in solution

through rock containing fossils. Fossils may also be deformed by metamorphic processes, often millions of years after burial and diagenetic alteration (see Box 1.5).

The calcium carbonate in shells occurs in four forms: aragonite, calcite (in two varieties – high-magnesium (Mg) calcite and low-Mg calcite), and combinations of aragonite + calcite. The commonest diagenetic process is the conversion of aragonite to calcite. After burial, pore fluids within the sediment may be undersaturated in $CaCo_3$, and the aragonite dissolves completely, leaving a void representing the original shell shape. Later, pore fluids that are supersaturated in $CaCO_3$ allow calcite to crystallize within the void, thus producing a perfect replica of the original shell. This process of replacement of aragonite by calcite occurs commonly, and may be detected by the change of the crystalline structure of the shell (Figure 1.14g). The regular layers of aragonite needles have given way to large irregular calcite crystals (sparry calcite) or tiny irregular calcite crystals (micrite).

A common diagenetic phenomenon is the formation of carbonate concretions, bodies that form within sediment and concentrate $CaCo_3$ (calcite) or $FeCO_3$ (siderite).

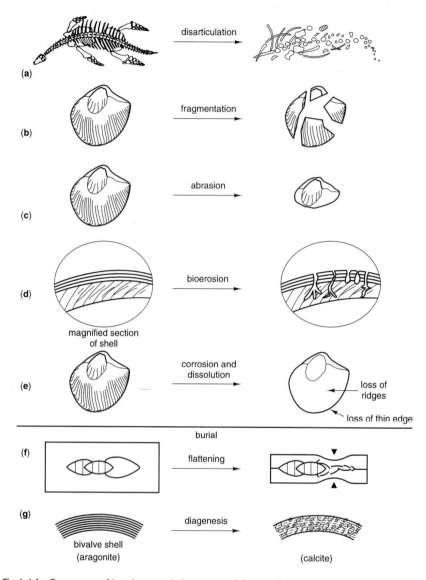

Fig 1.14 Processes of breakage and diagenesis of fossils. Dead organisms may be disartic-ulated (a) or fragmented (b) by scavenging or transport, abraded (c) by physical movement, bio-eroded (d) by borers, or corroded and dissolved (e) by solutions in the sediment. After burial, specimens may be flattened (f) by the weight of sediment above, or various forms of chemical diagenesis, such as the replacement of aragonite by calcite (g), may take place.

Carbonate concretions generally form early during the burial process, and this is demonstrated by the fact that enclosed fossils are uncrushed, having been protected from compaction by the formation of the concretion. Carbonate concretions form typically in black shales, i.e. sediments deposited in the sea in anaerobic conditions. Black shales contain abundant organic carbon, and when this is buried, bacterial processes of anaerobic decay begin. These decay processes reduce oxides in the sediment, and produce bicarbonate ions which may combine with any calcium or iron ions to generate carbonate and siderite concentrations. Such concentrations may grow rapidly to form concretions around the source of calcium and iron ions, usually the remains of an organism.

Box 1.5 Deformed fossils

Some fossils may be heavily deformed or distorted, so that they do not retain their original shapes. These distortions may be the result of collapse or diagenesis, but they may indicate metamorphism, i.e. processes connected with tectonic activity, faulting, folding and mountain building. If a mudstone is folded and, under high pressure, is changed into a slate, any contained fossils are likely to be stretched and distorted. With symmetrical fossils (e.g. Figure 1.15), it is possible to restore the original shape of the fossil and, indeed, to use that information to work out the nature of the tectonic forces that acted. Deformed fossils become commoner the further back in time one goes, simply because of the greater likelihood that any particular fossiliferous sediment has undergone metamorphism.

Fig 1.15 Deformation of numerous brachiopods, *Eoplectodonta,* from the Silurian of Ireland.

may combine with calcium ions to form apatite which can entirely replace dissolved calcareous shells. In other cases, the microbial processes enable soft tissues, and entirely soft-bodied organisms, to be replaced by phosphate. Coprolites (fossil dung) may also be phosphatized. In these cases, apatite has been liberated from the organisms themselves, and from surrounding concentrations of organic matter, and the replacement destroys most, or all, of the original skeletal structures.

Further reading

Briggs, D. E. G. (1991) Extraordinary fossils. *American Scientist,* **79**, 130–141.

Briggs, D. E. G. and Crowther, P. R. (eds) (1990) *Palaeobiology: a synthesis.* Blackwell Scientific Publications, Oxford.

Buffetaut, E. (1987) *A short history of vertebrate palaeontology.* Croom Helm, London.

Donovan, S. K. (1991) *The processes of fossilization.* Belhaven Press, London.

Mayr, E. (1991) *One long argument: Charles Darwin and the genesis of modern evolutionary thought.* Penguin, London.

Rudwick, M. J. S. (1972) *The meaning of fossils.* Science History Publications, New York.

References

Benton, M. J. and Storrs, G. W. (1994) Testing the quality of the fossil record: paleontological knowledge is improving. *Geology,* **22**, 111–114.

Gould, S. J. (1989) *Wonderful life. The Burgess Shale and the nature of history.* Hutchinson Radius, London.

Metcalf, S. J., Vaughan, R. F., Benton, M. J., Cole, J., Simms, M. J. and Dartnall, D. L. (1993) A new Bathonian (Middle Jurassic) microvertebrate site, within the Chipping Norton Limestone Formation at Hornsleasow Quarry, Gloucestershire. *Proceedings of the Geologists' Association,* **103**, 321–342.

Ryan, P. D., Harper, D. A. T., and Whalley, J. S. (1994) *The PALSTAT package.* Chapman & Hall, London.

Whittington, H. B. (1985) *The Burgess Shale.* Yale University Press, New Haven.

Another early diagenetic mineral which occurs in anaerobic marine sediments is pyrite (FeS_2). It is also produced as a by-product of anaerobic processes of microbial reduction within shallow buried sediments. Pyrite may replace soft tissues such as muscle in cases of rapid burial, and replaces hard tissues under appropriate chemical conditions. Wood, for example, may be pyritized, and dissolved aragonite or calcite shells may be entirely replaced by pyrite. In both cases, the original skeletal structures are lost.

Phosphate is a primary constituent of vertebrate bone and other skeletal elements. In some cases, masses of organic phosphates are modified by microbial decay, which releases phosphate ions into the sediment. These

2 Fossils in time and space

Key Points

- Fossil organisms allow the reconstruction of ancient animal and plant communities.
- Lithostratigraphy is the rock framework for geological studies; lithostratigraphical units are displayed on maps and measured sections.
- Biostratigraphy, using zone fossils, is the basis for correlation and can be investigated by graphical correlation and seriation.
- Chronostratigraphy, global standard stratigraphy, is the division of geological time into workable intervals with reference to type sections measured in terms of rock.
- Palaeoautecology analyses the operation of a single fossil organism whereas palaeosynecology investigates fossil communities.
- The palaeoecology of fossil organisms can be described in terms of their life strategies and trophic modes together with their habitats.
- Palaeocommunities are analyzed by a range of statistical techniques; the structure and composition of palaeocommunities have changed through time.
- Palaeobiogeography has provided basic data to suggest and test plate tectonic models.
- Changes in palaeogeography have promoted the interchange and migration of faunas and floras together with the radiation and extinction of taxa.
- Fossils from fold belts constrain the age and origin of tectonic events; fossil data also provide estimates of finite strain and thermal maturation.

Introduction

Fossil organisms were intimately associated with their surrounding environments, controlled by many biological, chemical and physical agencies. Environmental changes involving the compositions of the atmosphere and the oceans had an influential effect on evolving faunas and floras. The distributions of fossil organisms through time and space are thus very much part of the direction and shape of the history of life itself. But although fossils are the definitive evidence of evolving life on the planet they also hold a vast database of geological information: the age and environments of enclosing sediments and their palaeogeography, together with the finite strain and thermal maturation of rocks, can all be determined with reference to fossils.

Frameworks

Before the distributions of fossils in time and space can be studied, analyzed and interpreted, fossil animals and plants must be first described in context. A rock stratigraphy is the essential framework that geologists and particularly palaeontologists use to address and locate fossil collections accurately.

Background to stratigraphy

Pioneer work by Nicolaus Steno in northern Italy, during the late 17th century, established the simple fact that older rocks are overlain by younger rocks if the sequence has not been inverted. His principle of superposition of strata is fundamental to all stratigraphical studies. Moreover, about a century later Giovanni Arduino recognized, using superposition, three basically different rock suites in the Italian part of the Alpine belt. A crystalline basement of mainly Variscan rocks was unconformably overlain by mainly Mesozoic limestones deformed during the Alpine orogeny; these in turn were unconformably overlain by poorly consolidated clastic rocks, mainly conglomerates. These three units constituted his primary, secondary and tertiary systems (Figure 1a); the first two terms have been discarded, but the last term has, in fact, been retained and formalized for the period of geological time succeeding the Cretaceous.

On the ground: lithostratigraphy

Virtually all aspects of stratigraphy start from the rocks themselves. Basic stratigraphical data are first assembled and mapped through the definition of a lithostratigraphic scheme at a local and regional level. Lithostratigraphical units are recognized on the basis of rock type. The formation, a rock unit which can be mapped across country, irrespective of thickness, is the basic lithostratigraphical category. A formation may comprise one or several related lithologies, different from units above and below, and usually given a local geographic term. A member is a more local lithological development, usually part of a formation, whereas a succession of contiguous formations with some common characteristics are often defined as a group; groups themselves may comprise a supergroup. All stratigraphical units must be defined at a reference or type section, analogous to the holotype of a fossil and in a specified type area. Unfortunately, the entire outcrop of many lithostratigraphic units is rarely exposed; instead the bases of units are routinely defined in basal stratotype sections. Since the base of the succeeding unit defined the top of the underlying unit, only basal stratotypes need be defined.

A stratigraphy illustrated in terms of a map and measured sections is the basic framework required to monitor biological and geological changes through time and underpins the whole basis of Earth history. By the late 18th century, James Hutton and others had begun to appreciate that the Earth had evolved over immense periods of time, often called deep time. The spectacular unconformity at Siccar Point, Berwickshire, southern Scotland, where near-horizontal Old Red Sandstone (Devonian) strata overlie steeply dipping Silurian greywackes, convinced Hutton of the 'inconceivably long' time required for the operation of geological processes. Beneath the unconformity Hutton recognized the 'ruins of an earlier world', establishing the immensity of geological time, while also paving the way for our present concept of the Earth as a dynamic and changing system.

Use of fossils: discovery of biostratigraphy

Early in the 19th century, William Smith in the course of his work as a canal engineer, began to appreciate that different rocks units were characterized by distinctive assemblages of fossils. Since life has evolved throughout geological time, clearly each interval of geological time can be characterized by its own fossil taxa and commu-

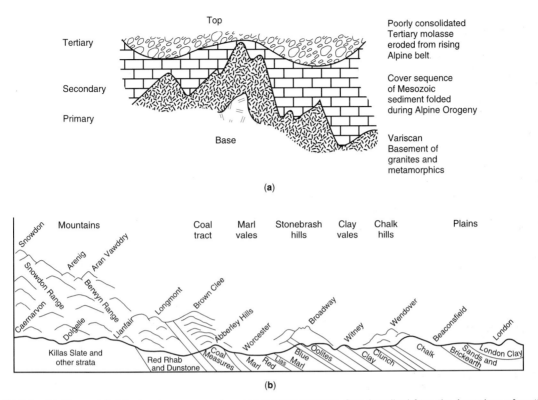

Fig 2.1 (a) Giovanni Arduino's primary, secondary and tertiary systems, first described from the Apennines of northern Italy in 1760; these divisions were built on the basis of Steno's 'Law of superposition of strata'. (b). Idealized sketch of William Smith's geological traverse from London to Wales; this traverse formed the basis for the first geological map of England and Wales. Data assembled during this survey were instrumental in the formulation of the 'Law of correlation by fossils'.

nities of taxa. In broad terms the Palaeozoic is dominated by brachiopods, trilobites and graptolites, whereas the Mesozoic assemblages have ammonites, belemnites, marine reptiles and dinosaurs as important components, and the Cenozoic is dominated by mammals and molluscan groups, such as the bivalves and the gastropods. Microevolutionary changes at the species and subspecies level can be tracked along phylogenetic lineages (see Chapter 3). Very accurate correlation is now possible using a wide variety of fossil organisms (see below). In a traverse from London to Wales, Smith encountered successively older groups of rocks, and he documented the change from the molluscan faunas of the Tertiary strata of the London Basin down through thick Mesozoic and Upper Palaeozoic sequences to the trilobite-dominated assemblages of the Lower Palaeozoic of Wales (Figure 2.1b). In France, a little later, the noted anatomist Georges Cuvier ordered and correlated Tertiary strata in the Paris Basin using series of mainly terrestrial vertebrate faunas, occurring in sequences separated by supposed catastrophes.

Dividing up geological time: chronostratigraphy

Throughout Europe geologists were describing units of rock with distinctive fossil assemblages using the concepts established by Steno, Arduino, Smith and Cuvier. For example, during the 1830s, while Roderick Murchison and Adam Sedgwick were claiming parts of Wales and the Welsh Borderlands for their respective Cambrian and Silurian systems, on the European continent Von Alberti was establishing the Triassic System based on sections in southern Germany. From the late 18th century to the mid 19th century, virtually all the geological systems had been named. The Ordovician was suggested later by Charles Lapworth in 1879 as a compromise to contain the contentious middle ground claimed by both Sedgwick and Murchison for their Cambrian and Silurian systems. Ironically the Ordovician is one of the longest and most diverse of the geological systems. The stratigraphical column itself was gradually assembled as new systems were named and placed in stratigraphical order.

Nevertheless, many of the geological systems, conventionally, were separated from each other by unconformities. Unconformities provided a convenient break between systems for investigators and, more importantly, it satisfied current thinking that major chunks of Earth history should be divided by global, catastrophic events. Unfortunately while many unconformities could be traced across Europe, evaluation of data from sections elsewhere in the world showed that many of these unconformities were apparently only regional breaks. The lower boundaries of most systems, as then understood, were represented by stratigraphical gaps, and gaps provided a very poor basis for the global correlation of systemic boundaries.

All the systemic boundaries have been or are currently being reinvestigated by working groups of the International Union of the Geological Sciences (IUGS). The potential of each base for international correlation must be maximized. Thus the traditional bases of these systems must be adjusted and placed within intervals of continuous sedimentation, with diverse and abundant faunas and floras in geographically and politically accessible areas that can be conserved and protected; ideally the sections should have escaped metamorphism and tectonism.

Chronostratigraphy (or global standard stratigraphy) is the most fundamental of all stratigraphical concepts. Everyday intervals of time, such as the second, minute and hour, are based on a universal time signal from an atomic clock. Units of geological time, such as the epoch and period, are much longer and of uneven lengths. The only standards available for the definition of these intervals are the rock successions themselves. Thus the rocks of the type section in the type area for the Silurian System act as an international standard for the Silurian Period, the time during which that system was deposited. The base of a chronostratigraphic interval is defined in a unique stratotype section, in a type area using the concept of a golden spike. All the usual criteria for a workable stratotype section must, of course, be satisfied. The Golden Spike, which represents a point in the rock section and an instant in geological time, is then driven into the section. In fact the spike is usually adjusted to coincide with the first appearance of a distinctive, recognizable fossil within a well-documented lineage (Figure 2.2). In practice, the ranges of all fossils occurring across the boundary are documented in detail as aids to correlating within the section and with sections elsewhere.

The Wenlock Epoch was one of the first intervals of geological time to be defined with reference to a stratotype section of the Wenlock Series (Figure 2.3). A lithostratigraphy was first established in the historical type area in terms of formations and members. On the basis of detailed collecting through the lithostratigraphy, a succession of biozones, based on the ranges of characteristic graptolite faunas, was then defined.

Sequence stratigraphy

More recently, North American geologists have emphasized the importance of unconformities in defining another form of stratigraphy based on the recognition and correlation of sequences with considerable application to hydrocarbon exploration. In the early 1960s, Sloss recognized that the Phanerozoic rocks of the North American craton could be split into five main sequences divided by unconformities (Figure 2.4). More minor

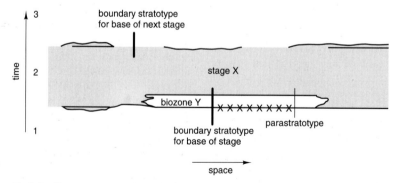

Fig 2.2   Key concepts in the definition of stratotypes and parastratotypes applicable to all stratigraphical units. The base of Stage X is defined at a suitable type section, coincident with the base of Biozone Y, which can be used to correlate the base of the stage. The type section is usually conserved and collecting across the interval is restricted to an adjacent parastratotype section. Base of stage: X X X X

Chronostratigraphy			Lithostratigraphy	Biostratigraphy
Series or epoch	Stage or age	Chronozone	Formations and member	Graptolite biozones
Wenlock	Homerian	Gleedon	Much Wenlock Limestone Formation	*ludensis*
			Farley Member of Coalbrookdale Formation	*nassa*
		Whitwell	Coalbrookdale Formation	*lundgreni*
	Sheinwoodian			*ellesae*
				linnarssoni
				rigidus
				riccartonensis
			Buildwas Formation	*murchisoni*
				centrifugus

Fig 2.3   Stratigraphical case study: description of the litho-, bio- and chronostratigraphy of the stratotype section, along Wenlock Edge in Shropshire, UK, of the type Wenlock Series. This section is the international standard for the Wenlock Epoch.

sequences could be recognized within these major cycles. The fact that sedimentary rocks can be described as packets of strata, presumably deposited during transgressive events, divided by periods of nondeposition during regressions, forms the basis for sequence stratigraphy. Moreover, various types and degrees of unconformities can be recognized on seismic profiles. Whereas most major sequence boundaries are probably due to global eustatic changes in sea level associated with climatic change or fluctuations in sea-floor spreading processes, sequences can also be generated by more local tectonic controls. Research teams in the Exxon Corporation have expanded the concept of sequence stratigraphy to build global sea-level curves for the entire Phanerozoic. Description of successions in terms of unconformity-bound sequences has proved valuable in hydrocarbon exploration, where sequence boundaries can be recognized at depth using seismic geophysics.

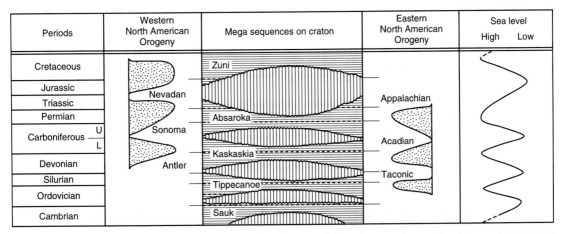

Fig 2.4   North American Phanerozoic sequences: the recognition of these large packages of rock or megasequences formed the basis for the modern discipline of sequence stratigraphy, established by Peter Vail and his colleagues at Exxon.

Biostratigraphy

Biostratigraphy is the establishment of fossil-based successions and their use in stratigraphical correlation. As noted above, measurements of the stratigraphical ranges of fossils, or assemblages of fossils, form the basis for the definition of biozones, the main operational unit of a biostratigraphy. Critics have argued that there are difficulties with the identification of many organisms flagged as zone fossils; and, moreover, it may be impossible to determine the entire range of a fossil or a fossil assemblage. None the less, to date, the use of fossils in biostratigraphy is the best and often the most accurate routine means of correlating and establishing the relative ages of strata.

There are several types of range zone (Figure 2.5). The concept of the range zone is based on the work of Albert Oppel during the mid 19th century. Oppel characterized

successive lithological units by unique associations of species; his zones were based on the consistent and exclusive occurrence of mainly ammonite species, and through Jurassic sections across Europe he recognized 33 zones in comparison with the 60 or so known today. His zonal scheme could be meshed with d'Orbigny's stage classification of the Jurassic, based on local sections with geographical terms. Although William Smith had recognized the significance of fossils almost fifty years previously, Oppel established a modern and rigorous methodology which now underpins modern biostratigraphy.

The range is thus drawn between the first and last appearance of the zone fossil in a given section. Clearly it is unlikely that the entire global vertical range of the zone fossil is represented in this section; nevertheless it is, in most cases, a workable approximation. This range, measured against the lithostratigraphy, is termed a biozone. It is the basic biostratigraphical unit, analogous to

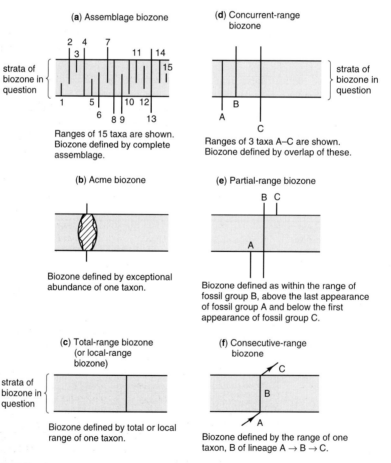

(a) Assemblage biozone

Ranges of 15 taxa are shown. Biozone defined by complete assemblage.

(b) Acme biozone

Biozone defined by exceptional abundance of one taxon.

(c) Total-range biozone (or local-range biozone)

Biozone defined by total or local range of one taxon.

(d) Concurrent-range biozone

Ranges of 3 taxa A–C are shown. Biozone defined by overlap of these.

(e) Partial-range biozone

Biozone defined as within the range of fossil group B, above the last appearance of fossil group A and below the first appearance of fossil group C.

(f) Consecutive-range biozone

Biozone defined by the range of one taxon, B of lineage A → B → C.

Fig 2.5   The main types of biozones, the operational units of a biostratigraphy.

Box 2.1 Zone fossils

The recognition and use of zone fossils is fundamental to biostratigraphical correlation. Fossil groups that are (1) rapidly evolving, (2) widespread across different facies and biogeographical provinces, (3) relatively common and (4) easy to identify, make the ideal zone fossils. In the early Palaeozoic macrofauna, graptolites are the closest to the ideal zone fossils whereas during the Mesozoic the ammonites are most useful. The use of efficient zone fossils ensures that relatively short intervals of geological time can be correlated, often with a precision of several hundred thousand years, over long distances through different facies belts around the world. In practice there are no ideal zone fossils. Most long-range correlations involve use of intermediate faunas with mixed facies. For example, in Ordovician rocks, deep-water facies are correlated with graptolites; these fossils are rare in shallow-water shelf deposits where trilobites and brachiopods are much more common. Nevertheless, biostratigraphical units based on both graptolite and shelly faunas may interdigitate in deep-shelf and slope sequences allowing correlation through these mixed facies from deep to shallow water.

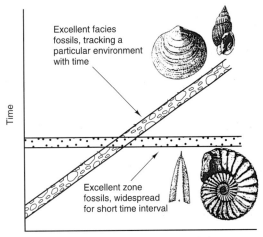

Fig 2.6 Behaviour of perfect zone and facies fossils through a hypothetical global stratigraphical section.

the lithostratigraphical formation. It is defined with reference to precise occurrences in the rock, and is defined again on the basis of a stratotype or basal stratotype section in a type area.

On a simple plot of space against time (Figure 2.6) an ideal zone fossil will represent a thin horizontal band reflecting a short duration but a widespread spatial distribution. In reality very few fossils approach the properties of an ideal zone fossil. The distribution of most is controlled to some degree by facies. A typical facies fossil is not constrained by time but appears to move spatially with time as it shadows the migration of a particular facies belt.

Many different animal and plant groups are used in biostratigraphical correlation (Figure 2.8). Graptolites and ammonites are arguably the best known and most reliable zone macrofossils. Amongst the vertebrates even lineages of fossil pigs have been used in biostratigraphy, particularly in the Quaternary rocks of East Africa where they have helped date horizons with important hominid

Box 2.2 Graphical correlation and seriation

A more rigorous, statistically based mode of correlation was developed initially for use in the petroleum industry. Graphical correlation by Shaw's method requires fossil range data from two or more measured sections (Figure 2.7a). Data of first and last occurrences of fossil species are plotted against a measured stratigraphical section; this is repeated for a second section. Usually only the more common taxa are plotted. A bivariate scattergram is then drawn with section 1 along the x-axis and section 2 along the y-axis. The first and last occurrences are then plotted as x,y coordinates; for example, the x coordinate represents the first appearance of species *a* along section 1 and the y coordinate its first appearance in section 2. A regression line is fitted to all the first and last appearance coordinates. This line of stratigraphical correlation can be used for interpolation, permitting the accurate correspondence of all levels in the two sections.

Biostratigraphers also use techniques established by archaeologists at the end of last century. Seriation is an ordering technique designed to analyse gradients. Usually the gradients are temporal, but biogeographical and environmental data have also been investigated by seriation. Biostratigraphers tend to enter the ranges of organisms on range charts in a fairly random manner. In simple terms, seriation shuffles the original data matrix until the stratigraphically higher taxa are on the left-hand side of the matrix and the stratigraphically lower taxa are on the right; any stratigraphical gradients in the data are then clearly visible (Figure 2.7b) and can be interpreted.

Box 2.2 (cont.)

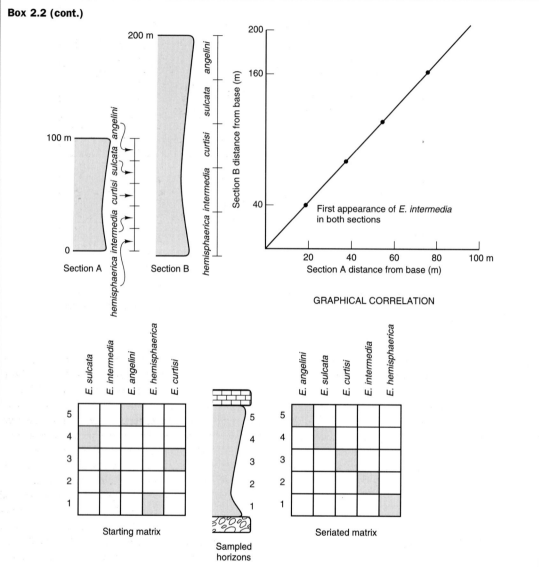

Fig 2.7 (a) Hypothetical and minimalist graphical correlation based on the stratigraphical distribution of five apparent chronospecies of the Silurian rhynchonellide brachiopod *Eocoelia*, in ascending order *E. hemisphaerica, E. intermedia, E. curtisi, E. sulcata* and *E. angelini*; the first four range through the Middle and Upper Llandovery while the last is characteristic of the Lower Wenlock. The ranges of these species have been reported from two artifical sections; the first appearances of each species in both sections are plotted as x and y coordinates. The straight line fitted to the points allows precise correlation between each part of the two sections. In this simple example all the points lie on a straight line; usually a regression line must be fitted to a scatter of data points. (b) Seriation of biostratigraphical data. The five species of *Eocoelia* were collected from five horizons in a stratigraphical section; the data were collected randomly and plotted as a range-chart matrix. The starting matrix is a structureless array of the data collected. Seriation seeks to establish any structure, usually gradients, within the data by maximizing entries in the leading diagonal. The seriated matrix reveals a stratigraphical sequence of *Eocoelia* species which is widely used for the correlation of lower Silurian strata. Most seriations are applied to much larger data matrices where any structure, if present, is initially far from obvious.

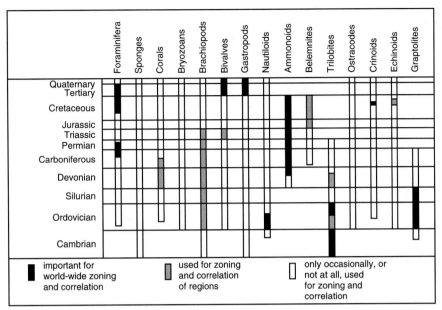

Fig 2.8 Approximate stratigraphical distribution through time of the main biostratigraphically useful invertebrate fossil groups.

remains. Microfossil groups, such as the conodonts, foraminiferans and plant spores, are now widely used, particularly in geological exploration (see Chapter 11). Microfossils are ideal zone fossils since they are usually common in small samples, such as drill cores and chippings, of most sedimentary lithologies, and many groups are widespread and rapidly evolving. The only drawback is that some techniques used in their extraction from rocks and sediments are rather specialized.

Palaeoecology

Palaeoecology is the study of the life and times of fossil organisms. This relatively new discipline is concerned with the lifestyles of individuals together with their relationships to each other and their surrounding environment. More recently, palaeoecology has developed a much wider importance: ecological data through time form the basis for models of the planet's evolving ecosystems. James Hutton suggested over two centuries ago that Earth itself can be modelled as a superorganism, and much more recently James Lovelock has described the planet as a living organism, Gaia, capable of regulating its environment through a careful balance of biological, chemical and physical processes.

There are, however, a number of constraints on virtually all palaeoecological investigations. Fossil animals and plants commonly are not preserved in their life environments; soft parts and soft-bodied organisms are usually removed by scavengers, whereas hard parts may have been transported elsewhere or eroded during exposure (see Chapter 1). Moreover, some environments are more likely to be preserved than others. Most importantly, palaeoecological studies must have a reliable and sound taxonomic basis. Although much palaeoecological deduction is based on uniformitarianism, some environments have changed through geological time, as have the lifestyles and habitats of many organisms. Nevertheless, a few basic principles hold true. Organisms are adapted for, and limited to, a particular environment however broad or restricted; moreover most are adapted for a particular lifestyle and all have some form of direct or indirect dependence on other organisms.

There are two main areas of palaeoecology: palaeoautecology is the study of the ecology of a single organism, whereas palaeosynecology looks at communities or associations of organisms. For example, autecology may investigate the detailed functions and life of a coral species but synecology will analyze the growth and structure of an entire coral reef, including the mutual relationships between species and their relationships to the surrounding environment. The autecology of individual groups is discussed in the taxonomic chapters. In most studies the functions of fossil animals or plants are established through analogies or homologies with living organisms or structures or by a series of experimental and modelling techniques. Geological evidence, however, remains the main test of these comparisons and models.

Taphonomic constraints

The vast majority of fossil assemblages have been intensely modified by taphonomic processes. The degradation of animal and plant communities after death results in the loss of soft-bodied organisms, while decay removes soft tissue with the disintegration of multi-plated skeletal taxa (see Chapter 1); moreover, transport and compaction add to the overall loss of information during the processes of fossilization. Nevertheless, areas occupied by dead communities may be recolonized and animal and plant debris may be supplemented by material washed in from elsewhere. These processes of time-averaging can thus artificially enhance the diversity of an assemblage.

In a series of detailed studies of the living and dead faunas of Copana Bay and the Laguna Madre along the Texas coast, Staff *et al.* (1986) have discussed the palaeoecological significance of the taphonomy of a variety of nearshore communities. Most taxa in living communities are not usually preserved; nevertheless the majority of taxa with a preservation potential (mainly shelled organisms) would in fact be fossilized. More taxa were actually located in death rather than living assemblages, where the effects of time-averaging were clearly significant. Suspension feeders and infaunal organisms were the most likely to be preserved. Measurements of biomass rather than those of numerical abundance and diversity are the best estimates of the composition of communities, and counts of the more stable adult populations are the most realistic monitors of community structure (Figure 2.9).

Another method to estimate taphonomic loss involves a census of an extraordinarily preserved Lagerstätte deposit. Whittington's (1980) detailed reinvestigation of the mid Cambrian Burgess Shale fauna revealed a community dominated by soft-bodied animals with very few of the more familiar skeletal components of post-Cambrian faunas such as brachiopods, bryozoans, gastropods, bivalves, cephalopods, corals and echinoderms. More importantly, the deep-water Burgess fauna is quite different from more typical coeval assemblages with phosphatic brachiopods and trilobites. Although the

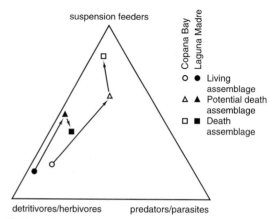

Fig 2.9 The transition from a living assemblage to a death assemblage. Relative proportions of different types of organism change in two living marine assemblages off the Texan coast. Living assemblages are dominated numerically by detritivores and herbivores, death assemblages by suspension feeders.

Burgess fauna has many other peculiarities (see Chapter 8), the high proportion of, for example, annelid and priapulid worms adds a different dimension to the more typical reconstructions of mid Cambrian communities (Figure 2.10).

These important taphonomic constraints must be addressed and built into any palaeoecological analysis and may be partly countered by a careful selection of sampling methods. A variety of methods involving the study of size–frequency histograms (see Chapter 1), the

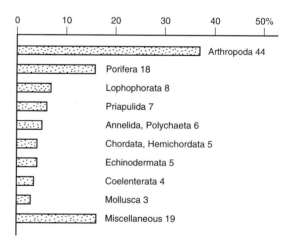

Fig 2.10 Census of organisms preserved in the Middle Cambrian Burgess Shale. Many groups, such as the priapulid and annelid worms, together with the diverse arthropod biota, are rarely represented in more typical mid Cambrian faunas, dominated by phosphatic brachiopods and trilobites.

degree of breakage, disarticulation and fragmentation of individuals, together with the attitude of fossils in sediments, generate useful criteria for the identification of autochthonous assemblages.

Habitats and niches

All modern and fossil organisms can be classified in terms of where they live (their habitat) or with reference to their lifestyle (their niche). Modern organisms occupy a range of environments from the top of Mount Everest at heights of nearly 9 km to depths of over 10 km in the Marianas Trench. Nearly fifty physical, chemical and biological factors may characterize an organism's environment and control its ecological niche; unfortunately, few can be recognized in the fossil record.

Some of the most abundant and diverse communities inhabit the littoral zone, where rocky shores hold some of the most varied and extensively studied faunas. For example, nearly 2000 individual organisms have been recorded from a 250 mm^2 quadrat on an exposed wave-battered platform around the Scottish island of Oronsay. Unfortunately, few rocky coasts have been recorded from the geological record, although Johnson (1988) has reviewed this environment through geological time, commenting on its neglect by palaeontologists.

The majority of animal fossils have been found in marine sediments, representing a wide range of depths and conditions. The distribution of the marine benthos is controlled principally by depth of water, oxygenation and temperature. The main depth zones and pelagic environments are illustrated in Figure 2.12. In addition, the photic zone is the water depth penetrated by light; this can vary according to water purity and salinity but in optimum conditions it can extend down to about 100 m. Terrestrial environments are mainly governed by humidity and temperature; organisms inhabit a wide range of continental environments, ranging from the Arctic tundras to the lush forests of the tropics.

Marine environments host a variety of lifestyles (Figure 2.11). The upper surface waters are populated with floating plankton whereas nektonic organisms swim at various levels in the water column. Within the benthos, mobile nektobenthos move across the sea-floor and the fixed or sessile benthos are attached by a variety of structures; infaunal organisms live beneath the sediment–water interface, while epifauna live above it.

Members of most communities are involved in some form of competition for food, light and space resources. For example, the stratification of tropical rain forests reflects competition in the upper canopy for light, while vegetation adapted for damp, darker conditions is developed at lower levels. Similar stratification or tiering is a feature of most marine communities, and the tiering has become higher and more sophisticated through geological time (Figure 2.13). Low-level tiers were typically occupied by brachiopods during the Palaeozoic while the higher tiers were occupied by crinoids. The Mesozoic and Cenozoic faunas, however, are more molluscan-based, with the lower tiers occupied by bivalves and gastropods.

An infaunal tiering structure has been discovered fairly recently, evident in trace fossil associations (see Chapter 12). During the early Palaeozoic, burrowing reached depths of 10–400 mm, but depths greater than 1 m were commonly penetrated by many Mesozoic and Cenozoic organisms as they sought to escape predation and exploit ever deeper sources of food and nutrients.

Trophic structures

Food pyramids form the basis of most ecological systems, defining the energy flow through a chain of different organisms from extremely abundant primary producers to relatively few predators. Marine food chains have been documented, including those dominated by suspension feeders such as brachiopods, bryozoans and sponges. These fed mainly on phytoplankton and other organic detritus. Suspension feeding was particularly common in Palaeozoic benthos; the Mesozoic and Cenozoic faunas were more dominated by detritus feeders.

Case histories

Food chains have been described for a number of fossil assemblages.

One of the most spectacular fossil lake deposits, dominated by amphibians, is found in the Upper Carboniferous of the Czech Republic (Figure 2.14a). The lacustrine ecosystem recreated for the inhabitants of the Nýřany Lake complex has three main ecological associations: an open-water lacustrine association, dominated by fishes together with anthracosauroid and loxommatid amphibians; the shallow-water and swamp–lake association with amphibians, small fishes, land plants and other plant debris; and finally the terrestrial–marginal association with microsaur amphibians and primitive reptiles. The trophic structure has been reconstructed in some detail for these palaeocommunities. For example, in the open-water environments fishes, such as the spiny acanthodians, fed on plankton but were themselves attacked

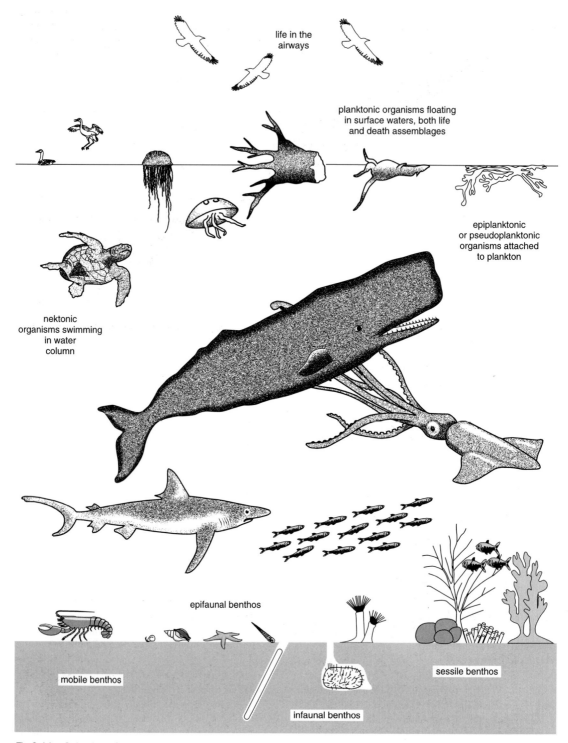

life in the airways

planktonic organisms floating in surface waters, both life and death assemblages

epiplanktonic or pseudoplanktonic organisms attached to plankton

nektonic organisms swimming in water column

epifaunal benthos

mobile benthos

sessile benthos

infaunal benthos

Fig 2.11 Selection of marine lifestyles above, at the surface, within and at the base of the water column.

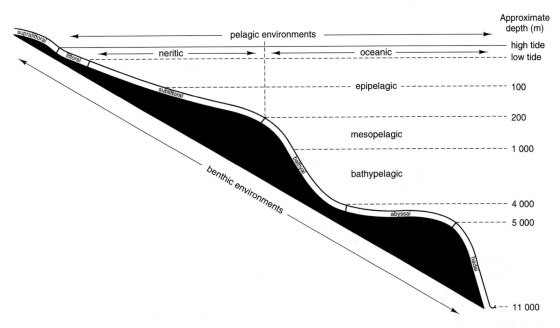

Fig 2.12   Review of modern marine environments and their depth ranges together with the approximate positions of the main benthic zones.

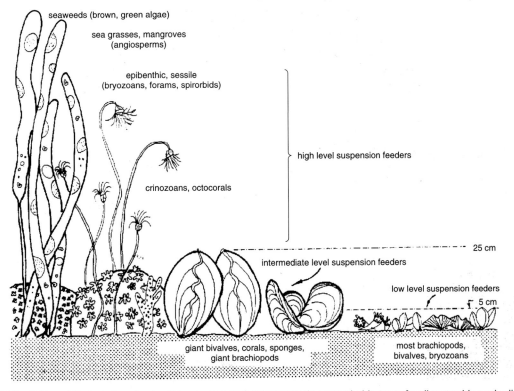

Fig 2.13   Epifaunal tiering of marine benthic communities; infaunal tiering recorded in trace fossil assemblages is discussed in Chapter 12.

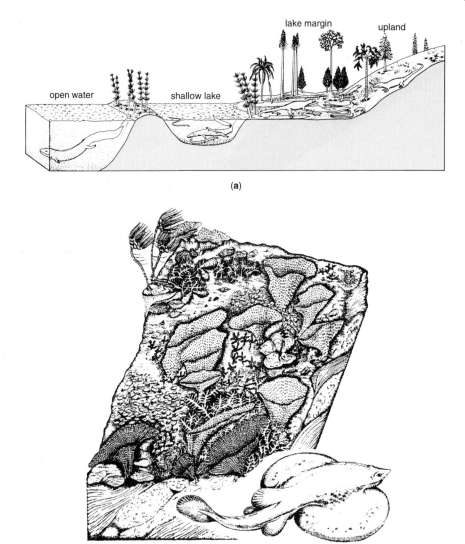

Fig 2.14   (a) Trophic structures in and around a late Carboniferous lake complex, Nýřany, Czech Republic. (b) Trophic structures in a late Permian reef complex, north-east England.

by the amphibians, presumably at the top of the food chain. In the associated terrestrial environments, plant material was consumed by a variety of invertebrates, including insects, millipedes, spiders, snails and worms; these provided food and nutrients for a range of amphibians, themselves prey for larger animals, mainly amphibians and reptiles.

Marine life in the Carboniferous was diverse, dominated by brachiopods, bryozoans, corals and crinoids together with the occasional shark. The marine ecosystem continued to flourish through the Permian with diverse and spectacular faunas in the Glass Mountains of

Texas and the Salt Ranges of Pakistan. The Zechstein Reef facies of northern Europe provides an important view of latest Palaeozoic marine communities and food chains prior to the end-Permian extinction event. A variety of late Permian marine communities occurs in the marginal Zechstein facies of north-east England (Figure 2.14b). The benthos was dominated by diverse associations of brachiopods overshadowed by the higher tiers of fan and vase-shaped bryozoans. Both groups were sessile filter feeders. Stalked echinoderms were rarer and occupied the highest tiers. Molluscs such as bivalves and gastropods were important deposit feeders and grazers.

One of the largest predators was *Janassa*, a benthic ray, equipped with a formidable battery of teeth capable of crushing the shells of the sedentary benthos.

Palaeocommunities

Palaeocommunities are recurrent groups of organisms related to some specific set of environmental conditions. Many of the concepts and techniques applied to marine fossil communities are based on the work of biologists such as Petersen, researching at the turn of the century. Petersen and his colleagues established a series of level-bottom benthic communities around the Scandinavian coasts; the major control on community distribution was water depth, although other factors such as the substrate were also influential. A few pioneer studies in the 1930s applied this approach to Carboniferous assemblages. In the mid 1960s, however, a series of six depth-related, mainly brachiopod-dominated communities was established in the lower Silurian rocks of Wales and the Welsh Borderlands (see also Chapter 6). These communities stretched from the intertidal zone to the deep shelf and continental slope. More recent studies, for example by Boucot (1975), have rationalized the scheme as a set of more widely applicable benthic assemblage zones (Figure 2.15). These zones are now defined on a wide range of faunal and sedimentological criteria and may be subdivided, internally, on the basis of, for example, substrate type and the degree of turbulence.

Describing the composition of palaeocommunities requires care and precision. The detailed faunal lists of older studies tended to give unequal weight to rarer species since all taxa were listed equally. More recent studies chart the absolute and relative abundance of each taxon, illustrated graphically with frequency histograms, and based on data derived from transects, quadrats or more commonly from bulk samples. The basic data can then be transformed into a more realistic picture of ancient communities populating past landscapes and seascapes.

Many statistical techniques have been used to analyse palaeocommunities and their distributions. Phenetic methods (see also Chapter 3) are based on the investigation of a similarity or distance matrix derived from a raw data matrix of the presence or absence or numerical abundance of taxa at each site. Cluster analysis is most commonly used in ecological studies and there is a wide range of both distance and similarity measures, together with clustering techniques, to choose from. R-mode analysis clusters the variables, in most palaeoecological studies the taxa, whereas Q-mode analysis clusters the cases, usually the localities or assemblages.

Communities and habitats through time

Most palaeoecological studies attempt to recreate the atmosphere and dynamism of past communities. Despite the significant loss of information during taphonomic processes, realistic reconstructions are possible, depicting the main components, their relationships to each other, and to the surrounding environment. McKerrow (1978) has reviewed, in broad terms, the development of communities throughout the Phanerozoic (see also Chapter 13). During the last 600 Ma both animals and plants, together with their communities, expanded and diversified. In marine environments increases in the height, complexity and stratification of benthic tiering were later matched by increases in the depth and sophistication of infaunal tiering as, particularly in Mesozoic and Cenozoic faunas, many more organisms adopted burrowing lifestyles and the benthos switched from filter- to deposit-feeding with significantly more predators. Terrestrial environments, initially dominated by small green plants, various arthropods and snails, together with diverse amphibian faunas in the mid–late Palaeozoic, changed significantly during the Mesozoic with the diversification of vegetation and eventually flowering plants, the rise of the dinosaurs and a huge radiation of various arthropod groups including the insects.

Case histories

Early Jurassic environments provide a wide range of communities and habitats showing the early stages of development of post-Palaeozoic faunas. A selection demonstrating environments, life modes and trophic strategies is illustrated in Figure 2.17. The Newark Supergroup and the Posidonia Shales provide important windows on life in continental and marine environments, respectively, during the early part of the Jurassic.

Major new finds in the Newark Supergroup and equivalent strata in eastern North America, have created a vivid picture of life on late Triassic and early Jurassic arid to humid landscapes of Laurentia, swept by occasional monsoons. Olsen *et al.* (1978, 1987) have described diverse dinosaur communities of both large and small carnivorous theropods, at the top of the food chain, together with large herbivorous sauropods and some early armoured forms. Most of the terrestrial tetrapods are preserved in volcaniclastic deposits, although adjacent fluviatile facies contain crocodiles. Lake facies have preserved diverse floras of conifers, cycads, ferns and lycopods. Fast-swimming holostean fishes patrolled the lakes and abundant insects of modern aspect, representing seven

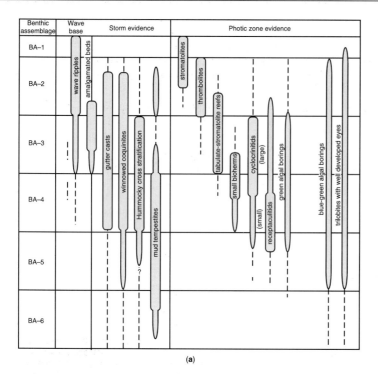

(a)

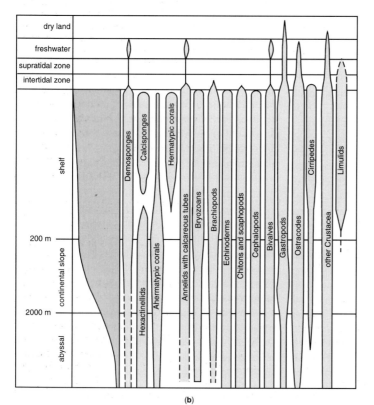

(b)

Fig 2.15 (a) Silurian marine benthic assemblages zones and identifying criteria. (b) Distributions of living marine organisms across depth gradient.

Box 2.3 Ecological statistics

A range of statistics has been used to describe aspects of fossil communities. Although the number of species collected from an assemblage provides a rough guide to the diversity of the association, obviously in most cases the larger the sample, the higher the diversity. Diversity measures are usually standardized against the sample size. Dominance measures have high values for communities with a few abundant elements and low values where species are more or less evenly represented; measures of evenness are usually the inverse of dominance. Examples include:

$$\text{Margalef Diversity} = S\text{-}1/\log N$$
$$\text{Dominance} = \Sigma \, (ni/N)^2$$
$$\text{Evenness} = 1/\Sigma \, (pi)^2$$

where S is the number of species, N is the number of specimens, ni is the number of the ith species, and pi is the relative frequency of ith species.

It is often difficult to assess the adequacy of a palaeoecological sample. Some authorities have suggested that samples of about 300 give a fairly accurate census of a fossil assemblage. Commonly, investigators plot rarefaction curves (Figure 2.16). These are produced simply by collecting samples of ten and identifying the number of species in each. For each sample of ten plotted along the x-axis, the cumulative number of species is plotted along the y-axis. The curve may level off at the point where no additional species are identified with additional collecting and this fixes the sample size adequate to count the majority of species present.

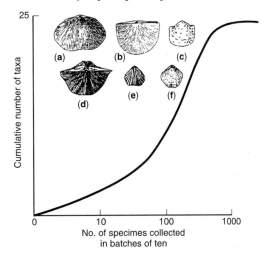

Fig 2.16 Construction of a rarefaction curve based on data collected from a mid Devonian brachiopod-dominated fauna, northern France. The main types of brachiopods are illustrated (a) *Schizophoria*; (b) *Douvillina*; (c) *Productella*; (d) *Cyrtospirifer*; (e) *Rhipidiorhynchus*; (f) *Athyris*: all approximately ×0.5 magnification. The curve levels off at about 300 specimens, suggesting this sample size is a sufficient census of the fauna.

Box 2.4 Ecological interactions

Animals and plants have participated in a wide range of relationships throughout geological time. Ecologists have classified these arrangements in terms of gain (+), loss (−) and neutrality (0). Antagonistic arrangements include antibiosis (−,0), exploitation (−,+) and competition (0,0) whereas symbiosis involves both commensalism (+,0) and mutualism (+,+). Antibiosis is difficult to demonstrate although mass mortalities of fishes (−) have been ascribed to dinoflagellate blooms (0); some palaeontologists believe the twisted skeleton of a late Cretaceous *Struthiomimus* (−) from Alberta may have died from strychnine poison (0).

Exploitation includes predation and parasitism. There are many records of bite marks, particularly by marine reptiles (+) on mollusc shells (−), while the stomach contents of Jurassic ichthyosaurs (+) have revealed a diet of fish (−). Moreover, a wide variety of nibble marks have been reported from fossil leaves. The relationship between the Devonian tabulate coral *Pleurodictyum* (−) and the worm *Hicetes* (+) fooled many palaeontologists.

Was this a bizarre compound organism? In fact the worm was probably a parasite; the association is common throughout the Rhenish magnafacies of Europe and virtually every specimen known of *Pleurodictyum* has a parasitic worm at its core. Others have suggested the relationship was mutualistic, with the worm dragging the coral around on its back.

Commensalism is one of the most common relationships apparent in the fossil record, where small epifauna use larger organisms for attachment and support. Small and immature productoid brachiopods (+) are often attached by clasping spines to crinoid stems (0), while *Spirobis* worms (+) are commonly attached near the exhalent currents of Carboniferous nonmarine bivalves (0). Some of the most spectacular examples have been reported from the shells of Devonian spiriferide brachiopods; Ager (see Chapter 6) reported a succession of epifauna (+), commencing with *Spirobis* followed by *Hederella* and *Paleschara* to finally the tabulate coral *Aulopora*, clustered near the inhalent current of the brachiopod (0).

(a)

10 cm

(b)

10 cm

(c)

10 cm

(d)

10 cm

(e)

10 cm

(f)

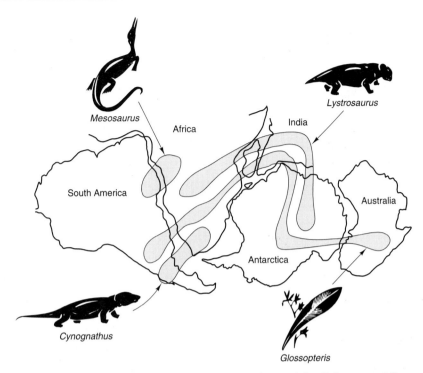

Fig 2.18 The tight fit of Gondwana; the correspondence of fossil faunas and floras across the southern continents suggested to Wegener and others, that South America, Africa, India, Antarctica and Australia had drifted apart since the Permo-Trias.

orders, populated the forests and shores or may have swum in the shallows together with crustaceans.

The Posidonia Shales crop out near the village of Holzmaden in the Swabian Alps. The shales are bituminous and packed with fossils, generally with echinoderms and vertebrates towards the base and cephalopods at the top. Seilacher *et al.* (1985) and his colleagues have recently reviewed the taphonomic aspects of this Lagerstätte and have emphasized the unique features of this stagnation deposit. Benthos was rare; however, encrusting and recumbent brachiopods, bivalves, crinoids and serpulids pursued a pseudoplanktonic life mode attached to a variety of hosts including ammonite shells, belemnite guards and driftwood. The dominant animals were nektonic ammonites and coleoids together with the superbly preserved ichthyosaurs and plesiosaurs, now displayed in many European museums. Some horizons are characterized by monotypic assemblages of small taxa such as diademoid echinoids and byssate bivalves, like *Posidonia* itself. These benthic colonization may have been promoted by storms providing more oxygenated conditions for short periods of time.

Palaeobiogeography

All living organisms have a defined geographical range; the ranges may be large or small, controlled by a variety of factors including mainly climate and physical barriers. By the middle of the 19th century both Charles Darwin and Alfred Wallace had recognized the reality of biogeographical provinces in their respective studies on the Galapagos Islands and in the East Indies. Discrete biogeographical provinces are defined by faunal and floral boundaries. Provinces are characterized by endemic or stenotopic taxa which have restricted ranges, in contrast to the more widespread cosmopolitan or eurytopic taxa. Continental configurations and positions have changed through time as have faunal and floral provinces. Faunal and floral provinces have changed through time as a consequence of moving continents; consequently palaeontological evidence has been valuable in charting the drift of wandering continents (Figure 2.18).

Fig 2.17 A cocktail of Jurassic environments. Early Jurassic: (a) sand, (b) muddy sand and (c) bituminous mud communities. Late Jurassic: (d) mud, (e) reef and (f) lagoonal communities.

Faunal and floral barriers

The origin and development of barriers can immediately fragment an existing province, with the propagation, eventually, of new biogeographically distinctive units with endemics. For example, rifting and basin formation can split and isolate into fragments many existing terrestrial and fringing shelf provinces, whereas the same effects in the sea may be induced by the formation of an isthmus.

George Gaylord Simpson distinguished three situations: corridors were open at all times, filters allowed restricted access, whereas sweepstake routes opened only occasionally. In continental settings the barriers may be mountain ranges, inland seas or even rain forests. Marine faunas may be separated by wide expanses of deep ocean, swift ocean currents or land. In general terms the endemicity of most marine faunas decreases with depth; the more cosmopolitan faunas are located in deep-shelf and slope environments. But in the deeper basins, populated by specialized taxa, faunas are again endemic.

In some situations, the development of a barrier for some organisms may provide a corridor for others. The emergence of the Isthmus of Panama has connected North and South America; however, it has isolated the Atlantic and Pacific oceans. South America was essentially isolated from North America for most of the past 70 million years, and was dominated by diverse specialized mammalian faunas consisting of marsupials, edentates, unique ungulates, and rodents. However, 3 million years ago the emergence of the Isthmus of Panama provided a land-bridge or corridor between the two continents and many terrestrial and freshwater taxa were free to move across the isthmus (Figure 2.19); the Great American Biotic Interchange (GABI) allowed the North American fauna to invade the south and essentially wipe out many of the continent's distinctive mammalian populations. South American mammals were equally successful in the north and some (such as the armadillo, opposum and porcupine) still survive in North America.

On the other hand, the emergence of the isthmus also promoted changes in the marine faunas of the Caribbean Sea. Jackson *et al.* (1993) documented marked increases in the diversity of Caribbean molluscs contrary to the extinctions expected in these faunas. The emergence of the terrestrial land-bridge and marine barrier initiated the upwelling of nutrients in the Caribbean area, with an increase in species diversity throughout the region. Valentine (1973) drew attention to a range of plate tectonic settings, including the evolution of spreading ridges, island arcs, subduction and transform zones, that may influence biological distributions. It is now recognized that in most cases provinciality develops through continental fragmentation and the controls of climatic or thermal gradients.

Island biogeography

Modern oceans are littered with islands. Most are transitory volcanic chains, developed above moving hot spots or at mid-oceanic ridges that will probably be subducted; some, however, are chunks of allochthonous continental crust often associated with volcanic arc systems. These lighter crustal complexes are usually amalgamated into mountain chains at active plate margins. The biogeography of islands is complex and has only recently been applied to ancient examples. Islands and archipelagos play a number of biological roles. Most islands are sufficiently isolated to act as sites for allopatric speciation through the Founder Principle. Moreover, island chains play an important part in migrations, acting as stepping stones. Finally, moving island complexes can allow cross-latitude transfer of organisms through time, acting as Noah's arks. In the longer term these complexes may function as 'Viking funeral ships' transporting exotic fossil assemblages to new locations.

Island biotas are often diverse, with many endemics and commonly a mixture of provincial signatures. Most oceanic islands are short-lived but nevertheless Pacific islands like Aldabra developed their own distinctive faunas and floras over relatively short periods of time. Unfortunately, as noted above, these assemblages end up in the world's orogenic belts where the rocks have been metamorphosed and tectonized. Many of these mountain belts form remote areas where few fossils have been preserved. Although these assemblages are potentially the most interesting, they are also the most difficult to study.

Geological and palaeontological implications

Despite the disbelief voiced over early palaeontology-based models for continental drift, modern plate tectonic and palaeogeographical reconstructions now rely heavily on biological data to both suggest and test plate distributions. Terrane models recognize that orogenic zones consist of a collage of tectonic units with unique geological histories mutually separated by tectonic structures such as faults. Virtually all orogens can now be interpreted in this way. One of the most famous is the North American Cordillera, although fossil data have been used to reconstruct more ancient ocean systems such as the early Palaeozoic Iapetus Ocean (Figure 2.20).

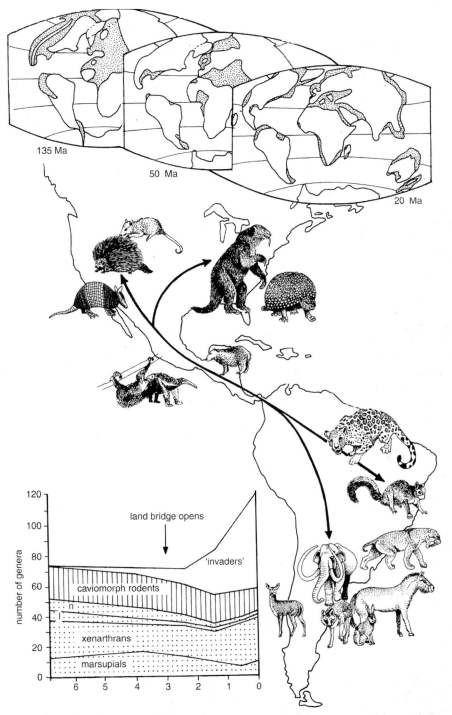

Fig 2.19 The emergence of the Isthmus of Panama promoting the Great American Biotic Interchange (GABI) between the North and South American terrestrial vertebrates together with the radiation of the marine benthos of the Caribbean Sea. Abbreviations: l, litopterns; n, notoungulates.

Two main types of biogeographic analysis are widely used, based on either phenetic or cladistic methods. Cladistic methods are based on the assumption that an original province has since fragmented with the creation of subprovinces characterized by new endemics, essentially analogous to apomorphies in taxonomic cladistics (see Chapter 3). The phenetic methods usually start from a similarity matrix between sites based on the presence and absence of taxa or more rarely the relative abundance of organisms across the sites (see also Chapter 3). There are a large number of distance and similarity measures to choose from. A few of the commoner coefficients are as follows:

$$\text{Dice coefficient} = 2A/2A+B+C$$
$$\text{Jaccard coefficient} = A/A+B+C$$
$$\text{Simple Matching coefficient} = A+D/A+B+C+D$$
$$\text{Simpson coefficient} = A/A+E$$

where A is the number of taxa common to any two samples, B is the number in sample 1, C is the number in sample 2, D is the number of taxa absent from both samples, and E is the smaller value of B or C.

On the basis of an intersite similarity or distance matrix, a dendrogram can be constructed linking first the sites with the highest similarities or the closest distances. When the distance or similarity matrix is recalculated to take into account the first clusters, additional sites or genera are clustered until all the data points are included in the dendrogram. Clearly the first clusters have the greatest significance and less importance is usually attached to later linkages.

Biogeography and climatic gradients are related to patterns of changing biodiversity. In broad terms, low latitudes support high-diversity faunas; biodiversity decreases away from the tropics towards the poles. Studies on modern bivalve, bryozoan, coral and foram faunas show marked increases in diversity towards the equator, and since many cool-water species breed later in life, individuals may be larger than their tropical counterparts. Diversity gradients established across early Jurassic ammonoid faunas have also aided reconstructions (Smith, 1990); Tethyan faunas, occupying tropical belts, were generally more diverse than counterparts at higher latitudes.

Many authors have suggested that changing plate configurations, oscillating between fragmentation and integration, have affected biodiversity through time

(Smith, 1990). For example, the huge early Ordovician radiation of marine skeletal faunas may be related to the break-up of Gondwana while the end-Permian extinction event coincides with the development of Pangaea; more recent diversifications have occurred during the late Mesozoic fragmentation of this supercontinent (Figure 2.21).

Fossils in fold belts

Fossils are difficult to study in mountain belt regions since they are usually poorly preserved, metamorphosed and tectonized; fossils in these orogenic zones are also rare and difficult to collect from often hazardous terrains. Fossils, however, can provide key age and geographic data for geologists disentangling these complex zones of the Earth's crust. Fossils in thrust belts helped recognize large-scale horizontal movements in the Swiss Alps, the Northwest Highlands of Scotland and in the Scandinavian Caledonides over a century ago. Not surprisingly, the eminent Alpine geologist Rudolf Trümpy once stated 'One bad fossil is worth a good working hypothesis'.

Fossil data have provided critical age constraints in many mountain belts; fossils cannot be reset by later thermal and tectonic events like radiometric clocks. Charles Lapworth's interpretation of the complex structure and stratigraphy of the Southern Uplands of Scotland, over a century ago, was based on sequence recognition of graptolite faunas. More recently in central Scotland reliable early Ordovician dates from the Highland Border Complex, previously included as part of the mainly Precambrian Dalradian Supergroup, suggest that these rocks were deposited in one of a series of basins along the margin of North America, quite separate from the Dalradian.

Fossils have also been critical in distinguishing exotic blocks of crust. Seaward of cratonic margins, oceanic terranes evolved unconstrained and are often termed suspect. Faunal and floral data have helped fingerprint terranes, while their changing provincial signals can track their movements across latitudes. The North American Cordillera contains a number of suspect terranes, identified by their faunal signatures (see above). In the older Caledonides, Harper and Parkes (1989) have described terranes located north of the Iapetus suture containing typically North American faunas, whereas terranes far to the south of the suture contain different faunas which developed marginal to the microcontinent of Avalonia. Some smaller terranes in central

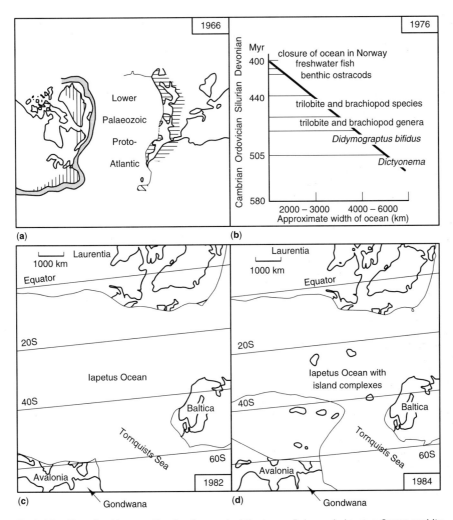

Fig 2.20 Changing ideas on the development of the Lower Palaeozoic Iapetus Ocean and its faunas: (a), (c) and (d) are palaeogeographic reconstructions; (b) indicates the mobility of organisms as the ocean closed.

Ireland almost certainly evolved within the Iapetus Ocean itself, with their own distinctive, often endemic, faunas. All these terranes are bounded by faults; the fossil evidence suggests there have been considerable displacements between the blocks since quite different faunas are now juxtaposed.

For many years structural geologists have used the deformed shapes of initially symmetrical fossils to determine the amount of finite strain recorded in tectonized rocks. For example, different shapes of a single genus of mid Devonian brachiopod, the Delabole butterfly, a distinctive wide-hinged spiriferide from Delabole town in Devon, commonly occur on single bedding planes; this variation convinced early workers

that a number of separate species of spiriferide was present. But the effects were caused by tectonic strain of a randomly orientated population (see also Chapter 1). Nevertheless useful palaeontological information can still be obtained from deformed fossils. Early this century, seven species of trilobite from the Kashmir region of the Himalayas were described and placed in three different superfamilies. Hughes and Jell (1992) restudied and unstrained the trilobites to reveal only one morphospecies bent and stretched into a wild variety of shapes. Moreover this single species indicates a precise mid Cambrian age for the fauna and firmly establishes its affinities with those from India and northern China.

Burial and tectonic activity generate heating and thermal gradients in the crust; during these processes some of the minerals in fossil skeletons may be altered. A number of groups of microfossils change colour with changing temperature (see also Chapter 11), functioning as palaeothermometers (Figure 2.22). The upper end of these thermally-induced colour ranges has even allowed geologists to map metamorphic zones in orogenic belts (see Box 2.6).

Box 2.6 Thermal maturation

Conodonts (see Chapter 11) in particular are useful thermal indicators. They change colour from grey to white, and eventually translucent, on a scale of Conodont Alteration Indices (CAI values) from 1 to 8, through a temperature range of about 60–600 °C. A range of chitinous organisms, including the graptolites, also show colour changes, as does vitrinite derived from plant material. These changes have also been documented in detail for acritarchs, where Acritarch Alteration Indices (AAI values) range from 1 to 5. Spores and pollen have spore colour indices (SCI values) ranging from 1 to 10, with colours ranging from colourless to pale yellow through to black. Other groups such as phosphatic microbrachiopods and chitinozoans show similar prospects, but their colour changes have yet to be calibrated with precise palaeotemperatures in detail.

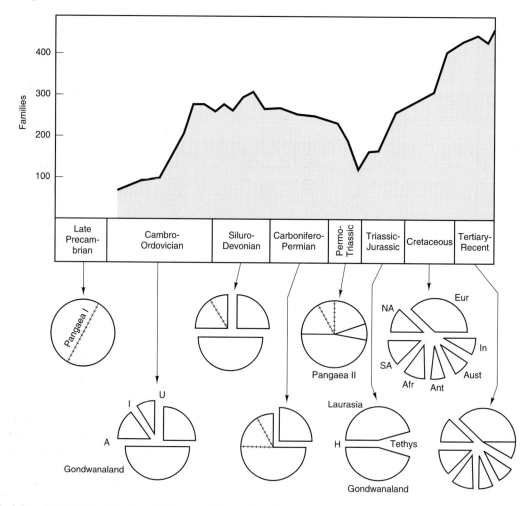

Fig 2.21 Relationship of continental fragmentation to changing global biodiversity. A, pre-Appalachian-Variscan Ocean; H, Hispanic Corridor; I, Iapetus Ocean; U, pre-Uralian Ocean.

CAI	Colour	Palaeotemperature °C	Mean temperature °C	Vitrinite reflectance	Palynomorph Trans index	Thermal alteration index approx.	Metamorphic grade	Metamorphic zones
1	pale yellow	50–80	65	0.8	1–5	1.5 2.0		
1.5	very pale brown	50–90	70	0.7–0.85	5-5ur		oil and gas window	Diagenetic zone
2	brown to dark brown	60–140	100	0.85–1.3	5–6	2.5		
2.5		85–180	135					
3	very dark grey brown	110–200	160	1.4–1.95	5ur-6	2.7 3.2		
3.5		150–260	205			3.5	dry gas	
4	light black	190–300	245	1.95–3.6	6	4.0		
4.5		230–340	285			5.0		
5	dense black	300–400	330	3.6	6ur-7	lower greenschist chlorite/muscovite		Anchizone
5.5	dark grey black	310–420	365					
6	grey	350–435	400			greenschist meta-argillite		Epizone
6.5	grey-white	425–500	460					
7	opaque white	480–610	550			upper greenschist biotite-garnet		
7.5	semi-translucent	>530						Mesozone
8	crystal clear	>600				garnet		

Fig 2.22   Colour change with temperature in microfossils.

Further reading

Ager, D. V. (1963) *Principles of paleoecology*. McGraw-Hill, New York.

Ager, D. V. (1993) *The nature of the stratigraphical record*, 3rd edition. John Wiley and Sons, Chichester.

Benton, M. J. (ed.) (1993) *Fossil Record 2*. Chapman and Hall, London.

Briggs, D. E. G and Crowther, P. R. (eds) (1990) *Palaeobiology – a synthesis*. Blackwell Scientific Publications, Oxford.

Bruton, D. L. and Harper, D. A. T. (eds) (1992) Fossils in fold belts. *Terra Nova*, **4**, special issue.

Copper, P. (1988) Paleoecology: paleoecosystems, paleocommunities. *Geoscience Canada* **15**, 199–208.

Cox, B. C. and Moore, P. D. (1993) *Biogeography. An ecological and evolutionary approach*. Blackwell Scientific Publications, Oxford.

Dodd, J. R. and Stanton, R .J. (1990) *Paleoecology, concepts and applications*, 2nd edition. Wiley, New York.

Goldring, R. (1991). *Fossils in the field. Information potential and analysis*. Longman Scientific and Technical, London.

References

Boucot, A. J. (1975) *Evolution and extinction rate controls*. Elsevier, Amsterdam.

Harper, D. A. T. and Parkes, M. A. (1989) Palaeontological constraints on the definition and development of Irish Caledonide terranes. *Journal of the Geological Society of London*, **146**, 413–415.

Hughes, N. C. and Jell, P. A. (1992) A statistical/computer-graphic technique for assessing variation in tectonically deformed fossils and its application to Cambrian trilobites from Kashmir. *Lethaia*, **25**, 317–33.

Jackson, J. B. C., Jung, P., Coates, A. G. and Collins, L. S. (1993) Diversity and extinction of tropical American mollusks and emergence of the Isthmus of Panama. *Science*, **260**, 1624–1626.

Johnson, M. E. (1988) Why are ancient rocky shores so uncommon? *Journal of Geology*, **96**, 469–480.

McKerrow, W. S. (ed.) (1978) *The ecology of fossils*. Duckworth, London.

Olsen, P. E., Remington, C. L., Cornet, B. and Thomson, K. S. (1978) Cyclic change in Late Triassic lacustrine communities. *Science*, **201**, 729–732.

Olsen, P. E., Shubin, N. H. and Anders, M. H. (1987) New Early Jurassic tetrapod assemblages constrain Triassic–Jurassic tetrapod extinction event. *Science*, **237**, 1025–1028.

Seilacher, A., Reif, W.-E. and Westphal, F. (1985) Sedimentological, ecological and temporal patterns of fossil Lagerstätten. *Philosophical Transactions of the Royal Society*, **B311**, 5–23.

Smith, P. L. (1990) Paleobiogeography and plate tectonics. *Geoscience Canada*, **15**, 261–279.

Staff, G. M., Stanton, R. J. Jr, Powell, E. N. and Cummins, H. (1986). Time-averaging, taphonomy, and their impact on paleocommunity reconstruction: death assemblages in Texas bays. *Bulletin of the Geological Society of America*, **97**, 428–443.

Valentine, J. W. (1973) *Evolutionary paleoecology of the marine biosphere*. Prentice-Hall, New Jersey.

Whittington, H. B. (1980) The significance of the fauna of the Burgess Shale, Middle Cambrian, British Columbia. *Proceedings of the Geologists' Association*, **91**, 127–148.

3 Macroevolution

Key Points

- The evolution of life may be represented by a single branching phylogenetic tree.
- Cladistics is a method of reconstructing phylogeny based on the identification of shared derived characters (homologies).
- Phylogenies may also be reconstructed by stratophenetics, for groups with an excellent fossil record, and by comparisons of protein and DNA sequences of living organisms.
- Speciation often occurs by the establishment of a barrier, and the isolation of part of a previously interbreeding population.
- It is still not clear how much evolution takes place within species lineages (phyletic gradualism), and how much at the time of speciation (punctuated equilibrium).
- There may be a process of species selection, acting independently of natural selection, but examples have been hard to find.
- Fossil species may show allometry, changes in relative proportions during growth.
- The development of an organism may give some evidence about phylogeny.
- Changes in developmental rates and timing (heterochrony) may affect evolution.

Phylogeny first

Fossils offer fundamental information on the history of life and on large-scale patterns of evolution. The focus of a great deal of evolutionary research today is on living organisms, their ecology, behaviour, genetics and molecular biology, and this has given rise to many remarkable new insights. There has also been a revolution in the ways in which palaeontologists interpret the evolutionary aspects of the fossil record. New techniques and fresh evidence have revealed a great deal about the shape of the history of life (phylogeny), the nature of speciation and lineage evolution, the relationships between growth and evolution, and the nature of large-scale radiations and extinctions. These topics are often grouped together as aspects of macroevolution, i.e. evolution on a large scale (generally at, or above, the species level). The first three of these macroevolutionary topics will be described here, and the last (radiations and extinctions) will feature in the closing chapter of the book.

Darwinian evolution and phylogeny

Charles Darwin's *On the Origin of Species* (1859) is usually remembered as the book that made the case for natural selection as the mechanism of evolution. The book was equally about a major aspect of evolution that is crucial for the palaeontologist, namely phylogeny. Darwin's idea was that life had become diverse, consisting of perhaps millions of species, as a result of continued splitting of species from a common stem (Figure 3.1). Indeed, he proposed the idea that all of life, modern and ancient, could be followed back down the phylogenetic tree to a single point of origin: modern evidence confirms this remarkable insight.

Darwin's branching diagram also explained for the first time the meaning of the natural hierarchy of life that Linnaeus had discovered (see p. 4). This natural inclusive branching hierarchy is the basis of cladistics, a codified method of determining the shape of phylogenies, and of expressing such phylogenies as classifications (see Box 3.1). Cladistic methods rely on detailed analysis of the characters of organisms, and an estimation of

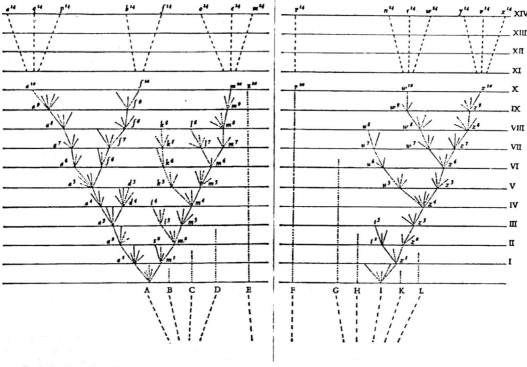

Fig 3.1   Branching diagram of phylogeny, the only illustration in Charles Darwin's *On the Origin of Species* (1859). It shows how two species, A and I, branch and radiate through time. The units I–XIV are time intervals of variable length, and the lower case letters (a, b, c) represent new species.

Box 3.1 Cladistics: reconstructing life's hierarchy

The key to the cladistic method is to carry out a detailed character analysis of the group of plants or animals of interest, and to seek characters that are phylogenetically informative. These are termed apomorphies, or derived characters. These are features that arose once only in evolution, and which, therefore, define all the descendants of that first organism to possess the new character.

Groups defined by apomorphies are said to be monophyletic (with a single common ancestor, and including all descendants of that common ancestor). Monophyletic groups, or clades, are distinguished from paraphyletic groups (with a single common ancestor, but not including all descendants) such as Reptilia, which excludes birds and mammals. Finally, polyphyletic groups are random assemblages of organisms that arose from more than one ancestor.

There are three steps in attempting to determine apomorphies. The first is to examine the definition of the characters. For example, detailed study of the feathers of birds as different as sparrows and ostriches shows that they share numerous similarities even in the finest details, and clearly are the same homologous structure. In evolutionary terms, feathers are a homology of all birds, and they are also an apomorphy that defines the Class Aves. Swimming limbs, on the other hand, differ from group to group (Figure 3.2): some show more than five fingers inside, others have the regular five; some show many finger joints, others a typical three or four. In detail, it can be shown that, anatomically, the paddles of ichthyosaurs and whales, plesiosaurs and seals, are not homologous; they are merely analogous. In evolutionary terms, analogues are structures that perform similar functions, and may look superficially similar, but which really had quite separate origins.

The second stage in testing whether shared characters are truly homologous, and appropriate at the level of testing, is outgroup comparison. In a study of the phylogeny of vertebrates, some of the apomorphies used to define the group Vertebrata might include possession of bone, possession of a skull and possession of a tail (Figure 3.3a). Anatomically, these are all rather complex features, and likely to be true homologues that arose once only in evolution: the question is, did they arise at the point of origin of Vertebrata (i.e. fishes plus the land vertebrates)? Outgroup comparison means searching for each of those three characters in the outgroup, i.e. all animals and plants that lie outside the ingroup, the group under study. Bone and skulls are present only in Vertebrata, and

hence define that group as apomorphies. A tail, on the other hand, is found in some other animals, such as sea squirt larvae and in amphioxus (see p. 196), and hence 'possession of a tail' becomes an apomorphy of a wider group, termed the Chordata (Figure 3.3b).

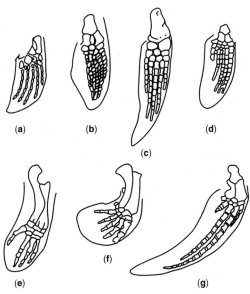

Fig 3.2   Swimming paddles of a variety of animals: forelimbs of: (a) *Archelon*, a Cretaceous marine turtle; (b) *Mixosaurus*, a Triassic ichthyosaur; (c) *Hydrothecrosaurus*, a Cretaceous plesiosaur; (d) *Plotosaurus*, a Cretaceous mosasaur; (e) *Dusisiren*, a Miocene seacow; (f) *Allodesmus*, a Miocene seal; (g) *Globicephalus*, a modern dolphin. (a)–(d) are reptiles, and (e)–(g) are mammals. The forelimbs are all homologous with each other, and with the wing of a bird and the arm of a human. However, as paddles, these are all analogues: each paddle shown here represents a separate evolution of the forelimb into a swimming structure.

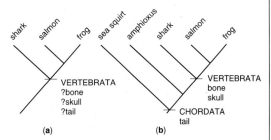

Fig 3.3   Reconstructing the phylogeny of vertebrates by cladistic methods. (a) Are the defining features of vertebrates the possession of bone, a skull and a tail? (b) The tail is found in a wider group, termed the Chordata, but the skull and bone define the Vertebrata.

Box 3.1 (cont.)

Table 3.1

Character	Shark	Salmon	Frog	Lizard	Chicken	Mouse
1. Fins	1	1	0	0	0	0
2. Legs	0	0	1	1	1	1
3. Warm-bloodedness	0	0	0	0	1	1
4. Bone	0	1	1	1	1	1
5. Diapsid skull	0	0	0	1	1	0
6. Loss of larval stage	0	0	0	1	1	1
7. Lung or swim bladder	0	1	1	1	1	1
8. Amniote egg	0	0	0	1	1	1
9. Elongate neck vertebrae	0	0	0	1	1	0
10. Marginal teeth	0	1	1	1	1	1

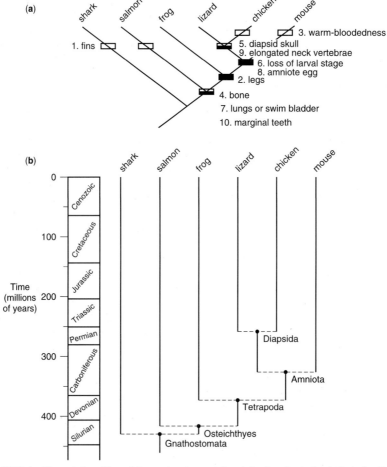

Fig 3.4 The relationships of the major groups of vertebrates, tested using six familiar animals. (a) Postulated relationships, based on analysis of characters in text. (b)
Phylogenetic tree, showing the cladogram from (a) set against a time-scale, and bas-
ing the dating of branching points on oldest known fossil representatives of each
group.

Box 3.1 (cont.)

The third stage can be illustrated by attempts to subdivide the vertebrates and to discover their true phylogeny. A first step might be simply to try to find the relationships among the following typical vertebrates: shark, salmon, frog, lizard, chicken and mouse. In comparing these animals, it is clear that the shark and the salmon have fins, while the others have legs. The chicken and mouse are warm-blooded, while the others are cold-blooded. Outgroup comparison of these two sets of characters suggests that warm-bloodedness might be an apomorphy (since most members of the outgroup, such as clams, oak trees and bugs, are cold-blooded), but it is harder to tell whether fins or legs are apomorphies or not. The characters may be coded with binary digits (0 or 1), where 0 means possession of the primitive state, and 1 means possession of the apomorphy. A number of features are tabulated in Table 3.1.

Scanning over the data in this table, it is clear that some groupings are indicated by several synapomorphies, but there are contradictions. For example, the diapsid skull (see p.204) supports a pairing of lizard and chicken, but warm-bloodedness suggests a pairing of chicken and mammal. Both pairings are not possible, and one of these synapomorphies must be wrongly interpreted. The method of testing at this point is to seek the most parsimonious pattern of relationships, i.e. the one that explains most of the data and implies least mismatch, or incongruence. The data may be run through a computer program, such as PAUP (Swofford, 1997), which extracts the most parsimonious tree of relationships, or cladogram (Figure 3.4a), and highlights the incongruent (i.e. probably misinterpreted) characters. The cladogram is of course a best effort, and further study of the specimens, and the discovery of new characters, can confirm or refute it.

The cladogram shows the best current evidence of relationships, and it can be made into a phylogeny by the addition of a time-scale (Figure 3.4b). Here, the fossil evidence for dates of origin of the various groups is used to give a picture of the true shape of this part of the phylogeny of life.

the point of appearance of particular unique features during the course of evolution.

Stratophenetics and molecules in phylogeny reconstruction

In practice, cladistic analysis is not always possible. Many plants and animals have simple forms, and there are not enough characters to analyse. In cases of this sort, stratophenetic techniques may be used to reconstruct phylogeny. Here, the fossils are first arranged in order of geological occurrence, with the oldest at the bottom and the youngest at the top, and phylogenetic links are sketched in, based on overall similarity in form. The weakness of this approach is that there is no final test of whether the discovered patterns of phylogeny are real. Phenetics has other useful applications, however, in determining the properties of members of populations (see Box 3.2).

A third approach to phylogeny reconstruction is based on comparison of molecules. With the birth of molecular biology in the 1950s and 1960s, it became clear that homologous proteins share similar structures in different organisms. The key is that proteins vary in small ways in the sequences of amino acids that make them up, and the more distantly related any pair of species is, the greater the amount of difference. The amount of these differences can be calibrated against the geological time-scale, based on cases where fossils of near-ancestral forms are well dated, and then the dates of branching of species with poorer fossil records can be assessed. The basis for correlating the amount of molecular difference with geological time is the molecular clock hypothesis, the idea that there is a certain regularity to the rate of molecular evolution.

Advances in laboratory techniques, and in particular the introduction of the polymerase chain reaction (PCR) method, has permitted molecular biologists to analyse tiny samples, which are cloned or multiplied before study. The PCR method has now opened up the real possibility of analysing the DNA of extinct organisms (see Box 3.3).

The origin of species

Speciation

Species consist of many highly variable individuals, often divided into geographically restricted populations and races. All human beings belong to a single species,

Box 3.2 Phenetics and variation within populations

Frequently, palaeontologists are faced with problems that require the simplification of a great mass of measurements. For example, a palaeontologist may have a large sample of brachiopods from a single rock horizon and may wish to determine whether these represent one or more species. It might be sufficient to plot univariate frequency histograms (see p.10) of particular measures, such as width, length and depth of the shells, as well as the hinge width, the diameter of the pedicle foramen, and the length and width of internal muscle scars. In addition, bivariate plots could be prepared, in which various measures are plotted against each other. However, it might still be difficult to differentiate clusters of points, and this approach means the palaeontologist has many separate graphs to compare.

Multivariate techniques can help solve these problems, by considering all the measured variates together. Two common techniques are cluster analysis and principal components analysis (PCA). In PCA, the maximum direction of variation is determined from the table of measurements of many characters, and this direction is termed eigenvector one. Further eigenvectors are then plotted in sequence perpendicular to the first. The first eigenvector reflects growth-related or size-dependent variation, and it is usually ignored in taxonomic studies. Species are usually plotted against the second and third eigenvectors, and tests can then be applied to determine whether there are separate clusters of points.

As an example, a comparison may be made between specimens of two species of brachiopod, *Dicoelosia biloba* from the Early Silurian of Sweden, and *D. hibernica* from rocks of the same age in Ireland. Four measurements were made on samples of both species and a PCA was performed. Both species were then plotted against the second and third eigenvectors (Figure 3.5). Although both samples overlap, in general, the Irish specimens have lower scores on eigenvector 2, showing that *D. hibernica* is wider and less deep than *D. biloba*.

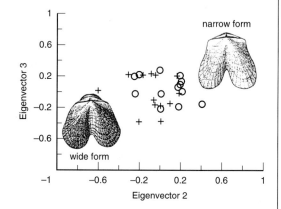

Fig 3.5   Variation in the brachiopod species *Dicoelosia biloba* (o) from the early Silurian of Sweden and *D. hibernica* (+) from Ireland, based upon numerous measurements. A principal components analysis plot separates wide and narrow forms along eigenvector 2, so there may truly be two species, although there is considerable overlap between the two.

Homo sapiens, and yet every person is different. The range of genetic and physical variation among humans is enormous, and much of it appears to be associated with geographic distribution. There has also been variation through time, with races of *Homo sapiens*, like *H. s. neanderthalensis*, the neanderthals, being stocky and heavily built, possibly as an adaptation to the cold Ice Age conditions of Europe 30 000 years ago (see p. 221). All species show geographic variation and, where the fossil record is good enough, temporal variation too.

Local populations may be to a great extent autonomous, i.e. isolated from other populations of the same species, and with a subtly different gene pool, the overall array of genetic material in all the individuals within the population. The cohesion of a species is maintained over its natural range by processes of gene flow, the occasional wandering of individuals from one area to another, which interbreed with members of neighbouring populations. These processes can be thought of as occurring on many different scales, ranging from the whole Earth for humans, to a tiny patch of forest for some insect species.

A convincing model for speciation was proposed in the 1940s by Ernst Mayr, based on the establishment of geographic barriers. Mayr suggested that the most likely explanation for species splitting was the separation of populations, and prevention of gene flow, by a barrier, such as a new strip of water, a new mountain chain, or even the building of a major road: anything that stops free genetic mixing among populations. Genetic and phenotypic (external appearance) divergence would occur for two reasons:

1. each population, or set of populations, would start out with a different gene pool, simply because part of the former genetic range of the intact species has now been separated off; and
2. selection pressures would be different, perhaps only subtly, on either side of the barrier.

Box 3.3: Fossil proteins: the real Jurassic Park?

Proteins were extracted from fossils in the 1960s and 1970s, but most of these are decay materials, the proteins of bacteria that decomposed the original tissues. Even in cases of exceptional preservation where soft tissues are preserved (see pp. 13–15), the proteins have usually long vanished. Until 1985, the oldest DNA, recovered in tiny quantities, came from Egyptian mummies, 2400 years old. Even well-preserved mammoths, some 11 000 years old, usually retain no original DNA – evidence of the rapid decay of that protein.

One possible scenario in which proteins might just survive unchanged is by preservation in amber, and this was the idea exploited by Michael Crichton in his book, *Jurassic Park*. Amber is ideal for two reasons: the insects are trapped instantly, usually overwhelmed by the sticky resin, and no decay takes place; the amber excludes oxygen and water so that no physical or chemical changes should occur during subsequent millenia. In 1992, original DNA was recovered from a termite in Oligocene–Miocene amber, and the record was pushed back to the Mesozoic by the report of original DNA extracted from an early Cretaceous weevil in amber from the Lebanon (Figure 3.6). In

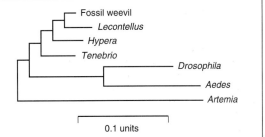

Fig 3.6   Sequencing the DNA of an ancient weevil from Cretaceous amber, 120–135 Ma. Phylogenetic tree based on comparisons of sequences from the ribosomal RNA of the extinct weevil, and selected modern insect species, the modern weevils *Lecontellus* and *Hypera*, the mealworm *Tenebrio*, the fruit fly *Drosophila*, the mosquito *Aedes*, and the brine shrimp *Artemia*. The extinct weevil shares most molecular similarity with the modern weevils.

1994, DNA was reported from dinosaur bone. These results have been disputed, and critics have suggested that the supposed fossil DNA is contamination in the laboratories from modern animals. Could it ever be possible to extract dinosaur DNA from blood in the belly of a Cretaceous mosquito? Blood-sucking insects are known from the Cretaceous, so who knows?

The allopatric ('other homeland'), or geographic, model of speciation may take two main forms. The process may be symmetrical (Figure 3.7a), with the ancestral species being divided roughly down the middle of its geographic range, and the two daughter species starting out with similar-sized populations. More dramatic effects may be seen when the split is asymmetrical (Figure 3.7b). Here, a small population, perhaps isolated on an island, evolves independently of the parent species, which may continue roughly unchanged. The smaller population may show unusual and rapid evolution because of what Mayr called the founder effect: the fact that its gene pool is a small sample of the overall gene pool, and that new environmental pressures and opportunities may occur.

Speciation and evolution in the fossil record

It was assumed by many palaeontologists that the fossil record showed how species had evolved gradually over vast spans of time, and that from time to time new species branched off. In this interpretation, most evolutionary change took place within gradually changing species lineages, and the origin of new species was really a side-effect, and not a focal point of evolution.

A forthright challenge to this gradualistic model of evolution was made by Eldredge and Gould (1972), who proposed an alternative, which they termed the punctuated equilibrium model. They argued that the fossil record does not show evolution occurring in species lineages: in fact, they argued, most species lineages show stasis, or no change, over long spans of time. Change occurs at the time of speciation. Eldredge and Gould contrasted their model of evolution by punctuated equilibria with the classic phyletic gradualism model in terms of the shape of a phylogeny:

1. in the phyletic gradualism model (Figure 3.8a), with sloping branches, most evolution takes place within species lineages, and speciation events involve no special additional amount of evolution;
2. in the punctuated equilibrium model (Figure 3.8b), with rectangular branches, almost no evolution takes place within species lineages (they show stasis), and evolution is concentrated in the speciation events, which coincide with major sideways shifts.

The two models of evolution seem so distinctive, both in the shape of phylogenies, and in their interpretation, that it would seem clearly possible to test between them by observations from the fossil record.

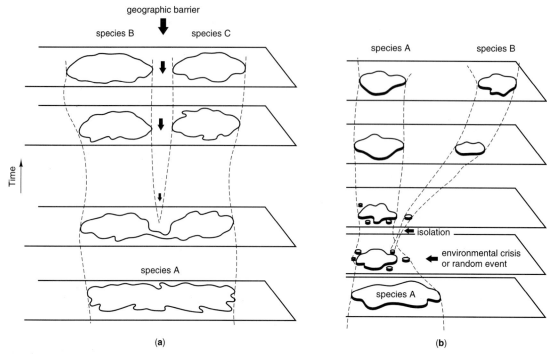

Fig 3.7 Allopatric speciation models, occurring either symmetrically (a), where the parent species is divided into two roughly equal halves by a geographic barrier, or asymmetrically (b), where a small peripheral population is isolated by a barrier. In the first case, two new species may arise; in the second, the parent species may continue unaltered, and the peripheral population may evolve rapidly into a new species.

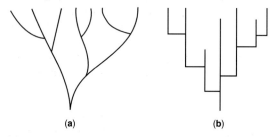

Fig 3.8 Two models of speciation and lineage evolution. (a) Phyletic gradualism, where evolution takes place in the lineages, and speciation is a side-effect of that evolution. (b) Punctuated equilibria, where most evolution is associated with speciation events, and lineages show little evolution (stasis).

Testing punctuated equilibrium

Eldredge and Gould argued that many test cases of the pattern of evolution at the species level could be studied from the fossil record. These should have the following features:

1. abundant specimens;
2. fossils with living representatives, so that species could be identified clearly;
3. information on geographic variation, so that rapid speciation events (punctuations) could be distinguished from migrations in or out of the area; and
4. good stratigraphic control, in terms of long continuous sequences of rocks without gaps, abundant fossils throughout, and good dating.

The problems in testing became evident early on, since sampling was generally not extensive enough. Williamson (1981) attempted to counter this problem in one of the most enormous sampling exercises ever. He studied millions of specimens of snails and bivalves in sediments deposited in the Lake Turkana area of Kenya, from 1.3 to 4.5 Ma (Figure 3.9). Lake Turkana lies in the East African Rift Valley, on a tectonically active line where the continent of Africa is unzipping to form two major plates. Lake muds and sands accumulated in thick deposits as the rift opened, and volcanic ash (tuff) beds occur sporadically throughout the sequence.

Williamson recorded changes in 19 species lineages,

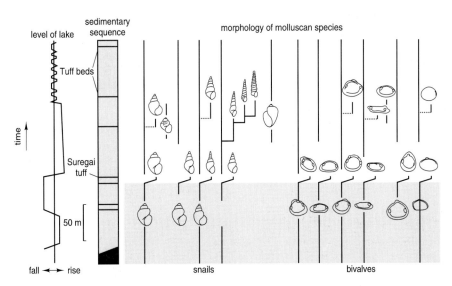

Fig 3.9 Fine-scale evolution in fresh-water snails and bivalves in Lake Turkana, Kenya, through the last 4 Myr. The volcanic tuff beds allow accurate dating of the sequence. Major speciation events seem to take place at times of lake-level change: are these examples of punctuational speciation, or merely ecophenotypic shifts?

and found that stasis was the normal state of affairs, but that rapid morphological shifts had taken place three times, two of which corresponded to substantial lake- level rises (Figure 3.9). He interpreted this as evidence for the punctuated equilibrium model, arguing that rapid environmental changes had caused evolutionary shifts and speciation events. The new species were short-lived, he argued, because the parental stock had survived in neighbouring unstressed lakes, and returned to colonize Lake Turkana after the lake-level changes had taken place.

However, even this enormous study aroused controversy. Critics pointed out that the sequence of sediments was not complete enough to be sure that all fossils had been found: there were gaps of 1000 years or more, and a great deal of gradualistic evolution could take place in that time. Secondly, Williamson's critics argued that the environmental stress of lake-level change induced short-term changes in shell shape, but when the stress was over, the shell shapes reverted to normal. Hence, they proposed that speciation had not taken place, but that the shells had changed shape ecophenotypically. This means that the changes happened during the animals' lifetimes, and were not genetically coded, and hence were not evolutionary.

Another similarly detailed study of evolutionary modes in fossil lineages has shown evidence for gradualistic change. Sheldon (1987) sampled thousands of trilobites belonging to eight lineages in an Ordovician

sequence in central Wales which spanned 3 Ma. He showed that gradualistic evolution had taken place in at least six of the lineages, with clear increases through time in the numbers of pygidial (tail segment) ribs (Figure 3.10). In detail, the changes zig-zagged back and forwards, but there appeared to be clear net changes. In other words, lineages do not always show stasis, and one 'species' is seen to evolve into another morphologically distinct 'species' in several of the lineages. Unfortunately, splitting of lineages is not observed in this example, so the rate of cladistic species formation cannot be assessed.

A number of conclusions may be drawn from these palaeontological studies of species change. Firstly, testing has been difficult because of problems of precision of dating, completeness of sampling, geographic movements of species, and distinguishing real evolutionary change from ecophenotypic effects. Secondly, both stasis and gradualistic change within lineages have been demonstrated. Thirdly, some speciation events at least appear to be geologically rapid, perhaps taking place within 1000–2000 years, but such events are hard to document because they usually take place in small isolated populations. Perhaps all extremes of evolutionary pattern illustrated in Figure 3.12 take place in nature. If punctuated equilibrium is a correct view of some evolution at least, it has been said that this threatens normal Darwinian interpretations of evolution.

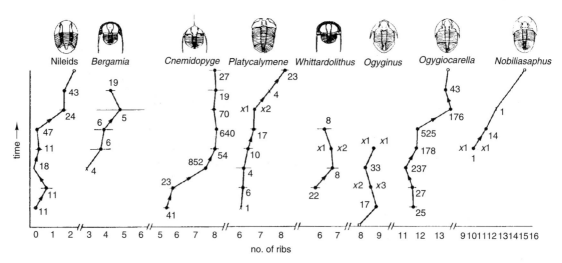

Fig 3.10   Evolution of trilobites in an Ordovician sequence from central Wales, representing a total time span of about 3 Myr. In each genus, there seems to be gradualistic evolution in the number of ribs in the pygidium, or tail segment (statistically significant shifts between consecutive samples are indicated by arrows). Numbers indicate sample sizes, and horizontal lines 95% confidence intervals for each sample.

Hierarchy, and expansion of Darwinian evolution

Steven Stanley argued that a punctuational model of evolution would imply a different kind of process, termed by him species selection. This would be analogous to natural selection, but it would act on species instead of individuals. A process such as species selection could be seen as a challenge to the standard Darwinian view of evolution in which all change is caused by natural selection.

Species selection means that, among a number of similar species, one or more survive since they possess some characters that give them an evolutionary advantage. The key to understanding species selection lies in these species-level characters: they must be quite distinct from individual-level characters. It is not enough to say that, among African large cats, lions might survive certain kinds of competitive situations because they are larger than the other hunters. Being large is an individual-level character, and selection for size is natural selection. Species-level characters must be irreducible to the individual level.

Possible species-level characters include the size of the geographic range of a species, the pattern of populations within the overall species range, characteristic levels of gene flow among the populations of a species, and mean species durations. Some studies have suggested that species-level characters of these kinds may play a part in evolution. Geographically widespread species of gastropods, for example, tend to have longer durations than more localized species, and hence can be said to survive longer because of a species-level character. If species selection is a real force in evolution, then Darwinian evolution would have to be expanded to incorporate a hierarchy, or multi-level array, of processes.

A possible resolution of this issue is the effect hypothesis of Vrba (1984). She argued that some species-level characters may be reducible indirectly to the individual level. That is to say, a broadly based feature of the species actually depends on some other character that is under the influence of natural selection. She gave an example from her own work on the evolution of antelope over the past 6 Myr (Figure 3.11). About 5 Ma the antelope phylogeny split into two, giving rise to one group of long-lived species that never became diverse, and a second group of shorter-lived species that radiated widely. Species duration in the first group was 2–3 Myr and total species diversity through the Plio-Pleistocene was two, while in the second group species duration was 0.25–3 Myr and 32 species evolved. Surely here species selection was taking place: the character of short species duration in the second group permitted great success, as measured by overall species diversity. Vrba noted, however, that the long-lived antelope had wide ecological preferences, while those in the second group were specialists.

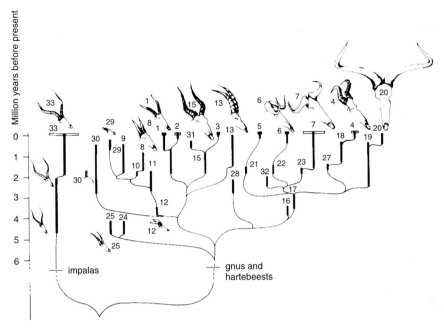

Fig 3.11 Reconstructed phylogeny of African antelopes. Two lineages diverged around 6–7 Ma: the slowly evolving impalas and the rapidly speciating gnus and hartebeests. The second group could be said to be evolutionarily more successful than the first, and this might be interpreted as a result of species selection of the species-level character, rate of speciation. However, the gnus and hartebeests have more specialized ecological preferences than do the species of impalas. Perhaps selection has occurred at the individual level (natural selection), and this has had an effect at the species level.

Hence, the whole pattern could be explained by natural selection at the level of individual antelope, where their ecological tolerances determine their evolutionary rates, and produce a superficial appearance of species selection.

Is evolution hierarchical? And, if so, was Darwin wrong? The case has been overstated by critics: evolution occurs by natural selection, as Darwin said in 1859. Many proposed examples of species selection can be explained by natural selection, coupled with rapid asymmetric geographic speciation and the effect hypothesis. None the less, species selection is a possibility, and real examples of it may be found.

Growth and form

Recognizing fossil species

Palaeontologists must interpret fossil species, and their ranges of variation, solely from the morphology, or external shape, of the specimens. There are problems in

deciding where one species ends and another begins. When there are close living relatives, it may be possible to compare the modern species with the fossils. But how are palaeontologists to decide just what is a species of dinosaur or trilobite?

For modern plants and animals, systematists apply the biological species concept. This defines a species as a group of individual plants or animals that are capable of interbreeding and of producing viable offspring. The production of viable offspring is important, since members of related species may sometimes cross-breed – horses (*Equus caballus*) and asses (*E. africanus*) for example – but the offspring are sterile.

In practice, most decisions on species of living plants and animals are based on assessments of the morphologies of dead specimens in museums: it is impossible to carry out extensive cross-breeding tests with living organisms. Thus, fossils are not in another league from modern taxa when it comes to determining species: the problems usually arise from the added dimension of time. If a palaeontologist finds a long evolving lineage, where should the dividing line be drawn between one species and the next? Decisions are often made easier by gaps in the fossil record which create artificial divisions within evolving lineages. Where gaps are not present, splitting events clearly mark off new species. If there are few of these, an evolving lineage is divided somewhat arbitarily into chronospecies ('time species'), each being defined by particular morphological features.

Variations in form within species

Within a species, there may be a range of morphologies; think of the variation among humans, or more dramatically, among domestic dogs. In naturally occurring species, morphology may vary as a result of several factors:

1. geographic variation, and physical differences between populations or subspecies in different parts of the overall species range;
2. sexual dimorphism, in which males and females may show different sizes, and different specialized features (e.g. horns, antlers, tail feathers);
3. growth stages, where there may be quite different larval and adult stages, or where body form alters during growth;
4. ecophenotypic effects, where local ecological

conditions affect the form of an organism during its lifetime (see p. 53).

The first three of these are genetically coded, and the last is not. Geographic variation may be substantial among members of modern species, particularly those distributed over wide ranges. Sexual dimorphism is also commonly observed among living animals, particularly in those where males engage in ritualized displays, or where females have special reproductive activities. Sexual dimorphism is also common in fossils, and it has often caused serious problems of identification where males and females look very different. For example, many ammonites show sexual dimorphism, where the postulated females are much larger than the males, and the males possess unusual lappets on either side of the aperture (Figure 3.12).

Allometry

Changes in form during growth are common. Think of human growth: babies have relatively large heads and eyes, and small limbs. Similar features are found in fossil cases too. Juvenile vertebrates, not just humans, usually have large eyes and heads in proportion to overall

(a)

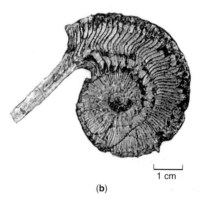

1 cm

(b)

Fig 3.12 Sexual dimorphism in ammonites, the Jurassic *Kosmoceras*. The larger shell (a) was probably the female, the smaller (b), the male.

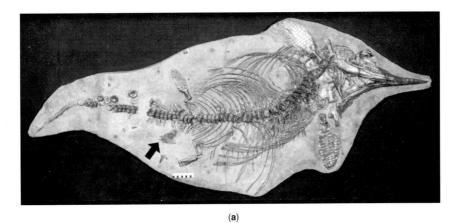

(a)

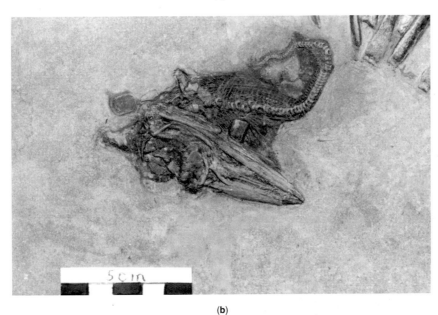

(b)

Fig 3.13 Adult female *Ichthyosaurus* (a) from the early Jurassic of Somerset, England, show-ing an embryo that has just been born (arrow), and detail of the curled embryo (b).

body size. A tiny embryo of an ichthyosaur (Figure 3.13) shows just these features. If measurements of the variable parts (e.g. eye diameter, head length) are scaled against a standard measure of the animal (e.g. total body length), it is evident that the proportions change as the animal grows older. In the case of the ichthyosaur, the ratio of eye diameter to body length diminishes as the animal approaches adulthood. This is an example of allometric ('different measure') growth. If there is no change in proportions during growth, the feature is said to show isometric ('same measure') growth.

Once the nature of any allometric change of parts or

organs has been established quantitatively, it is possible to investigate why such changes might occur. The large head of human babies is said to reflect the fact that they must have most of the large brain fully developed at birth. The large eyes and small noses of babies are said to make them look cute so their parents will look after them, and feed them.

Ichthyosaurs (Figures 3.13 and 3.14) were born live underwater, not from eggs laid onshore, as is the case with most other marine reptiles. Their large head at birth would have allowed them to feed on fish and ammonites as soon as they were born. The large eyes were perhaps

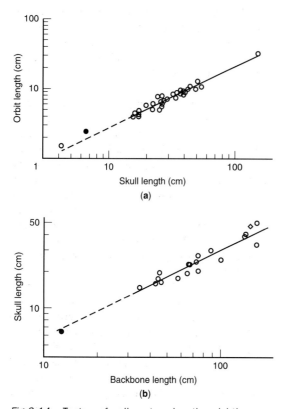

Fig 3.14 Tests of allometry in the ichthyosaur *Ichthyosaurus*. (a) Plot of orbit length against skull length; (b) plot of skull length against backbone length. The Somerset embryo (Figure 3.13) is indicated by a solid circle. Both graphs show negative allometry (orbit diameter = 0.355 skull length$^{0.987}$; skull length = 1.162 backbone length$^{0.933}$), confirming that embryos and juveniles had relatively large heads and eyes.

necessary also for hunting in murky water, and had to be near-adult size from the start. Or, perhaps, it made them look cute, and encouraged parental care!

Ontogeny and phylogeny

Biologists have long sought a link between ontogeny (development) and phylogeny (evolutionary history). In 1866, Ernst Haeckel, a German evolutionist, announced his Biogenetic Law, that 'ontogeny recapitulates phylogeny'. His idea was that the sequence of embryonic stages mimicked the past evolutionary history of an animal. So, in humans, he argued, the earliest embryonic stages were rather fish-like, with gill pouches in the neck region. Next, he argued, was an 'amphibian' stage and a 'reptile' stage, when the human embryo retained a tail

and had a small head, and finally came the 'mammal' stage, with growth of a large brain and a pelt of fine hair.

Haeckel's view was attractive at the time, but too simple. A more appropriate interpretation is Von Baer's Law (1828). Von Baer interpreted the embryology of vertebrates as showing that 'general characters appear first in ontogeny, special characters later'. Early embryos are virtually indistinguishable: they all have a backbone, a head and a tail (vertebrate characters). A little later, fins appear in the fish embryo, legs in the tetrapods. More specialized characters appear later: fin rays in the fish, beak and feather buds in the chick, snout and hooves in the calf, and large brain and tail loss in the human embryo.

'General characters appearing before special characters' has taken on a new meaning with the development of a strictly cladistic view of phylogeny. Von Baer's Law draws a parallel between the sequence of development, and the structure of a cladogram. In human development, the embryo passes through the major nodes of the cladogram of vertebrates. The synapomorphies of vertebrates appear first, then those of tetrapods, then those of amniotes, then those of mammals, of primates, and of the species *Homo sapiens* last.

Three other aspects of development throw light on phylogeny. Certain developmental abnormalities called atavisms, or throw-backs, show former stages of evolution, such as human babies with small tails or excessive hair, or horses with extra side toes (Figure 3.15a), showing how earlier horses had five, four or three toes, compared to the modern one.

Vestigial structures tell similar phylogenetic stories. These are structures retained in living organisms that have no clear function, and may simply be there because they represent something that was once used. So, modern whales have, deep within their bodies, small bones in the hip region that are remnants of their hind legs (Figure 3.15b). Whales last had functioning hind legs over 50 Ma in the Eocene, and the vestigial remnants are still there, even though they serve no known purpose.

The third aspect of development that forms links with phylogeny is the observation that ontogenetic patterns themselves have evolved. In particular the timing and rate of developmental events has varied between ancestors and descendants, often with profound effects. This phenomenon is termed heterochrony.

Heterochrony: are human adults juvenile apes?

Heterochrony means 'different time', and includes all aspects of changes of timing and rates of development.

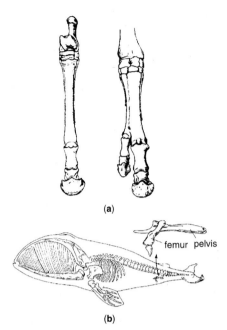

(a)

femur pelvis

(b)

Fig 3.15   Hints of ancestry in modern animals. (a) Extra toes in a horse: an example of an atavistic abnormality in development, or a throw-back, to earlier horses which had more than one toe; normal horse leg (left), extra toe (right). (b) The vestigial hip girdle and hindlimb of a whale: the rudimentary limb is the rudiment of a hindlimb that functioned around 50 Ma.

There are two forms of heterochronic change: paedomorphosis ('juvenile formation'), or sexual maturity in a juvenile body and peramorphosis ('over-development'), where sexual maturity occurs relatively late. These changes can each occur in three ways, by variation in timing of the beginning of body growth, the timing of sexual maturation, or the rate of morphological development (Table 3.2).

In studying heterochrony, it is necessary to have a good phylogeny of the organisms in question, a good fossil record of the group, and a good set of ontogenetic sequences for each species. This allows the palaeontologist to compare juveniles and adults throughout the phylogeny. A classic example is human evolution. It seems obvious that human adults look like juvenile apes, with their flat faces, large brains, and lack of body hair. These would imply a paedomorphic change in humans with respect to the human/ape ancestor. However, other characters do not fit this pattern. For example, developmental time in humans is far longer than in apes and ancestral forms, a feature of peramorphosis. Thus, heterochronic changes can occur in different directions in different characters, an example of mosaic evolution.

McNamara (1976) studied a brachiopod example where the phylogeny (Figure 3.16) suggested that species of *Tegulorhynchia* evolved into *Notosaria*. The main changes were a narrowing of the shell, a reduction in the number of ribs in the shell ornament, a smoothing of the lower margin, and an enlargement of the pedicle foramen (the opening through which a fleshy stalk attaches the animal to a rock). These changes related to a shift of habitats from deep to shallow high-energy waters: the large pedicle allowed the brachiopod to hold tight in rougher conditions, and the other changes helped stabilize the shell. The developmental sequence of the ancestral species, *Tegulorhynchia boongeroodaensis*, shows that its descendants are like the juvenile stage. Hence, paedomorphosis has taken place along a paedomorphocline ('child formation slope'). It is harder here to determine which type of paedomorphosis has taken place; perhaps it was neoteny.

A second example illustrates a peramorphic trend. Rhynchosaurs were a group of Triassic herbivorous reptiles. Later species had exceptionally broad skulls as adults, which gave them vast muscle power to chop tough vegetation. Juvenile examples of these late Triassic rhynchosaurs retain the rather narrower skulls of the ancestral adult forms (Figure 3.17). Hence, the evolution of the broad skull is an example of peramorphosis, along a peramorphocline ('over-development slope'). The heterochronic mode here was probably hypermorphosis, since the adult Late Triassic rhynchosaurs are larger than most earlier forms, which implies that sexual maturation was delayed while the body continued to grow.

Further reading

Briggs, D. E. G. and Crowther, P. (eds) (1990) *Palaeobiology, a synthesis.* Blackwell Scientific Publications, Oxford..

Forey, P. L. Humphries, C. J., Kitching, I. J., Scotland, R. W., Siebert, D. J. and Williams, D. M. (1993) *Cladistics: a practical course in systematics.* Oxford University Press, Oxford.

Patterson, C. (1987) *Molecules and morphology in evolution: conflict or compromise?* Cambridge University Press.

Ridley, M. (1996) *Evolution,* 2nd edition, Blackwell Scientific Publications, Oxford.

Skelton, P. (ed.) (1993) *Evolution; a biological and palaeontological perspective.* Addison Wesley, Wokingham.

Table 3.2 The processes of heterochrony: differences in the relative timing and rates of development

	Onset of growth	Sexual maturation	Rate of morphological development
Paedomorphosis			
Progenesis	—	early	—
Neoteny	—	—	reduced
Post-displacement	delayed	—	—
Peramorphosis			
Hypermorphosis	—	delayed	—
Acceleration	—	—	increased
Pre-displacement	early	—	—

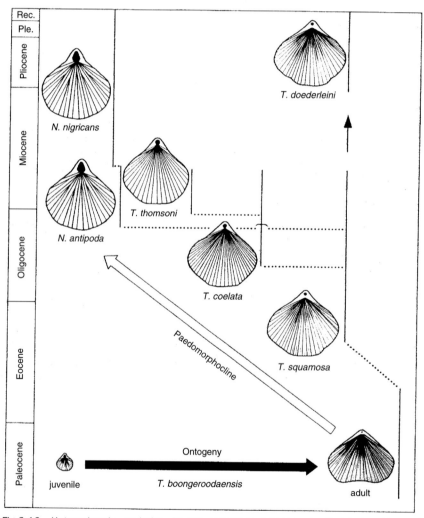

Fig 3.16 Heterochronic evolution in the Cenozoic brachiopods *Tegulorhynchia* and *Notosaria*. Adults of more recent species are like juveniles of the ancestor. Hence, paedomorphosis ('juvenile formation') is expressed in this example Ple., Pleistocene; Rec., Recent.

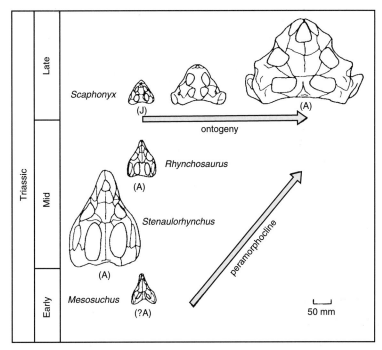

Fig 3.17   Heterochronic evolution in the Triassic rhynchosaurs. The skull of adult (A) late Triassic forms has developed beyond the size and shape limits seen in earlier Triassic adult forms. Here, the juveniles (J) of the descendants resemble the ancestral adults, and this is thus an example of peramorphosis ('beyond formation').

References

Crichton, M. (1991) *Jurassic Park*. Random House, London.

Eldredge, N. and Gould, S. J. (1972) Punctuated equilibria: an alternative to phyletic gradualism. In T. J. M. Schopf (ed.) *Models in paleobiology*. Freeman, Cooper and Co., San Francisco, pp. 82–115.

McNamara, K. J. (1976) The earliest *Tegulorhynchia* (Brachiopoda: Rhynchonellida) and its evolutionary significance. *Journal of Paleontology*, **57**, 461–473.

Sheldon, P. R. (1987) Parallel gradualistic evolution of Ordovician trilobites. *Nature*, **330**, 561–563.

Swofford, D. L. (1997) *PAUP, Phylogenetic Analysis Using Parsimony*, version 4.0. Documentation and software. Sinauer, New York.

Vrba, E. S. (1984) Patterns in the fossil record and evolutionary processes. In M.-W Ho. and P. T. Saunders (eds) *Beyond Neo-Darwinism: an introduction to the new evolutionary paradigm*. Academic Press, London, pp. 115–142.

Williamson, P. G. (1981) Palaeontological documentation of speciation in Cenozoic molluscs from the Turkana Basin. *Nature*, **293**, 437–443.

4 The origin of life

Key Points

- Life originated by fusion of organic molecules about one billion years after the formation of the Earth.
- The earliest forms of life were prokaryotes, cyanobacteria and bacteria, in rocks up to 3.5 billion years old. These existed as isolated cells, and also built stromatolites.
- The oldest eukaryotes, cells with a nucleus and organelles, date back 900 million years.
- Algae consist of a mixed assemblage of organisms, some primitive, and some related to green plants.

Introduction

Life arose on the Earth about 3500 million years ago. The first living organisms were simple single-celled prokaryotes similar to modern cyanobacteria (blue-green algae) and bacteria are often loosely called plants. More complex cells, eukaryotes, arose only much later, about 900 million years ago, and much later than that came the first true plants and animals. This means that the first three-quarters of the history of life passed by in the company of organisms that were neither plant nor animal.

The oldest organisms: prokaryotes

The origin of life

One of the key questions that people of all cultures have asked is: where did life come from? There have been four categories of idea used to explain the origin of life:

1. creation myths;
2. spontaneous generation;
3. inorganic model (see Box 4.1);
4. biochemical theory.

Creation myths are common to many religions, and they explain the origin of life by divine intervention. These ideas are not scientific since they cannot be tested.

Two scientific theories for the origin of life are spontaneous generation and the biochemical theory. Mediaeval scholars believed that many organisms sprang into life directly from non-living matter. For example, frogs were said to arise from the spring dew, and maggots were generated by rotting flesh. However, careful tests proved that there was no truth in these ideas. Louis Pasteur in 1861 enclosed pieces of meat in airtight containers, and maggots did not appear. Thus, he had disproved the theory of spontaneous generation, and he showed that flies laid their eggs on rotting meat, the eggs hatched as maggots, and the maggots then turned into flies.

The biochemical theory for the origin of life was developed in the 1920s independently by a Russian biochemist, A. I. Oparin, and a British evolutionary biologist, J. B. S. Haldane. They argued that life could have arisen through a series of organic chemical reactions which produced ever more complex biochemical structures (Figure 4.1). They proposed that common gases in the early Earth atmosphere combined to form simple

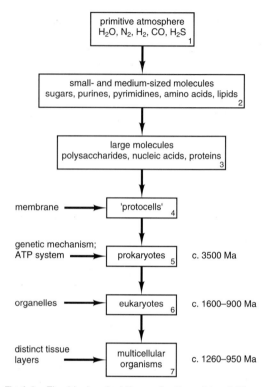

Fig 4.1   The biochemical theory for the origin of life, as proposed by I. A. Oparin and J. B. S. Haldane in the 1920s. Biochemists have achieved steps 1–3 in the laboratory, but scientists have so far failed to create life.

organic chemicals, and that these in turn combined to form more complex molecules. Then, the complex molecules became separated from the surrounding medium, and acquired some of the characters of living organisms. They became able to absorb nutrients, to grow, to divide (reproduce) and so on. These ideas have been tested successfully in the laboratory, but scientists have not created life yet (see Box 4.2). The biochemical theory was based on certain assumptions about early Precambrian atmospheres.

The inorganic model for the origin of life proposes that the first genetic material arose in association with clay minerals on the early Earth, and that organic molecules became involved only later (see Box 4.1), but this recent idea is highly controversial.

The early Precambrian world

The first two billion years of Earth history are known as the Archaean. This includes the time from the origin of the Earth, 4600 million years ago (Figure 4.2), to about

Box 4.1 Seven clues to an inorganic origin of life

Not all scientists accept an organic origin for life. Graham Cairns-Smith has suggested that the first genes were inorganic. The first genetic material was also 'naked', lacking the protection of a cell or organism, but nevertheless storing information and replicating in much the same way as DNA. Cairns-Smith (1985) presented his ideas as a detective story, drawing on seven clues from different areas of science and technology.

1. *Biology.* Since genetic information is the only part of an organism to evolve and be transmitted in a form rather than a substance, then the first genes may have been naked, and lacking an organism or phenotype.
2. *Biochemistry.* DNA and RNA are both complex molecules, far removed from the products of routine chemical reactions. They may have taken some time to evolve, and were probably late arrivals on the scene.
3. *Construction.* Most major structures are built within scaffoldings which themselves act as temporary structures. When a building is completed, the scaffolding is removed. This is proposed as an analogy for the construction of DNA and RNA on an inorganic template.
4. *Ropes.* None of the individual fibres stretches from one end of a rope to another. New fibres can be added and old fibres removed without changing the overall structure of the rope. Another analogy.
5. *History of technology.* Early machinery is often quite different in design from its advanced counterparts. For example, the first computer, the abacus, contrasts with today's PC, but in broad terms fulfilled the same function. Yet another analogy.
6. *Chemistry.* There are many simple substances that may have acted as low-tech genetic material. Crystal growth processes, as occur in clay minerals, can achieve precise replication.
7. Geology. Clay is abundant on the Earth's surface, providing a ready source of material for the first naked genes.

Cairns-Smith's hypothesis requires, firstly, a hypothetical minimal organism with no phenotype, a naked gene. The genetic material may have been inorganic crystals such as clay minerals, in which mutations arose through defects and twinning in the crystal lattice. Some of these clay minerals might have had the ability to produce a membrane that shielded them from environmental fluctuations in, say, water chemistry. The first low-tech organisms may have been sedimentary rocks where the growth of the 'genes' could change the behaviour or porosity and permeability of the rock. At some stage, these functions were taken over by organic materials with more securely held atoms and smaller molecules.

Box 4.2 Testing the biochemical theory

It took some years before biochemists were able to test the Oparin–Haldane theory for the origin of life (Figure 4.1). In 1953, Stanley Miller, then a student at the University of Chicago, made a model of the Precambrian ocean in a laboratory glass vessel. He exposed a mixture of water, nitrogen, carbon monoxide and nitrogen to electrical sparks, to mimic lightning, and found after a few days a brownish sludge in the bottle. This contained sugars, amino acids and nucleotides.

Further experiments in the 1950s and 1960s led to the production of polypeptides, polysaccharides and other larger organic molecules. Sidney Fox even succeeded in creating cell-like structures, in which a soup of organic molecules became enclosed in a membrane. His protocells seemed to feed and divide, but they did not survive for long. The crucial step (Figure 4.1) from molecules in solution to living cells has not been achieved in the laboratory.

2500 million years ago. At first, the Earth was a molten mass, and it cooled slowly, separating into an outer cool crust, and an inner molten mantle and core. Massive volcanic eruptions occurred throughout, and these produced great volumes of gases: carbon dioxide, nitrogen, water vapour and hydrogen sulphide. Impacts by comets also brought water vapour and carbon dioxide. Until about 3500 Ma, temperatures on the Earth were probably too high, and the crust was too unstable for any form of carbon-based life to exist. By 3500 Ma, things had changed: sedimentary rocks are found, and this proves that the crust had cooled and rivers were flowing and eroding rocks. Life also appeared.

Early Precambrian atmospheres contained the volcanic gases, but no oxygen. Oxygen levels are maintained in the atmosphere today by the photosynthesis of green plants and prokaryotes, and this was the source of the initial build-up of oxygen during the first part of the Precambrian.

The first organisms had anaerobic metabolisms, i.e. they operated in the absence of oxygen. Indeed the first prokaryotes would have been killed by oxygen. The

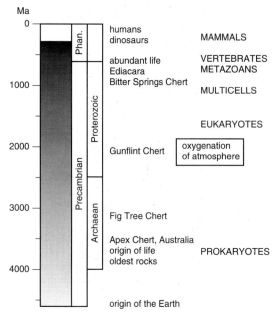

Ma

Fig 4.2 Time-scale showing major events in the history of the Earth and of life. Most of the time-scale is occupied by the Precambrian, whereas the well-known fossil record of the Phanerozoic (Phan.) accounts for only one-seventh of the history of life.

early diversification of life led to an increase in global atmospheric oxygen levels. By about 2 billion years ago, early in the Proterozoic eon (Figure 4.2), oxygen was being produced in sufficient quantites for aerobic organisms to originate. The presence of oxygen is proved by massive banded iron formations dating from 2.2 to 1.8 billion years old. The oxygen pumped out by early prokaryotes first entered the oceans, and then oxidized all the iron on the surface of the Earth, essentially producing vast quantities of rust. When all the available iron had been oxidized, oxygen then began to build up in the atmosphere. This marked the beginning of modern-style biogeochemical cycles, in which oxygen and carbon are exchanged continuously between living organisms and the Earth's crust.

Precambrian prokaryotes

The first traces of life occur in rocks dated about 3500 Ma. These include rare structures identified as possible stromatolites from various parts of the world, although the organic origin of such structures older than 3.0 Ma has been questioned. The Precambrian stromatolites were presumably constructed then, as now, by cyanobacteria and other prokaryotes (Figure 4.3). Cyanobacteria

live in shallow sea-water, and they require good light conditions to enable them to photosynthesize. The cyanobacteria form thin mats on the sea-floor in order to maximize their intake of sunlight, but from time to time, the mat is overwhelmed by sediment. The cyanobacteria migrate towards the light, and recolonize the top of the sediment layer, which may again be swamped by gentle seabed currents. Over time, extensive layered structures may build up. In fresh waters, and sometimes in the sea, stromatolites build up by precipitation of calcite. In most fossil examples, the cyanobacteria are not preserved, but the layered structure remains.

Some rare single-celled fossils are known from Australian and African cherts dated at 3500 Ma. These oldest prokaryote remains include spherical and filament-like chains of cells from Western Australia and South Africa. The best-corroborated of these earliest evidences of life are the 11 species of bacteria and cyanobacteria from the Apex Chert (Schopf, 1993). All specimens are filament-like microbes (Figure 4.4), ranging in length from 10 to 90 μm.

There is a long gap in time before the next diverse assemblage of prokaryote fossils come from the Gunflint Chert of Ontario, Canada, dated at 2000 Ma. The Gunflint micro-organisms include six distinctive forms, some shaped like filaments, others spherical, and some branched, or bearing an umbrella-like structure (Figure 4.5). These Precambrian unicells resemble in shape various modern prokaryotes, and some were found within stromatolites.

Records of Precambrian fossils have come to light only since the 1950s, and an enormous amount of effort in the 1980s and 1990s has produced dramatic new discoveries (Mendelson and Schopf, 1992). These show that at least 12 major groups of prokaryotes arose in the Precambrian, and survived to the present (see Box 4.3).

The classification of life

It is often assumed that everything living is either a plant or an animal. Traditionally, many simple organisms have been loosely classified as plants, but this is quite incorrect, as shown by a number of recent cladistic and molecular studies. Indeed, there were 50 or more major divisions on the evolutionary tree before the separation of multicelled plants and animals (metaphytes and metazoans).

Recent attempts to determine the relationships of the major groups of organisms, based on phylogenetic analysis of small subunit ribosomal RNA sequences, shows the complexity of early splitting events (Figure 4.6). The

Fig 4.3   Stromatolites, a Precambrian example from California, USA (×0.25).

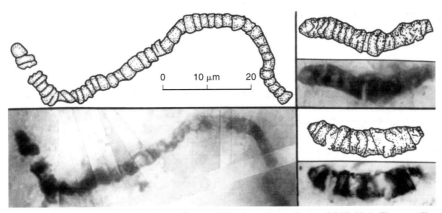

Fig 4.4   Prokaryotes from the Apex Chert of Western Australia (*c.* 3465 Ma). Filament-like microbes preserved as carbonaceous traces in thin sections; all are examples of the prokaryote cyanobacterium-like *Primaevifilum*, which measures 2–5 μm wide.

prokaryotes are a paraphyletic assemblage of bacteria (themselves paraphyletic) and cyanobacteria. The eukaryotes form a clade. The basal pattern of branching within Eukaryota produced a complex array of single-celled forms that are often lumped together as 'algae', another highly paraphyletic group. Among the 'algae', a crown-group clade includes *Acanthamoeba*, green algae, metaphytes, fungi, choanoflagellates and metazoans. The most startling observation from these results is that the fungi are more closely related to the multicellular animals than to the multicellular plants, and this has been confirmed in several analyses.

Life diversifies: eukaryotes

Eukaryote characters

Life until about 1 billion years ago consisted entirely of prokaryotes, simple microscopic organisms. One of the most important events in evolution took place when eukaryotes originated, since that marks the beginning of more complex life, real multicellularity and, in the end, larger plants and animals.

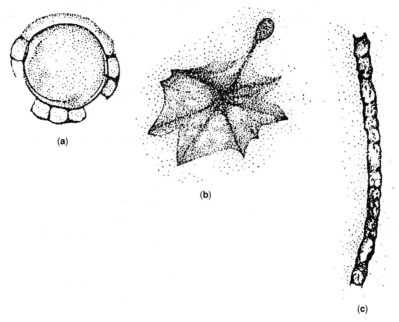

Fig 4.5 Prokaryote fossils from the Gunflint Chert of Ontario, Canada (*c.* 2000 Ma). (a) *Eosphaera*, (b) *Kakabekia*, (c) *Gunflintia*. Specimens are 0.5–10 μm in diameter.

Eukaryotes are distinguished from prokaryotes (Figure 4.7a, b) by having a nucleus containing their DNA in chromosomes (prokaryotes have no nucleus, and they have only a single strand of DNA) and cell organelles, i.e. specialized structures that perform key functions, such as mitochondria for energy transfer, flagella for movement, and chloroplasts in plants for photosynthesis. There are also many major biochemical differences between prokaryotes and eukaryotes.

The origin of eukaryotes is mysterious since they are in many ways so different from prokaryotes. The most attractive idea for their origin is the endosymbiotic theory, proposed by Lynn Margulis in the 1970s. According to this theory (Figure 4.7c), an amoeba-like prokaryote was invaded by, or consumed, some smaller energy-producing prokaryotes, and the two species evolved to live together in a mutually beneficial way. The small invader was protected by its large host, and the amoeba received supplies of sugars. These invaders became the mitochondria of modern eukaryote cells. Other invaders may have included worm-like swimming prokaryotes (spirochaetes), which became motile flagella, and photosynthesizing prokaryotes, which became the chloroplasts of plants.

The endosymbiotic theory cannot be tested by observation of fossils, but study of modern cells confirms it. For example, when one of your cells divides, the chromosomes arrange themselves carefully and duplicate within the nucleus. At the same time, the mitochondria set about duplicating and dividing independently, as if they are still invaders within the host.

Precambrian eukaryotes

Some of the oldest possible eukaryotes have been identified from the Bitter Springs Cherts of central Australia, dated at about 900 Ma. Some of the Bitter Springs cells apparently show nuclei (Figure 4.8), and if correctly interpreted, this would be clear evidence that these cells are eukaryotes. However, Knoll (1992) and others, consider that the dark areas may not represent nuclei at all, but merely condensations of the cell contents. Other putative early eukaryotes include specimens from Australia, China, India and North America in rocks dated between 1600 and 1000 Ma.

During the last 500 Ma of the Proterozoic, eukaryotes diversified, giving rise to diverse plant-like groups, usually grouped loosely as 'algae'. These included some close relatives of the metaphytes or true plants (see pp. 223–241). In addition, multicelled animals, or metazoans, also appeared later in the Proterozoic, and these included the complex Ediacaran animals (see pp. 76–79).

Box 4.3 Classification of the prokaryotes

Prokaryotes today occupy a range of habitats, from hot springs to Antarctic snow fields, from human digestive systems to many hundreds of metres down in the crust of the Earth. The subgroups of prokaryotes are distinguished by the nature of their metabolic activity, by aspects of their chemical composition, by their temperature preferences, and by the molecular structure of their RNA, DNA and proteins. There are two main sub-groups of prokaryotes, as indicated by studies of these characters, and by recent molecular phylogenetic analyses (Figure 4.6). The Kingdom Eubacteria includes the Cyanobacteria (blue-green algae) and most groups commonly called Bacteria. The Kingdom Archaebacteria, consisting of the Halobacteria (salt-digesters), Methanobacteria (methane producers) and Eocytes (heat-loving sulphur-metabolizing bacteria), is closer to the origin of more complex life.

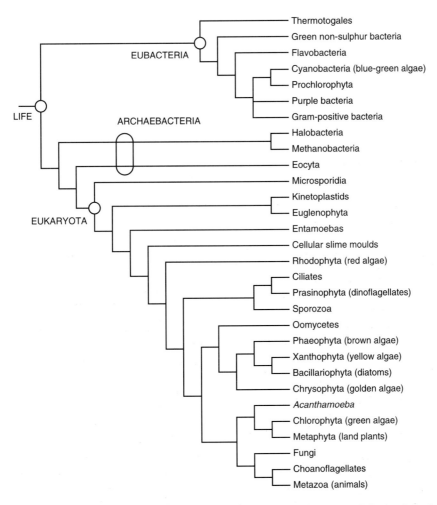

Fig 4.6 The relationships of basal organisms, based on recent molecular phylogenetic work. The major prokaryote groups are indicated (Eubacteria + Archaebacteria), as well as the major subdivisions of Eukaryota. Archaebacteria incudes Halobacteria, Methanobacteria and Eocyta. Among eukaryotes, most of the groups indicated are traditionally referred to as 'algae', both single-celled and multicelled, and these are usually regarded as primitive plants. The metaphytes (land plants), fungi, and metazoans (animals) form part of a derived clade within Eukaryota, indicated here near the base of the diagram.

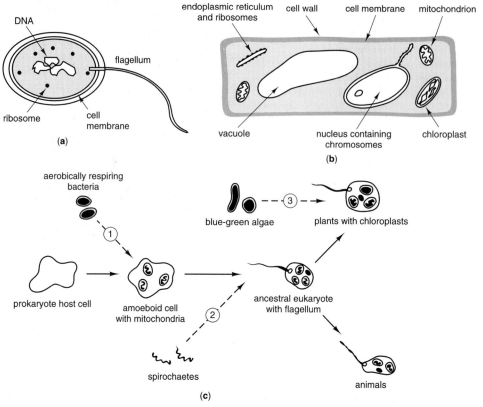

Fig 4.7 Eukaryote characters. A typical prokaryote cell (a) differs from a eukaryote plant cell (b) in the absence of a nucleus and of organelles. The endosymbiotic theory for the origin of eukaryotes (c) proposes that cell organelles arose by a process of mutually beneficial incorporation of smaller prokaryotes into an amoeba-like prokaryote (steps 1, 2, 3).

Diversification of multicellular algae

Prokaryotes are single-celled organisms, although some form filaments and loose 'colonial' aggregations. True multicellular organisms arose only among the eukaryotes. These are plants and animals that are composed of more than one cell, typically a long string of connected cells in early forms. Multicellularity had several important consequences, one of which was that it allowed plants and animals to become large (some giant seaweeds or kelp, forms of algae, reach lengths of tens of metres). Another consequence of multicellularity was that cells could specialize within an organism, some being adapted for feeding, others for reproduction, others for defence, and others for communication.

One of the oldest multicellular organisms is a bangiophyte red alga (rhodophyte) preserved in silicified carbonates of the Hunting Formation, eastern Canada, dated at 1260–950 Ma (Figure 4.9a). This simple strand of

50-µm-wide cells was neither very large, nor did its cells specialize. However, new modes of life became possible. In the Lakhanda Group of eastern Siberia, 1000–900 Ma, five or six metaphyte species have been found (Figure

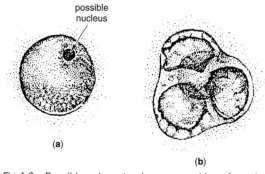

Fig 4.8 Possible eukaryotes in an assemblage from the Bitter Springs Cherts of central Australia (*c.* 900 Ma). (a) *Glenobotrydion*, a cell containing a possible nucleus. (b) *Eotetrahedrion*, a set of four cells supposedly preserved in the process of cell division, just after meiosis.

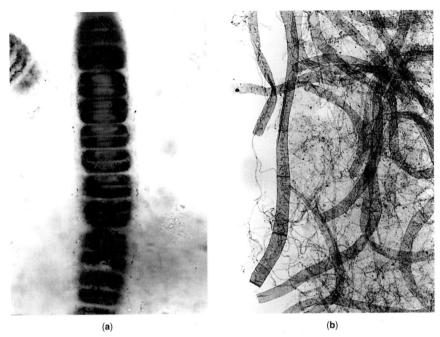

(a)

(b)

Fig 4.9   Some of the oldest algae. (a) A bangiophyte red alga from the Hunting Formation, of Canada (*c*. 950 Ma), 50 μm wide. (b) Filamentous alga from the Lakhanda Group, Siberia (c. 1000 Ma), 400 μm wide.

4.9b), as well as a colonial form that forms networks rather like a slime mould. The Svanbergfjellet Formation of Spitsbergen, *c*. 750 Ma, has yielded various multicellular green algae (chlorophytes).

The 'algae' are a paraphyletic assortment of single-celled and multicelled organisms, all of them eukaryotes, and many of them photosynthetic. The major groups are distinguished by their colour, morphology and biochemical properties. Molecular phylogenies (Figure 4.6) show that most of the basal eukaryotes have traditionally been termed 'algae'. Several algal groups now seem to be closely related to true plants (see p. 225). The fossil record of algae is patchy, but exceptions are the biostratigraphically useful acritarchs, dinoflagellates, coccoliths and diatoms, and calcareous algae such as dasycladaceans, charophytes and corallines (see pp. 244–252).

References

Cairns-Smith, G. (1985) *Seven clues to the origin of life: a scientific detective story*. Cambridge University Press, Cambridge.

Knoll, A. H. (1992) The early evolution of eukaryotes: a geological perspective. *Science*, **256**, 622–627.

Mendelson, C. V. and Schopf, J. W. (1992) Proterozoic and selected Early Cambrian microfossils and microfossil-like objects. In J. W. Schopf and C. Klein (eds) *The Proterozoic biosphere: a multidisciplinary study*. Cambridge University Press, New York, pp 865–951.

Schopf, J. W. (1993) Microfossils of the Early Archean Apex Chert: new evidence of the antiquity of life. *Science*, **260**, 640–646.

Further reading

Briggs, D. E. G. and Crowther, P. R. (eds) (1990) *Palaeobiology: a synthesis*. Blackwell Scientific Publications, Oxford.

Schopf, J. W. (1992) *Major events in the history of life*. Jones and Bartlett Publishers, Boston.

5 Early metazoans

Key Points

- Few basic body plans have appeared in the fossil record; most invertebrates have a triploblastic architecture.
- Two main groupings are recognized: echinoderm–hemichordate–chordate (radialian and deuterostomous) and mollusc–annelid–arthropod (spiralian and protostomous) lines.
- The Ediacara fauna was a soft-bodied assemblage of mainly probable cnidarians, reaching its acme during the late Proterozoic.
- The small shelly fauna (SSF) was the first skeletalized assemblage of metazoans.
- Worms and worm-like organisms are mainly soft-bodied and poorly represented in the fossil record and may contain the common origins of many invertebrate phyla.
- Parazoans are a grade of organization of multicellular complexes with few cell types and lacking variation in tissue or organs; the Phylum Porifera (the sponges) contains typical parazoans which lack a gut.
- Sponges are filter-feeding members of the sessile benthos; the group contains a variety of grades of functional organization. Sponge reefs were dominated by calcareous grades.
- Stromatoporoids are a poriferan grade of organization, important in reefs during the mid Palaeozoic and mid Mesozoic. Archaeocyathans were mainly solitary, Cambrian parazoans with a modular growth mode.

Introduction

The term 'early metazoans' is treated here as an operational category for animals that first appeared at and near the Precambrian–Cambrian boundary. The grouping thus contains some simple organisms such as the parazoan sponges, with cellular-level organization, together with the controversial cnidarian-type animals in the Ediacara fauna; however, undoubtedly more complex metazoans occur within the diverse micro-assemblages of the small shelly fauna. These organisms helped set the evolutionary agenda for Phanerozoic life.

Invertebrate body and skeletal plans

Life on our planet has been evolving for nearly 4 billion years (see Chapter 4), during which time as many as 35 phyla evolved. Despite the infinite theoretical possibilities for invertebrate body plans, relatively few basic types became established (Figure 5.1). These body plans are defined in terms of the number and type of enveloping walls of tissue together with the presence, absence or form of the coelom. The basic unicellular grade is typical of protozoan organisms and is probably ancestral to the entire animal kingdom. The parazoan body plan, seen in sponges, is characterized by groups of cells usually organized in two layers separated by jelly-like material, punctuated by so-called wandering cells or amoebocytes; the cell aggregates are not differentiated into tissue types or organs. The diploblastic grade or body plan, typical of cnidarians, has two layers – an outer ectoderm and an inner endoderm – which are separated by the inert, gelatinous mesogloea. The triploblastic plan, seen in most other animals, has three layers of tissues from the outside in: the ectoderm, the mesoderm and the endoderm. Acoelomate triploblasts, such as the platyhelminth and possibly nemertine worms, lack a coelom. Pairs of

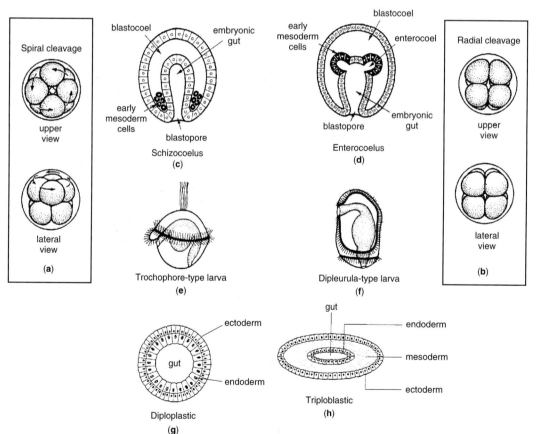

Fig 5.1   Main invertebrate body plans and larvae: upper and lateral views of spiral (a) and radial (b) patterns of cell cleavage, development of the mesoderm in the spiralians (c) and radialians (d), diploblastic (g) and triploblastic (h) body plans and the trochophore- (e) and dipleurula- (f) type larvae.

organs are arranged along the length of the animal. A blastocoel, a cavity between dividing cells during early ontogeny, can develop in many metazoans to be filled later by tissue. The triploblastic pseudocoelomates, such as the nematode worms, possess this false coelom during adult stages.

Most invertebrate groups, however, are variations on the triploblastic coelomate body plan, diversifying with the acquisition of skeletons at the base of the Cambrian. The metamerous coelomates (e.g. the annelid worms and the arthropods) have a segmented body; each segment possesses identical paired organs such as kidneys and gonads usually with appendages. Amerous coelomates have simple undivided coelomic cavities; they lack mineralized skeletons and many are burrowing forms. Pseudometamerous coelomates (e.g. the molluscs) have an undivided coelom and irregularly duplicated organs.

The oligomerous plan has a coelom divided longitidinally into two or three zones, each with different functions. Based around this plan the lophophorates are char-acterized by sac-like bodies with a specialized feeding and respiratory organ, the lophophore. However, many hemichordates possess a crown of tentacles and some have paired gill slits. The echinoderms have an elaborate water vascular system which drives feeding, locomotion and respiration.

The identification of invertebrate body plans is a useful method of grouping organisms according to their basic architecture; however, similarities between grades of construction unfortunately do not always mean a close taxonomic relationship. Certain body plans have clearly evolved more than once in different groups.

The skeleton is an integral part of the body plan of an animal, providing support, protection and attachment for muscles. Many animals such as the soft-bodied annelids possess a hydraulic skeleton with the movement of fluid providing support. Rigid skeletons, usually based on mineralized material, may be external (exoskeleton), in the case of most invertebrates; or internal (endoskeleton) structures, in the case of a few molluscs, echinoderms

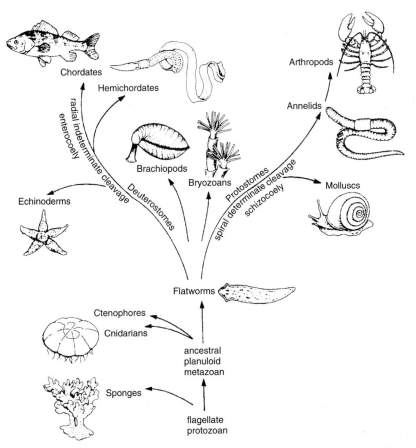

Fig 5.2 Phylogenetic relationships among the main invertebrate groups.

and the vertebrates. Growth is accommodated in a number of ways. Most invertebrate skeletons grow by the addition of new material, a process termed accretion; arthropods however grow by periodic bursts between intervals of ecdysis or moulting. Echinoderms grow by both accretion to existing material and by the appearance of new calcitic plates; the skeleton is constantly changing owing to a dynamic balance between growth and resorption.

Classification and relationships

There are two patterns of cell division in the early development of metazoans, and these define two groups: the spiralians, with an initial sequence of cell division following a spiral track with the early separation of cells, and the radialians in which cell cleavage promotes a radial pattern of new cells having a longer more flexible history. These styles of cell division, together with the mode of mouth development, has helped define two further major invertebrate categories. Most protostomes are spiralians where the mouth develops directly from an opening, the blastopore, resulting from cell growth and migration. The deuterostomes, however, are radialians with a mouth arising from a secondary opening; the true blastopore often develops as an anus. Not all phyla fit simply into these two major divisions, but by consensus based on comparative morphology, two main streams emerge: the echinoderm–hemichordate–chordate (radialian and deuterostomous) and the mollusc–annelid–arthropod (spiralian and protostomous) groupings (Figure 5.2). The cnidarians and the lophophorates do not fit readily into either lineage, although both are characterized by radial cleavage in the early embryo. Both protostome and deuterostome taxa characterize the Cambrian fauna. Divergence must have occurred at least 570 million years ago. In general, the deuterostome groups share more recent common ancestors than those of the protostomes.

The phylogeny of the earliest metazoans is studied by means of three lines of evidence: fossils (Conway Morris, 1993a), molecular phylogeny and larval stages. For example, the nauplius larva is most typical of crustaceans and the trochophore larva occurs in the molluscs; those groups with trochophores may, therefore, have shared a common ancestor. However, Williamson (1992) has suggested that larvae may be transferred across apparently unrelated groups and this would severely limit their use in phylogenetic reconstructions.

Ediacara fauna

Since the first impressions of soft-bodied animals were identified in the Upper Proterozoic rocks of Namibia and in the Pound Quartzite in the Ediacara Hills, north of Adelaide in southern Australia, 50 years ago, the remarkable Ediacara assemblage has now been documented from all continents. More than 100 species of these unique animals have been described on the basis of moulds usually preserved in shallow-water siliciclastic sediments, more rarely carbonates, or even turbidites. Although morphologically diverse, the Ediacarans have many features in common. All were soft-bodied, with high surface-to-volume ratios and marked radial or bilateral symmetries. These thin, sometimes ribbon-shaped, animals may have operated by direct diffusion processes, obviating the need for gills and other more complex internal organs. Most have been studied from environments within the photic zone; those collected from some deeper-water deposits were probably washed in. Provincialism among these Upper Proterozoic faunas was weak, and many later taxa had nearly world-wide distributions. It is possible that the flesh of the Ediacara animals littered areas of the late Precambrian sea-floor; predators and scavengers had yet to evolve in sufficient numbers to remove it.

Morphology and classification

The Ediacara fauna has been viewed from three different perspectives. Firstly, the Ediacara animals were traditionally assigned to a variety of Phanerozoic invertebrate groups on the basis of apparent morphological similarities; secondly, Fedonkin (1990) has developed a form classification for the group (Figure 5.3), while thirdly, Seilacher (1989) has removed them from the Metazoa altogether (Box 5.1). Fedonkin's classification (1990) includes the Radiata (radial animals) containing most colonial organisms in the fauna, e.g. *Charnia*, *Charniodiscus* and *Rangea*, assigned to coelenterates and part of the sessile benthos; many taxa have been compared with sea-pens or the pennatulaceans. The Bilateria (bilateral animals) includes both smooth and segmented forms. The smooth *Vladimissa* and *Platypholinia* may be turbellarians, types of platyhelminth worms; the segmented *Dickinsonia* may represent an early divergence from the radial forms, whereas *Spriggina*, although superficially similar to some annelids and arthropods, has a unique morphology.

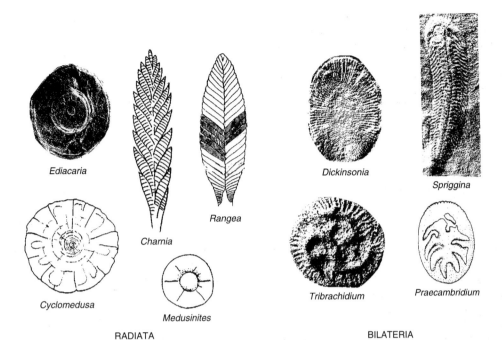

Ediacaria

Rangea

Charnia

Cyclomedusa

Medusinites

Dickinsonia

Spriggina

Tribrachidium

Praecambridium

RADIATA

BILATERIA

Fig 5.3   Some typical Ediacara fossils; those illustrated on the left, the Radiata, have been associated with the cnidarians, whereas those on the right, the Bilateria, may be related to the annelids and arthropods. *Ediacaria* (×0.3), *Charnia* (×0.3), *Rangea* (×0.3), *Cyclomedusa* (×0.3), *Medusinites* (×0.3), *Dickinsonia* (×0.6), *Spriggina* (×1.25), *Tribrachidium* (×0.9), and *Praecambridium* (×0.6)

Ecology

There is little doubt that the Ediacara biotas dominated the latest Precambrian marine ecosystem, occupying a range of ecological niches and pursuing varied life strategies probably within the photic zone. It is also possible that these flattened animals hosted photosymbiotic algae, maintaining an autotrophic existence in the tranquil 'Garden of Ediacara'; however this model does have its opponents. The ecosystem was dominated by medusoid pelagic animals and attached, sessile benthos, infaunal animals were sparse; food chains were probably short and the trophic structure was apparently dominated by suspension and deposit feeders (Figure 5.5).

Extinction of the Ediacarans

The Ediacara fauna, as a whole, became extinct about 550 million years ago; nevertheless in terms of longevity, the ecosystem was very successful. Moreover, the lobopod *Xenusion* from the lowermost Cambrian may be a relict of the Ediacara fauna; significantly this animal appears to be armoured with sets of spines. In addition,

recent reports (Conway Morris, 1993b) of Ediacara-like fossils from Lower Cambrian Burgess Shale equivalents in North America suggest that some relicts survived into the Phanerozoic and may not have been so different from contemporary metazoans.

The rise of predators and scavengers, together with an increase in atmospheric oxygen, may have at last prevented the routine preservation of soft parts and soft-bodied organisms. More importantly, the Ediacara body plan offered little defence against active predation. There is abundant evidence for Cambrian predators: damaged prey, actual predatory organisms and the appearance of defence structures, such as trilobite spines and multielement skeletons. All suggest the existence of a predatory life strategy which was probably established prior to the beginning of the Cambrian Period. The Proterozoic–Cambrian transition marked one of the largest faunal turnovers in the geological record, with a significant move from soft-bodied, possibly photo-autotrophic animals to heterotrophs relying on a variety of nutrient-gathering strategies. However, it is still uncertain whether a true extinction or the slamming shut of a taphonomic window accounted for the disappearance of the Ediacara fauna from the fossil record.

Box 5.1 Vendobionts or the first true metazoans

The extraordinary suggestion (Seilacher, 1989) that the Ediacara animals are not related to the familiar metazoans at all, but may have been an independent experiment in animal design that failed, has generated much debate. Not only are the fossils similarly preserved but they also share quilted pneu (rigid, hollow, airbed-like) structures with sometimes additional struts and supports together with a significant flexibility (Figure 5.4). If the Ediacara biota is in fact divorced from the true metazoans and indeed may be grouped together as a separate grade of organization – termed by Seilacher the Vendozoa – certain generalizations about their anatomy and behaviour, some speculative, may be made. Reproduction may have been by spores or gametes, and growth was achieved by both isometric and allometric modes. The skin or integument had to be flexible, although it could crease and fracture. Moreover, the skin must have acted as an interface for diffusion processes whilst providing a watertight seal to the animal. However, this stimulating and original view of the fauna remains controversial. The common frond-like morphs have been assigned to the pennatulaceans, a group of soft corals. Recently, Valentine (1992) has reinterpreted the distinctive *Dickinsonia* as a coral-like cnidarian possessing multiple polyps. Buss and Seilacher (1994) have suggested a compromise. Their Phylum Vendobionta includes cnidarian-like organisms, but lacking cnidae, the stinging apparatus typical of the cnidarians; vendobionts thus comprise a monophyletic sister group to the Eumetazoa. This interpretation requires the true cnidarians to acquire cnidae as an apomorphy for the phylum.

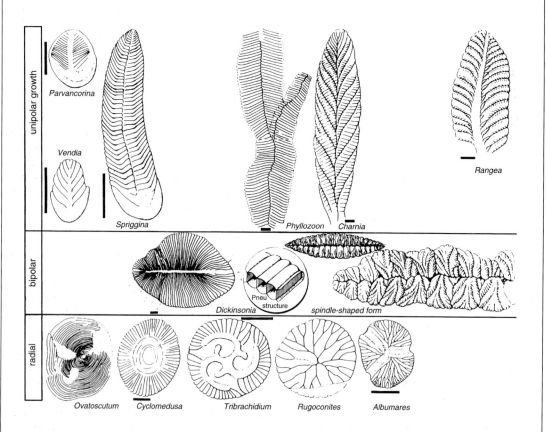

Fig 5.4   Vendozoan constructional morphology, recognizing unipolar, bipolar and radial growth modes within the Ediacara-type fauna. Scale bars are 10 mm.

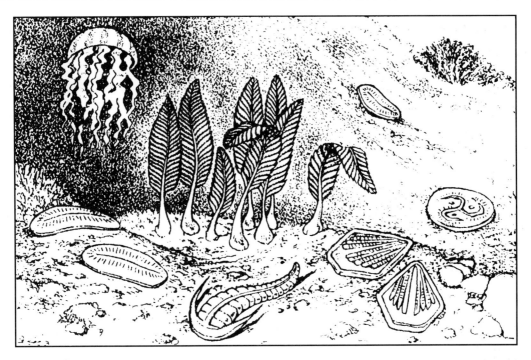

Fig 5.5 An Ediacara community including a fixed and mobile tiered benthos, together with nektonic and plank-tonic microcarnivores.

Small shelly fauna

A distinctive assemblage of small shelly fossils, tradi-tionally labelled the Tommotian fauna, has recently been documented in considerable detail from the Precambrian –Cambrian transition; the assemblage is most extrava-gantly developed in the lowest Cambrian stage of the Siberian Platform, the Tommotian, which gives its name to the fauna. A great deal is now known about the strati-graphical distribution and palaeobiogeography of these organisms through current interest in the definition of the base of the Cambrian System. Nevertheless, the biologi-cal affinities of many members of the fauna have yet to be established. The assemblage, although dominated by minute species, together with small sclerites of larger species, represents the first appearance of diverse skele-tal material in the fossil record, some 10 million years before the first trilobites evolved (see Chapter 8).

This type of fauna is not restricted to the Tommotian Stage; these fossils are also common in the overlying Adtabanian Stage. The less time-specific term, small shelly fossils (or SSF) was introduced by Matthews and Missarzhevsky (1975) to describe these assemblages. The fauna is now known to include a variety of phyla united by the minute size of their skeletal components and a sudden appearance at the base of the Cambrian. The small shelly fauna probably dominated the earliest Cambrian ecosystems where many metazoan phyla developed their own distinctive characteristics, initially at a very small scale.

Composition and morphology

Many of the small shelly skeletons (Figure 5.6) were retrieved from residues after the acid etching of lime-stones; thus there is a bias towards acid-resistant skeletal material in any census of the group as a whole. Moreover, it is not clear whether the acid-resistant skele-ton of the Tommotian animals was a primary construc-tion or a secondary replacement fabric. Nevertheless, the small shelly animals had skeletons composed of a vari-ety of materials. For example, *Cloudina* and the anabar-itids were tube-builders which secreted carbonate mater-ial, whereas *Mobergella* and the sclerites of *Lapworthella* were phosphatic; *Sabellidites* is an organ-ic-walled tube, possibly of an unsegmented worm.

Many of the small shelly fossils are form taxa since the biological relationships of most cannot be established

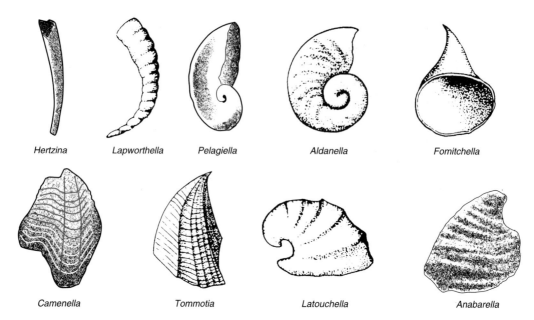

Fig 5.6   Elements of the Tommotian-type or small shelly fauna (all about ×20 magnification except *Fomitchella* which is about ×40).

and often there are few clues regarding the function and significance of each skeletal part. Most are short-lived and have no obvious modern analogues. Two groups are common: the hyolithelminthids have phosphatic tubes, open at both ends, whereas the tommotiids are phosphatic cone-shaped shells, usually occurring in pairs.

Recent discoveries of near-complete examples of *Microdictyon*-like animals, from the Lower Cambrian of China and of halkieriids from North Greenland (Conway Morris and Peel, 1990) have shown that some elements of the small shelly fauna were parts of chain-mail exoskeletons of appendiculate worms. These had round to oval plates arranged in pairs along the length of the body which may have provided a base for muscle attachment associated with locomotion; they were probably related to the onychophorans. As noted previously, many of the small shelly fossils are probably the sclerites of larger multiplated worm and worm-like animals.

Distribution and ecology

It is unclear whether many of the small shelly skeletons are isolated shells or sclerites forming part of an armour, and although the autecology of most groups is unknown, the assemblage is certainly the first attempt to establish a skeletalized benthos. Few of the small shelly skeletal parts exceed a centimetre; however, both mobile and fixed forms occurred together with archaeocyathans and nonarticulate brachiopods. The Meishucunian Stage in southern China has yielded some of the most diverse Tommotian-type assemblages in strata of Atdabanian age (Qian Yi and Bengtson, 1989). Nearly 40 genera belong to three, largely discrete, successive ecological assemblages. Many of these fossils are known from Lower Cambrian horizons elsewhere in the world, highlighting the global distribution of the fauna. Nevertheless the Chinese evidence emphasizes that this early fauna was possibly organized into a number of different 'community' types.

The micro-benthos of the Tommotian and Atdabanian was soon succeeded by the more typical Cambrian fauna, dominated by trilobites, nonarticulate brachiopods, monoplacophoran molluscs, primitive echinoderms, and archaeocyathans during the Atdabanian Stage (Figure 5.7).

Soft-bodied lower invertebrates

Of the 24 animal phyla documented by Margulis and Schwartz (1988), fewer than 35% have an adequate fossil record and many are only represented by a few taxa. There are, however, a number of larger phyla which have a poor fossil record because they lack a preservable

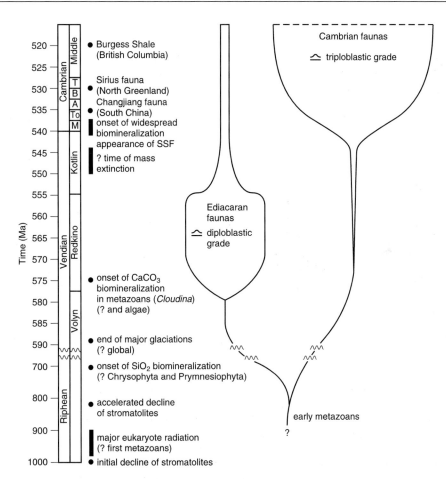

Fig 5.7 Stratigraphical distribution of late Precambrian and early Palaeozoic biotas, and the timing of some major environmental events. A, Atdabanian; B, Botomian; M, Manykian; T, Toyonian: To, Tommotian.

skeleton; a number of these soft-bodied forms are fortunately preserved in fossil Lagerstätten. Most are worms or worm-like organisms. But in spite of unspectacular fossil records there is considerable interest in these poorly represented invertebrates (Conway Morris, 1985). The origins of many higher taxa must be sought within the plexus of worm-like organisms (Figure 5.8). Moreover, many of these groups probably dominated marine palaeocommunities in terms of both numbers and biomass, and additionally contributed to associated trace fossil assemblages.

The platyhelminthes (i.e. the flatworms) are bilateral animals with organs composed of tissues arranged into systems. Most are parasites: but the turbellarians are free-living carnivores and scavengers. The Ediacaran animals *Dickinsonia* and *Palaeoplatoda* have been assigned to the turbellarian flatworms by some authors;

similarly, *Platydendron* from the Middle Cambrian Burgess Shale has been ascribed to the platyhelminthes.

The ribbon worms, or the nemertines, are characterized by a long anterior proboscis. The majority are marine, although some inhabit soil and freshwater. Although the bizarre *Amiskwia* from the Middle Cambrian Burgess Shale was assigned to this group, recent opinion favours morphological convergence rather than taxonomic affinity. The nematodes or roundworms are generally smooth and sac-like.

The priapulid worms are exclusively marine, short and broad with prosces covered in spines and warts. The Middle Cambrian Burgess Shale contains seven monospecific genera assigned to at least five families. The Burgess forms are all characterized by priapulid probosces, and most have little in common with modern forms. Nevertheless the most abundant taxon, *Ottoia*, is

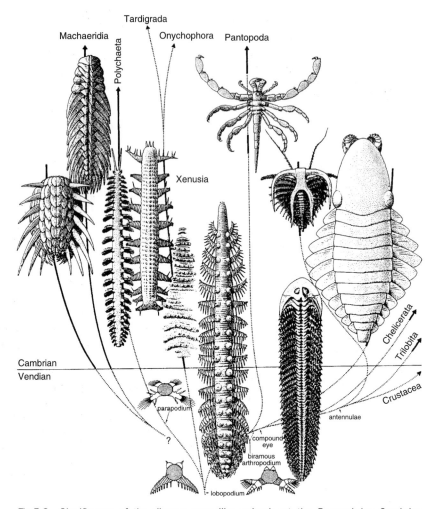

Fig 5.8 Significance of the diverse worm-like animals at the Precambrian–Cambrian boundary and the postulated origins of some major clades.

very similar to the living genus *Halicryptus*. Elsewhere in the fossil record, the Upper Carboniferous Mazon Creek fauna has yielded *Priapulites* which has a distinctly modern aspect.

The annelid worms have ring-like external segments coincident with internal partitions housing pairs of digestive and reproductive organs; the nervous system is well developed and the head has distinctive eyes. The annelid body is ornamented by bristles which aid locomotion and provide stability. Most are predators or scavengers living in burrows. The polychaetes, or paddle worms, have the most complete fossil record, which is enhanced by the relatively common preservation of elements of the chitinous jaw apparatus, called scolecodonts. Although some Ediacara animals, such as *Spriggina*, have been associat-

ed with the polychaetes, the first undoubted paddle worms are not known until the Cambrian. A diverse polychaete fauna has been described from the Burgess Shale; it even contains *Canada spinosa* which is very similar to some living polychaetes.

Parazoan strategy

The parazoan grade of organization is found in one or more phyla having aggregates of cells of only a few types, and which are not differentiated into tissues or organs. Parazoans are in some ways intermediate between the unicellular protozoans and the complex multicellular metazoans.

Nevertheless, the parazoan organization was probably an evolutionary dead end; it shared a common ancestor with the main metazoan lineage but certainly did not lead to it. The sponges, stromatoporoids and archaeocyathans successfully followed this simple strategy, often dominating reef ecosystems in a variety of marine environments throughout the Phanerozoic. They all probably operated as high-level filter feeders within a suspension-based food chain.

Porifera

The poriferans or sponges have a unique porous structure and a body plan based at the cellular level of organization. Most lack symmetry, differentiated tissues, and organs. There are over 10 000 species of sponge, all of which are aquatic, and the vast majority are marine. Sponges are part of the sessile benthos, pumping large volumes of water through their fixed but flexible bodies which act as filters for nutrients. The group has a remarkable range of morphologies; the more bizarre, stalked forms lived in deep-water environments. The classification of the phylum has recently undergone considerable revision (Box 5.2). Some well-established calcified groups (such as the 'chaetetids' and 'sphinctozoans') are probably polyphyletic, merely representing convergence towards common grades of organization.

Morphology

A typical sponge is sac-shaped with a central cavity or paragaster which opens externally at the top through the osculum (Figure 5.9). The sponge is densely perforated by ostia marking the entrances to minute canals through which pass the inhalent currents. There are three main cell types: flattened epithelial cells; collar cells or choanocytes which occupy the internal chambers and

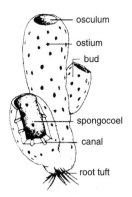

Fig 5.9   Basic sponge morphology.

move water along by beating their flagella; and amoeboid cells which have digestive, reproductive and skeletal functions. Nutrient-laden water is thus sucked through the ostia, flagellated by the choanocytes and processed by the amoeboids. Waste products and spent water are ejected upwards through the paragaster into the water column. Despite their flexibility, sponges are skeletal organisms. Skeletons are usually composed of a colloidal jelly or spongin, a horny organic material, together with calcareous or siliceous spicules; whereas some groups have a basal calcareous skeleton.

Three basic levels of organization have been recognized among the sponges (Figure 5.10). The simple ascon sponges are sacs with a single chamber lined by flagellate cells whereas the sycon grade have a number of simple chambers with a single central paragaster. The leucon grade, where a series of sycon chambers access a large central paragaster, is the most common.

Ecology

Sponges are part of the sedentary benthos, usually dominated by a large exhalent opening communicating upwards with the water column. The group is entirely

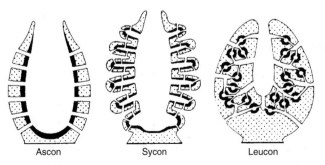

Fig 5.10   Main grades of sponges.

Box 5.2 Classification of the sponges

The Phylum Porifera was traditionally subdivided into four classes, the Demospongea, Sclerospongea, Calcarea and Hexactinellida, based mainly on the composition of the skeleton and type of spicules (Figure 5.11). Higher-level taxonomy is based exclusively on soft tissue morphology. Some workers have suggested the exclusion of the glass sponges from the Porifera. However, the sclerosponges, with additional calcareous skeletons, are now placed within the Demospongea. Thus three classes now comprise the phylum (see also Figure 5.12).

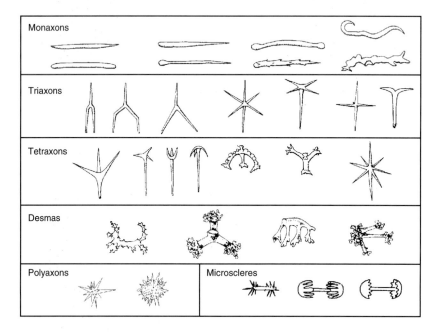

Fig 5.11 Main categories of spicule morphology (all about ×75 magnification except microscleres which is about ×750).

Class Calcarea (calcareous sponges)
Sponges with calcitic spicules, commonly simple, or porous calcareous walls without spicules. Some species have a basal calcareous skeleton. Cambrian–Recent.

Class Demospongea (common sponges)
Sponges with skeletons of spongin, a mix of spongin and siliceous spicules or only siliceous spicules. The spicules may be of two different sizes, the larger ones are typically monaxons and tetraxons. Cambrian–Recent. Living sponges previously assigned to the Sclerospongea (coralline sponges) – sponges with a compound skeleton of siliceous spicules, spongin and an additional basal layer of laminated fibrous aragonite or calcite – are now also included here. Basal calcareous skeletons are developed in a number of species spread across several orders.

Class Hexactinellida (siliceous sponges)
These are the glass sponges with complex siliceous spicules having six rays directed along three mutually perpendicular axes. Cambrian–Recent.

Box 5.2 (Cont.)

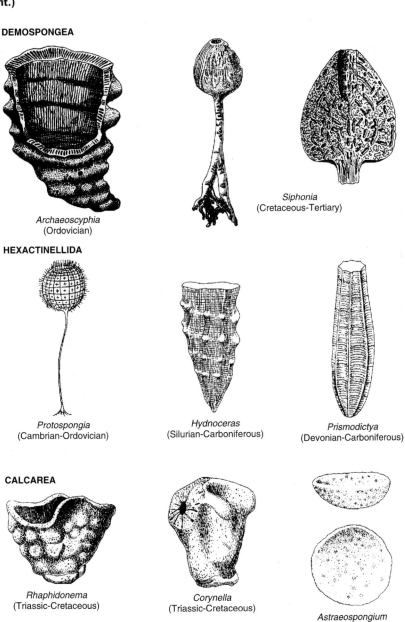

DEMOSPONGEA

Archaeoscyphia
(Ordovician)

Siphonia
(Cretaceous-Tertiary)

HEXACTINELLIDA

Protospongia
(Cambrian-Ordovician)

Hydnoceras
(Silurian-Carboniferous)

Prismodictya
(Devonian-Carboniferous)

CALCAREA

Rhaphidonema
(Triassic-Cretaceous)

Corynella
(Triassic-Cretaceous)

Astraeospongium
(Silurian-Devonian)

Fig 5.12 Some examples of the main groups of sponges. *Archaeoscyphia* (×0.25), *Siphonia* (×0.4 and 0.8), *Protospongia* (×0.4), *Hydnoceras* (×0.25), *Prismodictya* (×0.6), *Rhaphidonema* (×0.8), *Corynella* (×0.8), *and Astraeospongium* (×0.4).

aquatic, living attached in a range of environments from the abyssal depths of oceans to the moist barks of trees in the humid tropics. Most Palaeozoic and early Mesozoic forms inhabited mainly shallow water environments. Today, however, sponges apparently occupy a much wider range of environments. Modern hexactinellids prefer depths of 200–600 m, probably extending down into the abyssal depths of submarine trenches, whereas the calcareous sponges are most common in depths of less than 100 m. The modern calcified sponges are often cave-dwellers, hidden in the shadows of submarine crevices at depths of 5–200 m, mainly in the Caribbean, although the group occurs elsewhere including the Mediterranean. Few predators attack sponges, although some fishes, snails, starfish and turtles have been observed eating their soft tissues in the tropics.

Stratigraphical distribution and sponge reefs

The evolution of the sponges (Wood, 1991) is intimately related to their participation in reef ecosystems (Figure 5.13). Sponges can possess a rigid skeleton by the fusion of strong spicules or by the development of an additional basal calcareous skeleton, whereas particular grades of organization suited special environmental conditions. The first sponges probably appeared in the late Proterozoic as clusters of flagellate cells. The stratigraphical distribution of sponge abundance is strongly linked with major phases of reef development. The Cambrian sponge fauna, of thin-walled and weakly fused

spiculate demosponges, hexactinellids, and early calcisponges, is mainly cosmopolitan. In contrast, the Ordovician sponge faunas are characterized by the heavier, thick-walled demosponges which continued to dominate Silurian faunas.

The sponges were the dominant components of reefs during the mid to late Ordovician, where the stromatoporoid grades, possibly demosponges, formed the frameworks. The most impressive structures were developed during the mid Devonian. During the late Carboniferous the chaetetid calcified sponges were locally important. In the Permian and mid Triassic, structures constructed by sphinctozoans exist, and the mid to late Jurassic was marked by bioherms of lithistid demosponges. Jurassic sponge reefs dominated by hexactinellids and lithistids have been documented throughout the Alpine region. Cup-shaped and discoidal morphotypes dominate soft substrates and developed a substantial topography above the sea-floor; encrusting morphotypes favoured hard substrates.

As noted previously, the acquisition of a calcareous skeleton was not confined to any one class; the calcareous skeleton was developed a number of times, convergently, across the phylum, with a few basic plans superimposed on a pre-existing sponge morphology. Consequently, various groups have been recognized on the basis of the calcareous skeleton, but components of each group arose independently in different clades. In broad terms, the chaetetids and sphinctozoans, together with the archaeocyathans and the stromatoporoids, were important calcareous reef-builders. However, the decline

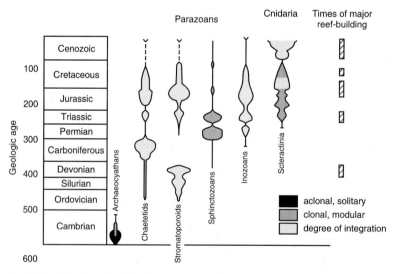

Fig 5.13 Stratigraphical distribution of reef-building sponges and related parazoans together with the scleractinian corals.

of the calcareous sponges in reef ecosystems, during the Mesozoic, is often correlated with the rise of the scleractinian corals equipped with symbiotic zooxanthellae (see Chapter 6).

Stromatoporoidea

The stromatoporoids were marine colonial organisms which appeared by the mid Ordovician. These animals were common components of Silurian and early to mid Devonian shallow-water marine communities, forming irregular mounds on the seabed, associated with calcareous algae and corals. They have a superficial resemblance to some tabulate corals. The group reached an acme during the mid Devonian but declined during the later Palaeozoic and Mesozoic. Although stromatoporoids have been classified with the cnidarians, their similarity to the modern calcified sponges supports inclusion within the poriferans. In common with a number of other poriferans, the group is polyphyletic, with stromatoporoid taxa showing gross morphological convergence towards a common body plan or grade of organization. Thin sections are usually required to describe and classify most stromatoporoid taxa.

Morphology and classification

Typical stromatoporoids have a calcareous skeleton with both horizontal and vertical structures and often a fibrous microstructure (Figure 5.14). The skeleton or coenosteum is constructed from undulating layers of calcareous laminae punctuated perpendicularly by vertical pillars. The surfaces of some forms are modified by small swellings or mamelons together with astrorhizae, radiating stellate canals, which are the traces of the exhalent current canal system. Siliceous spicules have been identified in some Carboniferous and Mesozoic taxa, suggesting that the original skeleton was in fact spiculate; the calcareous casing is secondary, with aragonite and calcite precipitated within a framework of spongin.

Ecology

Stromatoporoids were marine organisms usually associated with shallow-water carbonate facies often deposited in turbulent environments. Many genera were important constituents of reefs particularly during the Silurian and Devonian. The spectacular Silurian bioherms on the Swedish island of Gotland are characterized by a variety of stromatoporoid growth forms (Figure 5.15), and throughout North America and northern Europe, Devonian reefs are dominated by stromatoporoids. These animals had a multi-oscular, colonial or modular water system. Growth modes included columnar, dendroid, encrusting and hemispherical forms.

Stratigraphical distribution

Animals with a stromatoporoid grade of organization have been identified from rocks of Botomian age; however, these forms were apparently short-lived.

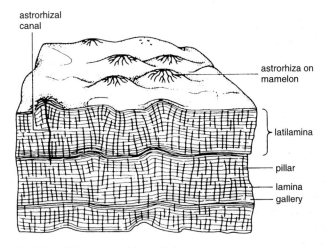

astrorhizal canal

astrorhiza on mamelon

latilamina

pillar

lamina

gallery

Fig 5.14 Stromatoporoid morphology.

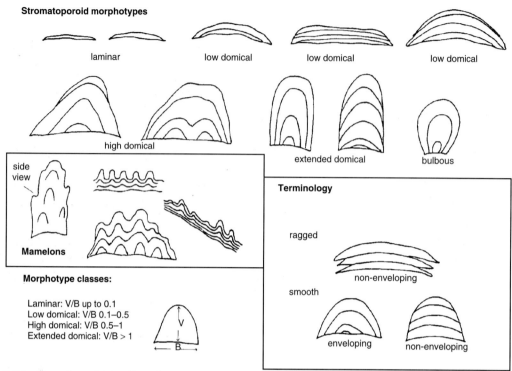

Stromatoporoid morphotypes

laminar   low domical   low domical   low domical

high domical   extended domical   bulbous

side view

Mamelons

Terminology

ragged

non-enveloping

smooth

enveloping   non-enveloping

Morphotype classes:

Laminar: V/B up to 0.1
Low domical: V/B 0.1–0.5
High domical: V/B 0.5–1
Extended domical: V/B > 1

Fig 5.15 Stromatoporoid growth modes.

Pseudostylodictyon from the Middle Ordovician of New York and Vermont may be the oldest true stromatoporoid, derived from a soft-bodied sponge-like ancestor in the Cambrian. Stromatoporoids were important reef-builders during the Silurian and Devonian, becoming largely extinct during the end-Frasnian (late Devonian) event. The grade returned in the mid and late Jurassic when stromatoporoids again participated in reef frameworks. Nevertheless, most groups disappeared at the end-Cretaceous extinction event.   Wood (1991) has emphasized however, that some living sponges have a stromatoporoid grade of organization: *Astrosclera* and *Calcifimbrospongia* are both calcified demosponges with stromatoporoid architecture.

Archaeocyatha

The Archaeocyatha or 'ancient cups' are one of only a few animal phyla that are entirely extinct. The group exploited calcium carbonate during the early part of the Cambrian radiation to construct porous cup or cone-like skeletons, usually growing together in clumps and often living with stromatolites to form reefs. The

Archaeocyatha dominated shallow-water marine environments usually in tropical palaeolatitudes. From an early Cambrian origin on the Siberian Platform, the group spread throughout the tropics, forming the first Palaeozoic reefs; however, by the end of the early Cambrian, and the start of the middle Cambrian, archaeocyathans are known only from Australia, the Urals and Siberia. They disappeared at the end of the Cambrian.

Morphology

Archaeocyathans are most commonly found in limestones, and details of their morphology are usually reconstructed from thin sections. The exoskeleton of the archaeocyathan animal is aspiculate and usually composed of a very porous, inverted cone composed of two nested concentric walls separated from each other by radially arranged, vertical septa (Figure 5.16). Both the inner and outer walls are densely perforated and together define the intervallum, partitioned into a number of segments (loculi) by the radial septa which are often less porous than the walls, or sometimes aporous. The inner wall circumscribes the central cavity, open at the top and closed at its base to form a tip; the apex of the skeleton

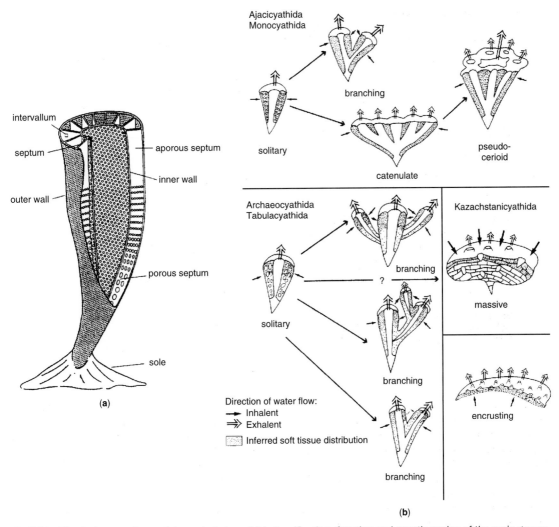

Fig 5.16   The Archaeocyathans: (a) morphology and (b) classification, function and growth modes of the main groups.

is usually buried in the sediment with a basal flange and roots or holdfasts adding anchorage and stability. In some taxa the intervallum is partitioned horizontally firstly by porous shelves or tabulae or secondly with aporous, convex dissepiments often extending into the adjacent central cavity (see also Figure 5.17).

Ecology

The archaeocyathans were exclusively marine, occurring most commonly at depths of 20–30 m on carbonate substrates. The phylum developed an innovative style of growth based on modular organization (Wood *et al.*, 1992). Such modularity permitted encrusting abilities and the possibility of secure attachment on a soft substrate; moreover, growth to a large size was encouraged, together with a greater facility for regeneration. The archaeocyathans were thus ideally suited to participate in some of the first reefal structures of the Phanerozoic (Figure 5.18), in intervals of high turbulence and rates of sedimentation, during the early Cambrian. However, although archaeocyathan reefs were probably not particularly impressive, usually up to 3 m thick and between 10 and 30 m in diameter, they were nevertheless the first animals to form biological frameworks and to establish this complex type of ecosystem. Most archaeocyathans, however, did not form frameworks.

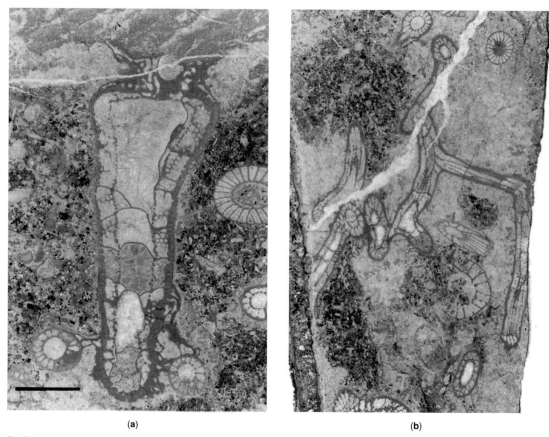

(a)   **(b)**

Fig 5.17   Some archaeocyathans, from the Lower Cambrian of Western Mongolia, in thin section: (a) cryptic, solitary individual of *Cambrocyathellus* showing holdfast structures (×7.5); (b) branching *Cambrocyathellus tuberculatus* with skeletal thickening between individuals associated with transverse sections of *Rotundocyathus lavigatus* (×5).

Box 5.3 Classification of the archaeocyathans

Two main groups have been defined within the phylum: the Class Regulares and the Class Irregulares. The regular forms have an aporous initial single-walled stage lacking dissepiments. The inner and outer walls are punctuated by septa and tabulae developed either singly or together. Most archaeocyathans belong to this class which includes the groups Monocyathida and Ajacicyathina; however, the apparent abundance of regular genera may be a result of excessive splitting. The irregulars have initial aporous single-walled stages with dissepiments. The twin walls have irregular pore structures, always dissepiments and the skeleton itself is characterized by an asymmetrical shape. The group is smaller but includes the orders Archaeocyathida and Kazachstanicyathida.

Stratigraphical and palaeogeographical distribution

The first archaeocyathans are known from the lowest Cambrian (Tommotian) rocks of the Siberian Platform and are represented by mainly solitary regulars (Figure 5.20). During the early Cambrian, the phylum diversified, migrating into areas of North Africa, the Altai Mountains of the former USSR, North America and South Australia. Archaeocyathans were most common in the mid-Early Cambrian (Botomian) when a number of distinct biogeographical provinces can be defined, but by the Lenian Stage the group was very much in decline; few genera have been recorded from the Middle Cambrian and only one is known from Upper Cambrian strata. Archaeocyathan history demonstrates a progressive move towards a more modular architecture in response to conditions of high turbulence. In general, solitary taxa domi-

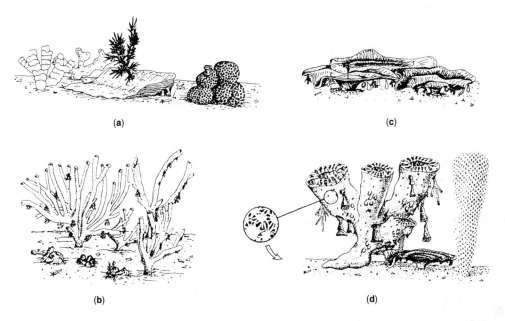

Fig 5.18   Archaeocyathan reef structures, which, when preserved, become (a) boundstones, (b) bafflestones, (c) bindstones to (d) bioherms.

Box 5.4 Functional morphology of the archaeocyathans

In an experimental biomechanical study, Savarese (1992) has described three morphotypes (aseptate, porous septate and aporous septate), constructed and subjected to currents of coloured liquid in a flume (Figure 5.19). The first morphotype, a theoretical reconstruction, performed badly, with fluid escaping through the intervallum while also leaking through the outer wall. The porous septate forms, however, suffered some slight leakage through the outer wall, but no fluid passed through the intervallum. The aporous septate form was most efficient, with no leakage through the outer walls and no flow through the intervallum. Significantly, in ontogenetic series, initially porous septate morphotypes develop an aporous condition in later life perhaps to avoid leakage through the outer wall.

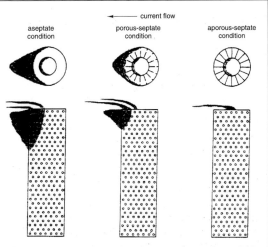

Fig 5.19   Modelling the functional morphology of the archaeocyathans.

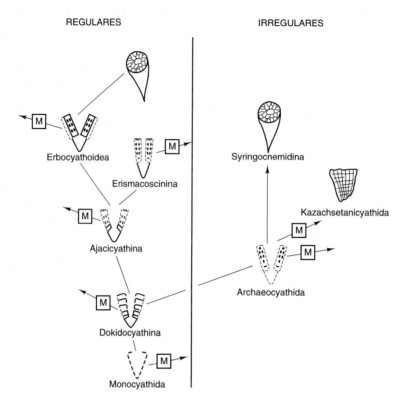

REGULARES   IRREGULARES

Erbocyathoidea

Erismacoscinina

Syringocnemidina

Ajacicyathina

Kazachsetanicyathida

Archaeocyathida

Dokidocyathina

Monocyathida

Fig 5.20 Evolutionary trends within the archaeocyathans; modular forms, appearing itera-tively, are indicated by M.

nated the early Cambrian; but following the late Botomian, modular morphotypes continued after the extinction of most solitary forms. Nevertheless, the abundance and diversity of the group in some parts of the world, particularly in Lower Cambrian rocks, has prompted its effective use in biostratigraphical correlation.

Fossil invertebrates. Blackwell Scientific Publications, pp. 107–115.

Rigby, J.K. and Scrutton, C.T. (1985) Sponges, chaetetids and stromatoporoids. In J.W. Murray, (ed.) *Atlas of invertebrate macrofossils.* Longman, London, pp 3–10.

Further reading

Clarkson, E.N.K. (1993) *Invertebrate palaeontology and evolution,* 3rd edition. Chapman and Hall, London.

Glaessner, M. (1984) *The dawn of animal life.* Cambridge University Press, Cambridge.

Rigby, J.K. (1987) Phylum Porifera. In R.S. Boardman, A.H. Cheetham and A.J. Rowell (eds) *Fossil invertebrates.* Blackwell Scientific Publications, pp. 116–139.

Rigby, J.K. and Gangloff, R.A. (1987) Phylum Archaeocyatha. In R.S. Boardman, A.H. Cheetham and A.J. Rowell (eds)

References

Buss, L.W. and Seilacher, A. (1994) The phylum Vendobionta: a sister group of the Eumetazoa? *Paleobiology* **20**, 1–4.

Conway Morris, S. (1985) Non-skeletalized lower invertebrate fossils: a review. In S. Conway Morris, J.D. George, R. Gibson, and H.M. Platt. (eds). *The origins and relationships of lower invertebrates. Special Volume of the Systematics Association,* **28**, 343–359.

Conway Morris, S. (1993a) The fossil record and the early evolution of the Metazoa. *Nature,* **361**, 219–225.

Conway Morris, S. (1993b) Ediacaran-like fossils in the Cambrian Burgess shale-type faunas of North America. *Palaeontology*, **36**, 593–635.

Conway Morris, S. and Peel, J.S. (1990) Articulated halkieriids from the Lower Cambrian of north Greenland. *Nature*, **345**, 802–805.

Fedonkin, M.A. (1990) Precambrian Metazoans. In D.E.G. Briggs and P.R. Crowther (eds) *Palaeobiology, a synthesis*. Palaeontological Association and Blackwell Scientific Publications, pp. 17–24.

Margulis, L. and Schwartz, K.V. (1988) *Five kingdoms. An illustrated guide to the phyla of life on Earth*. W.H. Freeman and Company, New York.

Matthews, S.C. and Missarzhevsky, V.V. (1975) Small shelly fossils of late Precambrian and early Cambrian age. *Journal of the Geological Society of London*, **131**, 289–304.

Qian Yi and Bengtson, S. (1989) Palaeontology and biostratigraphy of the Early Cambrian Meishucunian Stage in Yunnan Province, South China. *Fossils and Strata, **24***.

Savarese, M. (1992) Functional analysis of archaeocyathan skeletal morphology and its paleobiological implications. *Paleobiology*, **18**, 464–480.

Seilacher, A. (1989) Vendozoa: organismic construction in the Proterozoic biosphere. *Lethaia*, **22**, 229–239.

Valentine, J.W. (1992) *Dickinsonia* as a polypoid organism. *Paleobiology*, **18**, 378–382.

Williamson, D.I. (1992) *Larvae and evolution. Toward a new zoology*. Chapman and Hall, London.

Wood, R. (1991) Problematic reef-building sponges. In A.M. Simonetta and S. Conway Morris (eds) *The early evolution of Metazoa and the significance of problematic taxa*. Cambridge University Press, pp 113–124

Wood, R., Zhuravlev, A. Yu. and Debrenne, F. (1992) Functional biology and ecology of Archaeocyatha. *Palaios*, **7**, 131–156.

6 Radialians I: cnidarians and lophophorates

Key Points

- The cnidarians are the simplest of the true metazoans; they include the jellyfish, sea anemones and corals.
- The Palaeozoic rugose and tabulate corals displayed a wide range of growth modes often related to environments; neither group was a successful reef-builder.
- The scleractinians radiated during the Mesozoic, dominating biological reefs; scleractinian-like forms in Ordovician faunas arose independently.
- Lophophorates have filamentous feeding organs and include brachiopods, bryozoans and phoronids.
- Brachiopods are bivalved shellfish, with a lophophore and usually a pedicle, pursuing life strategies in the sessile benthos.
- The phylum is currently divided into the lingulates, with phosphatic shells, and the calciates, with calcareous shells.
- Brachiopods dominated the low-level filter-feeding benthos of the Palaeozoic but never recovered from the end-Permian extinction event.
- Bryozoans are well-integrated colonial lophophorates commonly displaying marked ecophenotypic variation across a wide range of environments.

Introduction

Cnidarians and lophophorates were conspicuous elements of Palaeozoic sea-floor life. The cnidarians include the corals, carnivorous members of the sessile benthos and important reef-building organisms from the Mesozoic onwards. The lophophorates, mainly the brachiopods and bryozoans, dominated the low-level filter-feeding benthos of the Palaeozoic, but brachiopods were much less diverse following the end-Permian extinction event. Whereas the coral body plan is diploblastic, based on a radiate arrangement, the lophophorates, although triploblastic, are not easily associated with either of the two main, protostomous and deuterostomous, invertebrate lineages.

Cnidarians

The cnidarians, or the sea-nettles, include the jellyfish, sea anemones and corals, and are the least advanced of the true metazoans, having cells organized into a few different tissue types in a radial plan (Figure 6.1). There are no specialized organs or tissues. In the distant past, the group was referred to the Coelenterata, but since that phylum also included the sponges and the gelatinous ctenophores or comb-jellies, the more restricted term Cnidaria is now preferred. Two basic life strategies occur: the polyps are usually sessile or attached, although some can jump and somersault; while the medusae swim, trailing their tentacles like the snakes adorning the head of Medusa herself. Many

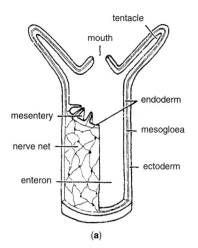

(a)

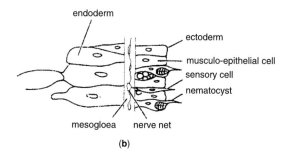

(b)

Fig 6.1   Morphology of *Hydra*; (a) general body plan, (b) detail of the body wall.

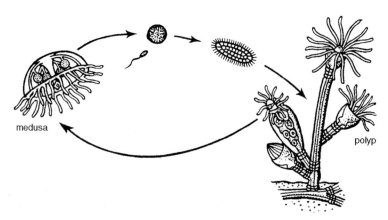

Fig 6.2   Cnidarian life cycle: generalized view of the life of the hydroid *Obelia*, alternating between the conspicuous polyp and medusa stages.

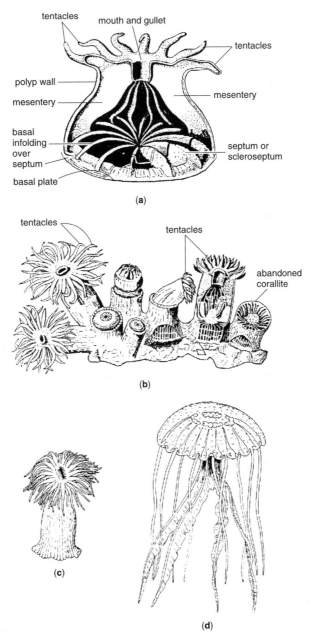

tentacles

mouth and gullet

tentacles

polyp wall

mesentery

mesentery

basal
infolding
over
septum

septum or
scleroseptum

basal plate

(a)

tentacles

tentacles

abandoned
corallite

(b)

(c)

(d)

Fig 6.3 Main cnidarian body plans: (a) generalized scleractinian polyp, (b) generalized part of scleractinian coral colony, (c)
living anemone, (d) living jellyfish.

cnidarians experience both styles during their polymorphic life cycles (Figure 6.2); others, however, predominantly or uniquely may opt for one life mode.
The Portuguese man-of-war is a spectacular colonial
form with a medusoid module for flotation and various
polypoid modules which help feeding, locomotion and
reproduction. As a whole, the group is carnivorous,
activating poisonous stinging cells (cnidoblasts) to
attack crustaceans, fishes, worms and even microscopic diatoms.

Basic cnidarian morphology

The cnidarians are multicellular, having a single body cavity or enteron; the opening at the top, surrounded by tentacles with stinging cells or cnidoblasts, functions both as a mouth and an anus. The body itself, although diploblastic, is composed of three layers: the inner endoderm and the outer ectoderm both consist of living cells, while the intervening mesogloea is a gelatinous non-living substance containing disparate cells. The outer layer of the body wall contains cnidoblast cells with primed stings or nematocysts, usually confined to the tentacles, whereas a primitive nerve net is embedded in the mesogloea. Flanges of endoderm commonly project into the enteron, forming radial partitions which increase digestive efficiency and nutrient absorption. In the case of the corals, the mesenteries control the secretion of calcium carbonate to form solid calcified vertical partitions or septa. The group is almost entirely marine.

Classification and design of main groups

The Phylum Cnidaria is usually split into three classes (see Box 6.1). The hydrozoans include freshwater and marine colonial forms together with the fire corals; there are over 3000 living species inhabiting water depths up to 8000 m, mainly in marine environments. Hydrozoans have been recorded from the late Precambrian Ediacara fauna, where genera such as *Eoporpita* and *Ovatoscutum* may be the oldest sessile and pelagic members of the phylum, respectively. They reproduce either sexually or by asexual budding. The scyphozoans are mainly free-swimming medusae or jellyfish often inhabiting open-ocean environments. Some of the first scyphozoans, e.g. *Conomedusites* and *Corumbella*, are probably represented in the late Precambrian Ediacara fauna; however, many of the best-preserved fossil forms have been collected from the late Jurassic Solnhofen Limestone of Bavaria. Living members of the group include *Aurelia*, the jellyfish, and the compass jellyfish, *Chrysaora*. Although the anthozoans include the sea anemones, sea fans, sea pens and sea pansies, the class also includes the soft and stony corals. Following a short mobile planula larval phase, all members of the group pursue a sessile life strategy as polyps.

The corals

The anthozoans are the most abundant fossil cnidarians, pursuing a polypoid lifestyle. The Class Anthozoa contains two subclasses with calcareous skeletons. Whereas the Octocorallia have calcified spicules and axes, the orders Rugosa, Tabulata and Scleractinia include the more familiar fossil coral groups. The Octocorallia or Alcyonaria have eight complete mesenteries and a ring of eight hollow tentacles; the skeleton lacks calcified septa, but calcareous or gorgonin spicules and axes comprise solid structures in the skeleton. Although the group is only sporadically represented in Permian, Cretaceous and Tertiary rocks, the octocorals are important reef-builders today (Figure 6.4). Some familiar genera include *Alcyonium* (dead men's fingers), *Gorgonia* (sea fan) and *Tubipora* (organ-pipe coral).

Box 6.1 Classification of the Cnidaria

The phylum is characterized by radiobilateral symmetry, with the ectoderm and endoderm separated by mesogloea; the enteron has a mouth surrounded by tentacles with stinging cells. The phylum ranges from Upper Precambrian to Recent. The putative medusoid *Brooksella*, which predates the Ediacara fauna, may in fact be a trace fossil.

Class Hydroidea
  This includes small, polymorphic forms. Each has an undivided enteron, solid tentacles and may form colonies. There are six main orders; the Chondrophora include some of the oldest cnidarians. Ediacarian–Recent.

Class Scyphozoa
  Mainly jellyfish, contained in the Scypho-medusae. The extinct Conulata is often included here since the group has a tetrameral symmetry and apparently tentacles. Their long conical shells (e.g. *Conularia*) are, however, composed of chitinophosphate; conulates appeared in the Cambrian and were extinct by the mid Triassic. Ediacarian–Recent.

Class Anthozoa
  The three subclasses – Ceriantipatharia, Octocorallia and Zoantharia (including the orders Rugosa, Tabulata and Scleractinia) – all lack medusoid stages, possess hollow tentacles and have the enteron divided, longitudinally, by vertical mesenteries. Both solitary and colonial forms occur. The class includes corals, sea anemones and sea pens. Ediacarian–Recent.

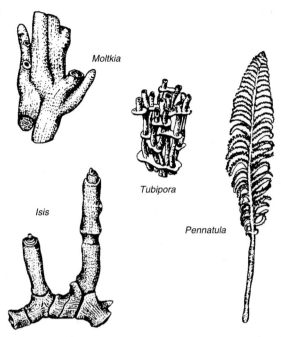

Fig 6.4　Some typical octocorals (all approximately ×1).

In the Zooantharia corals there are four main elements to the skeleton: radial and longitudinal structures, together with horizontal and axial elements. Following the planula larval stage, the scleractinian coral polyp initially rests on a basal plate or disc and begins secretion of a series of vertical partitions or septa in a radial arrangement. At the circumference, the septa are joined to the theca or skeletal wall which extends longitudinally from the apex of the corallum to the margin of the calice where the polyp is attached. During growth the polyp may secrete a series of horizontal sheets together with mounds of smaller plates or dissepiments. A columella may arise from the fusion of the axial edges of the septa, and may occupy the core region of the corallite. The rugose and tabulate corals usually start as small cones attached to clasts and develop either an outer epitheca or

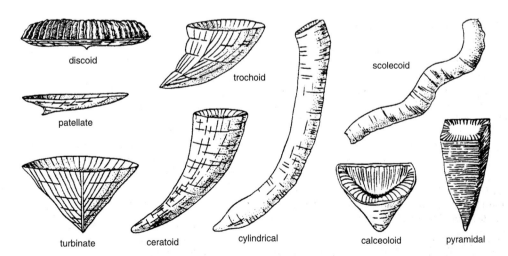

Fig 6.5　Terminology for the main modes of solitary growth in corals.

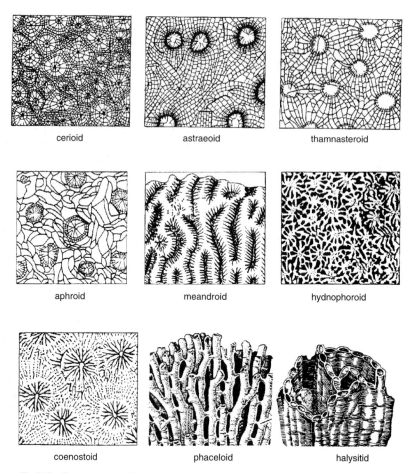

Fig 6.6   Terminology of the main modes of colonial growth in corals.

holotheca. The majority of rugose corals, and all tabulate corals, lack axial structures such as the columella. Despite the apparent simplicity of the skeletal elements of the group, there is considerable scope for variety in all these structures; this variety, together with a multiformity in both solitary and colonial growth programmes, has permitted some remarkable variations on the simple radiate body plan.

The three main orders of stony corals have both solitary and colonial or compound growth modes. There are a variety of solitary growth modes (Figure 6.5). Colonial corals with corallites have adopted either fasciculate, cateniform or massive growth modes. Fasciculate styles exhibit either dendroid or phaceloid strategies with either no or poor integration (Figure 6.6). The massive colonies are much more varied, with cerioid, astraeoid, aphroid, thamnasteroid, meandroid and coenostoid growth

modes. These growth modes are variably developed across the rugosans, tabulates and scleractinians through time (Figure 6.7), but the meandroid mode is restricted to the Mesozoic scleractinians.

Another method of colony analysis involves key measurements made on the colony and these are plotted on a ternary diagram (Figure 6.8). A series of fields can be mapped out within the triangle; for example, bulbous, columnar, domal, tabular and branching colonies are discriminated. These different growth strategies are usually ecophenotypic, commonly reflecting ambient environmental conditions.

Rugose corals

The rugose corals are generally robust, calcitic forms with both colonial and solitary life modes, more varied

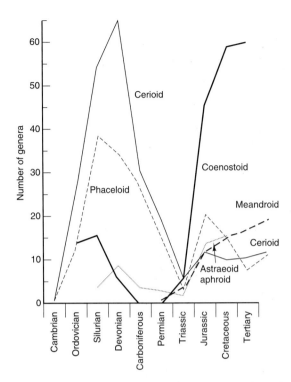

Fig 6.7 Schematic graph of the distribution of colonial growth modes through time.

than those of the tabulates (Figure 6.9). Rugosans had well-organized septal arrangements with six protosepta; metasepta are inserted in four spaces around the corallum – between the cardinal septa and the two alar septa and also between the alar and two counter lateral septa (Figure 6.10). Shorter minor septa are usually longer than the width of the dissepimentarium, where developed. Tabulae and dissepiments are also well developed across the order. Undoubted rugosans, such as *Streptelasma*, with short secondary septa but lacking a dissepimentarium, are not recorded until the mid Ordovician. By the Silurian, rugose faunas were well established, with the development of a wide variety of morphologies. The Silurian *Goniophyllum* was quadrangular with a deep calyx, whereas the Devonian *Calceola* was a slipper-shaped form with semicircular lid or operculum, and the compound *Phillipsastrea* had a massive, astraeoid growth mode.

Following near-extinction, diverse rugosan faunas developed during the Carboniferous Period (Figure 6.12). Solitary forms such as the large horn-shaped to cylindrical *Siphonophyllia*, the cylindrical *Dibunophyllum* with a marked axial complex, the long, cylindrical *Palaeosmilia* with its long septa, and the smaller, horn-shaped *Zaphrentis* are often conspicuous members of Carboniferous coral assemblages. Coral colonies (e.g. the fasciculate, phaceloid *Lonsdaleia*, and *Lithostrotion*, with its massive cerioid growth modes) are locally common. The order declined during the Permian until its eventual extinction at the close of the period.

Fig 6.8 Ternary plot of colonial growth modes based on the shape of the colonial coral.

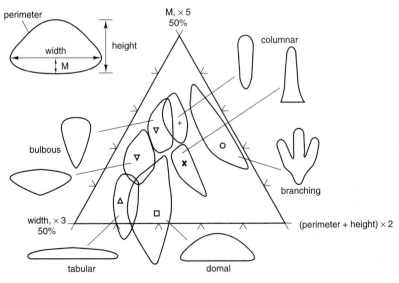

Box 6.2 Colonies and integration

Colonial integration usually involves a loss of individuality. Many organisms display a transition from solitary growth modes through morphologies with asexually budded modules to a fully integrated colony with the astogeny or growth of compound structure showing little variation across the individual corallites. The degree or level of integration of a colony is usually measured with reference to the amount of cohesion between the individual skeletal parts and soft tissues together with the development of polymorphism in the colony. Clearly there is a spectrum from phaceloid modes with little or no integration, to the thamnasteroid and meandroid modes with high levels of integration. Individual polyps were no longer separated by corallite walls and may have shared a common enteron and nervous system. Some colonies have several types of morph adapted for a variety of different functions. This suggests a high degree of integration where the colony approaches the body plan of a more advanced metazoan.

The high degree of integration associated with the scleractinian coral colonies has been correlated with the presence of symbiotic zooxanthellae. The relatively low levels of integration seen in the Rugosa and some Tabulata colonies, thus, suggest a lack of algal symbionts.

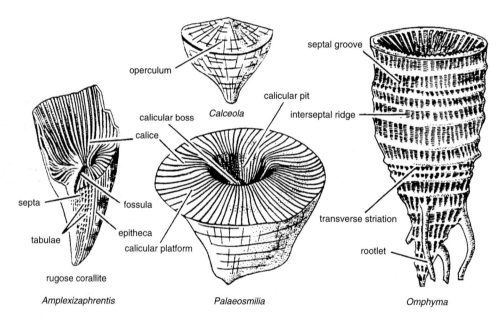

Fig 6.9   Rugose coral morphology: external morphology of a variety of solitary rugose corals.

Tabulate corals

As the name suggests, the tabulates have well-developed tabulae (Figure 6.13). In contrast, the septa are usually very much reduced to short spines or absent entirely. The group is varied with erect, massive, sheet-like and chain-like colonies; some authors have suggested that some tabulates, such as the heliolitids, may not, in fact, be cnidarians. Nevertheless, Copper (1985) has illustrated fossilized polyps in Silurian *Favosites*. Only colonial growth forms are known, usually with small, elongate corallites ranging from 0.5 to 5 mm in diameter

Commonly the corallite walls are perforated by minute holes or mural pores. The order first appeared in the early Ordovician, predating the first rugosans; forms such as *Lichenaria* have been recorded from Lower Ordovician rocks in the United States. Tabulates such as *Catenipora*, *Favosites* and *Heliolites* became widespread during the later Ordovician and Silurian.

Silurian coral assemblages (Fig. 6.2) were dominated by *Favosites* with massive, cerioid corallites, *Halysites*, the chain coral, with a series of linked long cylindrical corallites of elliptical cross-section, and *Heliolites*, the

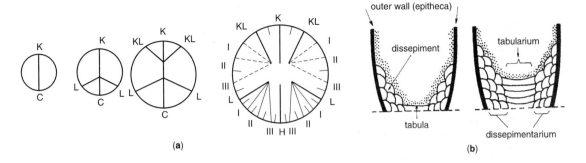

Fig 6.10   Septal and tabular development in solitary rugose corals with (a) details of vertical partitions: cardinal (C), counter cardinal (K), alar (L) and counter lateral (KL) septa shown; and (b) details of horizontal structures.

Box 6.3 Rugosan life strategies

Despite the apparent simplicity of rugosan architecture, these corals may have pursued a number of different life strategies, listed in decreasing order of importance (Figure 6.11).

1. A number of strongly curved rugosans (e.g. *Aulophyllum*) probably lay within the substrate, concave upwards. Successive increments of growth were directed more or less vertically, giving the coral exterior a stepped or rugose appearance. Many other solitary corals exhibit a similar terraced theca which may result from changes in growth direction associated with adjustments following toppling of the corallum during turbulence or storms.

2. *Grewingkia* pursued a fixosessile life mode, cemented to areas of hard substrate.

3. Liberosessile taxa (e.g. *Holophragma*) were initially attached to a clast, but subsequently toppled over to rest on the seabed.

4. The small discoidal *Palaeocyclus* may have been mobile, creeping over the substrate on its tentacles.

5. *Dokophyllum*, with a rhizosessile lifestyle, probably faced upwards in the sediment rooted by fine holdfasts extending from the epitheca.

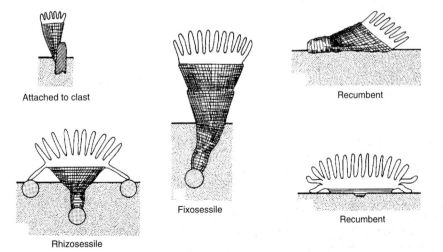

Attached to clast

Recumbent

Rhizosessile

Fixosessile

Recumbent

Fig 6.11   Rugose solitary life strategies displaying attached, fixosessile, rhizosessile and recumbent life modes.

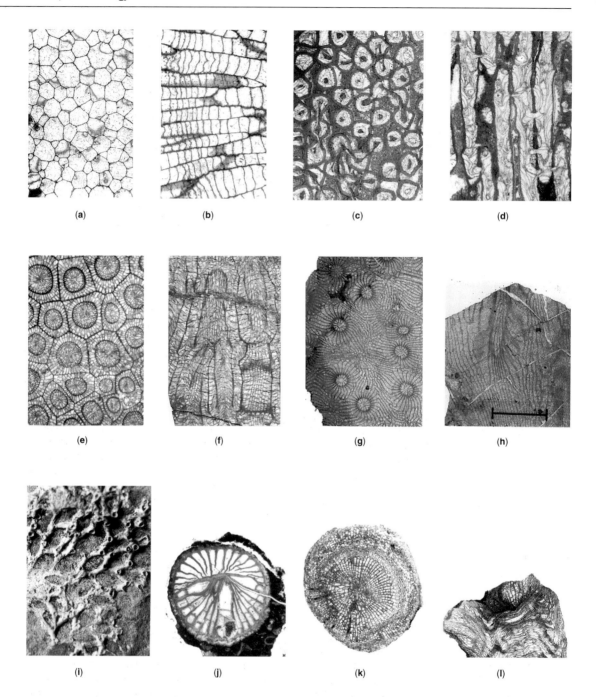

Fig 6.12 Some rugose and tabulate corals: (a), (b) cross and longitudinal sections of *Favosites* (Silurian); (c), (d) cross and longitudinal sections of *Syringopora* (Carboniferous); (e), (f) cross and longitudinal sections of *Acervularia* (Silurian); (g), (h) cross and longitudinal sections of *Phillipsastrea* (Devonian); (i) *Aulopora* (Silurian); (j) *Amplexizaphrentis* (Carboniferous); (k), (l) cross and longitudinal sections of *Palaeosmilia* (Carboniferous). (a)–(d) and (i) are tabulates at approximately ×2 magnification; (e)–(h), (j)–(l) are rugosans at approximately ×2 (e)–(h), ×3 (j) and ×1 (k)–(l) magnifications. Note: here and elsewhere, age assignments refer to the specimen figured, and not to the entire stratigraphical range of the taxon.

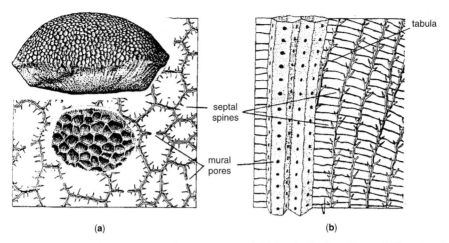

Fig 6.13 Tabulate morphology: (a) transverse and (b) longitudinal sections of *Favosites*; insets on (a), lateral and upper surfaces of entire *Favosites* colony.

sun coral, with corallites set in common colonial tissue. Similarly distinctive tabulates were characteristic of the Devonian; *Aulopora* usually formed branching, encrusting colonies, similar to the bryozoan *Stomatopora*, while the extraordinary *Pleurodictyum* with large mural pores and thorn-like septa was virtually always associated with the commensal worm *Hicetes*.

Carboniferous tabulates, such as *Michelinia*, formed small colonies possessing very large, massive thick-walled corallites, and the phaceloid *Syringopora* with long, thin cylindrical corallites characterize the coral faunas of the period. By the late Permian, the group was very much in decline following a long period of deterioration after the Frasnian extinctions; only five families survived to the end of the period.

Scleractinian corals

The scleractinians are elegant zoantharian corals composed of aragonite, usually with relatively light porous skeletons. Both solitary and colonial modes exist, with even more varied architectures than those of the rugosans (Figure 6.15). Metasepta are inserted in all six spaces between the six protosepta. Tabulae, in the strict sense, are absent although dissepiments and dissepimentaria are developed. Moreover, the scleractinian skeleton, although relatively light and porous, has the stability of a basal plate which initially aids anchorage on the substrate. Additionally, the scleractinian polyp can secrete aragonite on the exterior of the corallite often in the form of attachment structures. Both adaptations provided a much greater potential for reef building than the less stable rugose and tabulate corals of the Palaeozoic (Figure

6.6). Finally, scleractinians have a distinctive ultrastructure composed of aragonite and a widespread development of coenosarc. The group first appeared in the mid Triassic, with forms such as *Thamnasteria* quickly becoming widespread throughout Europe.

Montlivaltia is a small, cup-shaped coral common during the early Jurassic to Cretaceous, whereas *Thecosmilia* is a small, dendroid to phaceloid colonial form with corallites similar to the corallum of *Montlivaltia*, ranging from middle Jurassic to Cretaceous horizons; the massive, cerioid *Isastraea* has a similar range. Scleractinians are now the dominant reef-building animals in modern seas and oceans where they form biological structures in a variety of settings, usually within the tropics.

Coral ecology and reefs

Two ecological groups have been recognized among Recent scleractinians. Hermatypic corals are associated with zooxanthellae and are restricted to the photic zone to maintain this symbiosis. Symbiosis between zooxanthellate algae and cnidarians is widespread among the living representatives of the phylum, with algae associating not only with corals, but also anemones and gorgonians. The zooxanthellae are endosymbionts living in the tentacles and mouth of the cnidarian where they recycle nutrients, accelerate the rate of skeletal deposition and convey oxygen, organic carbon and nitrogen to the cnidarian in return for support.

Hermatypic corals are commonly multiserial forms, with small corallites displaying a high degree of inte-

Box 6.4 Computer reconstruction of colonies

The colonial tabulate *Aulopora* has a long geological history and mainly occupied an encrusting niche, coating brachiopods, stromatoporoids and other, larger corals. *Aulopora* grew by dichotomous branching, pursuing a creeping or reptant life mode, efficiently siting its corallites adjacent to potential sources of food at, for example, the inhalent currents through brachiopod commissures. Scrutton (1989) has reconstructed the colonies of free-living *Aulopora* in three dimensions using a computer-based technique (Figure 6.14). Serial sections of the colony were digitized and assembled on a micro-VAX mainframe with software routinely used for building up 3D views of diseased kidneys. Both the ontogeny of the procorallites and the astogeny of the colony as a whole were established in considerable detail by these techniques. The patterns of colonial development of the free-living *Aulopora* suggest that the group is an unlikely link between the rugose and tabulate corals. Moreover, to date, no early Ordovician auloporids have been reported.

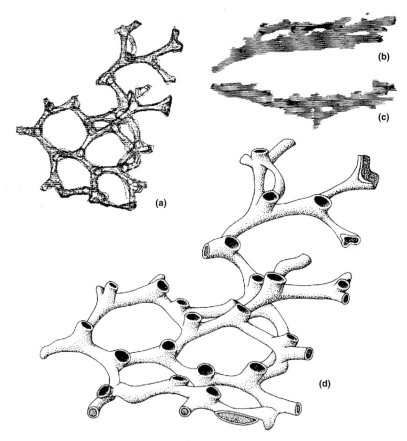

Fig 6.14 *Aulopora* morphology: computer-generated reconstructions of (a) the plan, (b) the lower side, and (c) the direction of the procorallite; (d) reconstruction of colony.

gration; ahermatypic forms lacking algal symbionts are commonly solitary or uniserial compounds with large, poorly integrated corallites and generally occur in deeper water. Coates and Jackson (1987) have suggested that coral morphology may help predict the presence of symbionts in fossil coral communities. It is probable that many tabulates were zooxanthellate whereas the rugosans were not. Nevertheless, in broad terms, platform and basin associations of rugose and tabulate corals have been recognized during the Palaeozoic which may be matched by the distinction between the hermatypic and ahermatypic scleractinian

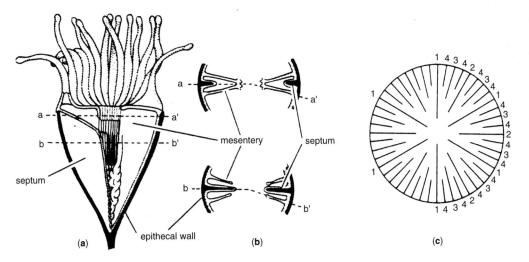

Fig 6.15   Scleractinian morphology; longitudinal (a) and transverse sections (b), and mode of septal insertion (c).

| Coral group

feature	Rugosa	Tabulata	Scleractinia
Growth mode	colonial and solitary	colonial	colonial and solitary
Septa	6 prosepta; later septa in only 4 spaces	septa weak or absent	6 prosepta; later septa in all 6 spaces
Tabulae	usual	well developed	absent
Skeletal material	calcite	calcite	aragonite
Stability	poor	poor	good with basal plate
Range	Ordovician – Permian	Ordovician – Permian	Triassic – Recent

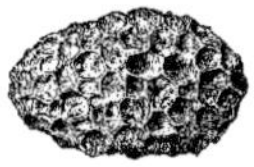

Fig 6.16   Features of the main coral groups.

corals during the Mesozoic and Cenozoic.

Reefs are biological frameworks with significant topography. Most coral reefs are restricted to the tropics, occurring within 30° latitude of the equator. Three main types of biological structure occur: fringing reefs develop directly adjacent to land areas; barrier reefs have an intervening lagoon; whereas atolls completely surround islands, usually of volcanic origin. The last will continue to grow as the volcanic island subsides until eventually only the atoll, enclosing a lagoon, remains.

Palaeozoic corals were not particularly successful reef-builders; many preferred soft substrates and lacked

Box 6.5 Kilbuchophyllida and iterative skeletalization

A remarkable coral genus, *Kilbuchophyllia*, has recently been described from the Middle Ordovician rocks of the Southern Uplands of Scotland (Scrutton and Clarkson, 1990). Mid Ordovician rugose and tabulate corals are not uncommon, but *Kilbuchophyllia* has patterns of septal insertion and a microstructure identical with those of the modern scleractinians (Figure 6.17). This genus, from debris-flow deposits with rich shelf and slope faunas, forms the basis for a separate order of coral within the Subclass Zoantharia, the Kilbuchophyllida. The discovery of these calcified scleractinian-like morphs is of considerable significance in cnidarian evolution. Skeletalization of an early Palaeozoic 'scleractinian' polyp was local and probably, in this instance, short-lived. It is unlikely that *Kilbuchophyllia* was the stem group for the scleractinians; however, clearly other groups of soft-bodied anemones with the potential for skeletalization were available early in the history of the group. Following the end-Permian extinction event when the rugose and tabulate corals finally disappeared, calcification of other scleractinian-type morphs during the Triassic marked the start of another highly successful calcified coral group.

Fig 6.17   Kilbuchophyllia – an Ordovician scleractiniomorph coral (approximately ×10).

structures that allowed anchorage and aided stability; calcareous algae and stromatoporoids were usually more important. Nevertheless, frameworks dominated by colonial tabulates and to a lesser extent rugosans do occur, particularly during the mid Palaeozoic. Pioneer and climax communities have been described from a number of Silurian and Devonian successions (Figure 6.18). The scleractinians, however, gradually became the dominant reef-builders during the Mesozoic and Cainozoic. Modern coral reef associations have been documented in detail from eastern Australia, the eastern Pacific and the Caribbean.

Corals and correlation

Around the turn of the century, detailed studies of Lower Carboniferous corals in Belgium and Britain, particularly in the Avon Gorge (Figure 6.20), provided a considerable prospect for their use in Carboniferous biostratigraphy. Corals are very common, often widespread, generally distinctive and well preserved in the Dinantian rocks of Europe. Unfortunately, more modern studies on

Carboniferous biostratigraphy using microfossils such as conodonts and foraminiferans, have shown that the occurrence of various coral groups is in fact diachronous, often controlled by facies. Nevertheless, many corals are still very useful for correlation.

Evolution

Although a growing number of coral-like forms have been described from the Cambrian, most are unlikely ancestors for Ordovician forms. *Cothonion*, for example, with poorly integrated corallite-like clusters and opercula, was probably a Cambrian experiment with coralization. The first tabulates appeared during the early Ordovician with cerioid growth modes (Figure 6.22); tabulae were rare and septa and mural pores absent. Nevertheless, by the mid and late Ordovician, when they dominated coral faunas, the more typical characters of the Tabulata had evolved. Some workers have removed the heliolitids, with individual corallites mutually separated by extensive coenosteum, from the Tabulata, into a distinct order. The group was common until the early

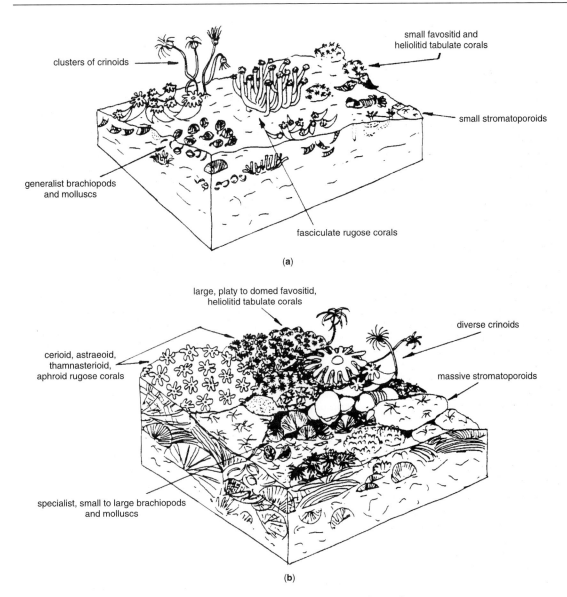

Fig 6.18 Pioneer (a) and climax (b) reef communities in Silurian and Devonian reef systems.

Silurian, when the more open structures of the favositids with massive cerioid colonies began to dominate.

The Rugosa appeared during the mid Ordovician. Many of the evolutionary trends across the order have been repeated many times in different families. The trend towards more complex, robust skeletons was dramatically reset by the late Devonian extinction event prior to the extinction of the order at the end of the Permian.

The first scleractinians were established by the mid

Triassic, derived from an ancestor within the sea anemones. The Triassic taxa were probably hermatypic, forming patch reefs in parts of the Tethyan belt. However, the group expanded significantly during the Jurassic, with the radiation of both hermatypic and ahermatypic groups in shallow- and deep-water environments respectively. Scleractinian evolution was marked by a number of morphological trends: solitary life strategies were eventually superseded by a dominance of colonial forms which display transitions from

Box 6.6 The Great Barrier Reef and Devonian reefs of the Canning Basin

The Great Barrier Reef occurs on the continental shelf of eastern Australia; it is the largest coral structure on Earth, approaching 3000 km long and up to 300 km wide. The reef extends from 9° to 25° S and comprises many multicoloured scleractinian corals, together with many other invertebrates and calcareous algae. The fore-reef deposits tumble eastwards into the western Pacific; landward back-reef lagoons, acting as major shipping lanes, are developed against eastern Australia. In Western Australia the Upper Devonian rocks of the Canning Basin contain comparable fossil barrier reefs dominated by calcareous algae, and stromato-poroids together with tabulate and rugose corals (Figure 6.19). The reef and its associated facies can be mapped in considerable detail, as the Windjana Gorge dissects the near-horizontal strata of the northern margin of the Canning Basin. An unbedded core of calcareous algae, stromatoporoids and corals sheltered a back reef and lagoonal environment packed with calcareous algae, stromatoporoids, corals and crinoids, together with brachiopods, bivalves, cephalopods and gastropods. In front, the fore-reef was steep and littered by reef talus. However, during the late Devonian extinction event, at the end of the Frasnian, associations dominated by stromatoporoids, together with rugose and tabulate corals disappeared; this type of reef ecosystem never recovered.

(a)

(b)

Fig 6.19   Devonian reefs of the Canning Basin Australia, (a) main face; (b) Windjana Gorge, forereef slope in the foreground with large blocks of unbedded reef material in the background; the reef is prograding over the forereef towards the viewer.

low levels of integration in phaceloid growth modes to higher levels in meandroid styles, common in modern reefs.

Lophophorates

Three superficially different phyla, the Phoronida, Brachiopoda and Bryozoa, are united in the Superphylum Lophophorata; all possess a complex feeding organ, the lophophore, and similar coelomic systems. All undoubtedly shared a common ancestor prior to the mineralization of the brachiopod and bryozoan skeletons, probably during the earliest Cambrian and earliest Ordovician, respectively.

The phoronids are tube-dwelling vermiform lophophorates, with the ten or so described species divided between two genera, *Phoronis* and *Phoronopsis*. These animals lack a mineralized skeleton and they pursue burrowing or boring life strategies with near-cosmopolitan distributions. The phylum probably has a long geological history, as some authors suggest that Precambrian and Lower Palaeozoic records of the vertical burrow *Skolithos* may, in fact, be the work of phoronids. The ichnogenus *Talpina* manifest as borings in both Cretaceous belemnite rostra and Tertiary mollusc shells, may have been constructed by more recent phoronids.

Brachiopods

The brachiopods are one of the most successful invertebrate phyla. The group was a conspicuous component of

Stages		Vaughan 1905-1906 (Avon Gorge)	
Dinantian	Brigantian	Horizon ε ------- *Dibunophyllum* Zone	Kidwellian = Upper Avonian = Visean
	Asbian		
	Holkerian	*Seminula* Zone	
	Arundian	*Syringothyris* (Lower *Caninia*) Zone	Clevedonian = Lower Avonian = Tournaisian
	Chadian		
	Courceyan	*Zaphrentis* Zone	
		Cleistopora Zone	
		Modiola Zone	
		Old Red Sandstone facies	

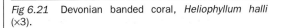

Fig 6.20   Dinantian coral sequence established in Avon Gorge, Bristol area, UK.

early Cambrian marine communities, diversifying throughout the Palaeozoic to dominate the low-level suspension-feeding benthos; a wide range of shell morphologies and sizes characterize the phylum, from the microscopic acrotretides to the massive gigantoproductids. Although only about 120 genera of brachiopods are present today, they occupy a wide range of habitats, from the intertidal zone to the abyssal depths. The brachiopods, or lampshells, are entirely marine bilaterally symmetrical animals with a ciliate feeding organ or lophophore contained within a pair of shells or valves. Internal structures such as hinge teeth and sockets, cardinal process and various muscle scars are all associated with the opening and closing of the two valves during feeding cycles.

Brachiopods have featured in many palaeoecological studies of Palaeozoic faunas, when they dominated the benthos. Their use in palaeobiogeographic analysis is well documented (see also Chapter 2). Nevertheless, brachiopods have also been widely used in regional biostratigraphy, and during the Silurian, in particular, a number of orthide, pentameride and rhynchonellide lineages show good prospects for international correlation.

Brachiopod morphologies

The basic plan of the brachiopod animal includes a pair of hinged shells or valves, opened and closed by a variety of muscles, commonly attached to the seabed by a fleshy stalk or pedicle, extending through a pedicle fora-

---

Box 6.7 Corals and the earth's rotation

Detailed analysis of the growth bands on coral epithecae has proved an interesting tool for the investigation of the changing speed of rotation of the Earth. Well-preserved corals often display fine growth lines, grouped together into thicker bands; the former are thought to reflect daily growth while the latter bands are monthly growth cycles, controlled by the lunar orbit (Figure 6.21). A set of more widely spaced bands may represent yearly growth. In a classic study, Wells (1963) documented the growth lines on a variety of Devonian corals, and suggested that the Devonian year had about 400 days; the implication that Devonian days were shorter suggests that the Earth's rate of rotation is decreasing.

Fig 6.21   Devonian banded coral, *Heliophyllum halli* (×3).

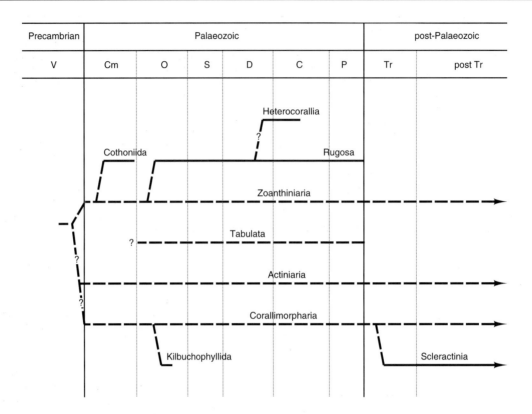

Fig 6.22   Stratigraphical ranges of the main coral groups.

men, and feeding with an extendable lophophore (Figure 6.24). The pedicle varies from thick, strong fleshy stalks, for attachment to large areas of substrate, to thread-like strands which can anchor shells like those of the abyssal genus *Cryptopora* in fine mud. Some brachiopods lost their pedicle during ontogeny; many Palaeozoic strophomenides followed a free-living quasi-infaunal lifestyle, while some modern terebratulides have been reported pursuing unattached life modes in sandy and muddy facies.

Brachiopods evolved a number of different lophophore styles. The earliest growth stage, the trocholophe, is an incomplete ring of filaments; the minute interstitial brachiopod *Gwynia* has kept this type of lophophore into adulthood presumably by paedomorphosis. The schizolophe stage develops a bilobed outline and may have characterized many small Lower Palaeozoic taxa such as the phosphatic acrotretides. From this simple pattern the plectolophe (typical of most living terebratulides) developed through a zygolophe stage (probably typical of early terebratulides). The spirolophe has a pair of spiral arms and is characteristic

of living inarticulates and rhynchonellides; moreover, the majority of the spire-bearing fossil brachiopods, such as the atrypides, athyrides and the spiriferides probably possessed a spirolophe. The more complexly lobate ptycholophe is developed in only a few living genera. There have been many attempts to relate these types of lophophore to the lophophore supports in fossil specimens but with varying degrees of success.

In contrast to the bivalve molluscs, where the right valve is the mirror image of the left valve, the plane of symmetry bisects both valves of the brachiopod perpendicular to the plane of the commissure. There is no general agreement on the terms used to define each of the two valves. Although traditionally the lower and larger of the two valves was termed the ventral valve and the smaller upper valve, the dorsal, there was uncertainty regarding the true orientation of the animal. Although many authors have preferred the terms pedicle and brachial valves for ventral and dorsal valves, the revised brachiopod Treatise is promoting the latter set of terms.

Brachiopods have a wide variety of shell outlines, profiles and surface ornament. The outlines range from cir-

Box 6.8 Brachiopod classification

The basic division of the phylum into Inarticulata, without articulating structures, and Articulata with, has been stable since its introduction by Thomas Huxley over 100 years ago. But there are undoubted anomalies highlighted by recent, cladistic-based analysis of the zoological features of living members of the main articulate and inarticulate groups (Figure 6.23). This analysis suggests a need for a reorganization of the higher-level taxonomy of the phylum. Currently a new bipartite division in terms of the chitinophosphatic lingulates and the calcareous calciates has been adopted. Nevertheless, there have been challenges to the accepted monophyletic origin of the phylum (e.g. Wright, 1979).

Class Lingulata
All brachiopods with phosphatic shells, comprising five orders.

Order Lingulida
Spatulate valves with pedicle emerging posteriorly between both valves. Examples: *Lingula*, *Glottidia* and *Obolus*. Cambrian–Recent.

Order Acrotretida
Mainly micromorphic forms with conical ventral valves and dorsal valves, commonly with complex internal structures. Examples: *Acrotreta*, *Numericoma* and *Scaphelasma*. Cambrian–Devonian.

Order Discinida
Subcircular shells with conical ventral valve and distinctive pedicle opening, often partly covered. Examples: *Discina*, *Orbiculoidea* and *Trematis*. Ordovician–Recent.

Order Siphonotretida
Biconvex subcircular shells bearing spines and having a large, commonly elongate pedicle opening. Examples: *Siphonotreta*, *Multispinula* and *Schizambon*. Cambrian–Ordovician.

Order Paterinida
Strophic shells with variably developed interareas. Examples: *Dictyonina*, *Micromitra* and *Paterina*. Cambrian–Ordovician.

Class Calciata
All brachiopods with calcareous shells, comprising three nonarticulate orders, two intermediate orders and eight articulate orders. The first three have been assigned to the Subclass Craniformea.

Order Craniida
Usually attached by poorly calcified lower, ventral valve. Upper, dorsal valve subconical with quadripartite muscle scars. Examples: *Crania*, *Neocrania* and *Petrocrania*. Ordovician–Recent.

Order Craniopsida
Small oval valves with marked concentric growth lines and internal platforms. Example: *Craniops*. Ordovician–Carboniferous.

Order Trimerellida
Typically giant aragonitic shells with platforms, umbonal cavities and marked interareas. Examples: *Dinobolus*, *Gasconsia* and *Trimerella*. Ordovician–Silurian.

Order Obolellida
Oval calcareous valves with primitive articulation. Example: *Obolella*. Cambrian.

Order Kutorginida
Calcareous valves with straight hinge line and interareas; lacking articulatory structures. Example: *Kutorgina*. Cambrian.

Order Orthida
Variably biconvex valves with straight hinge lines, open delthyria and notothyria and commonly simple linear cardinal process. Both punctate and impunctate forms. Examples: *Orthis*, *Enteletes*, *Clitambonites* and *Triplesia*. Cambrian–Permian.

Order Strophomenida
Commonly concavoconvex valves with straight hinge lines and bilobed cardinal process. Pedicle foramen atrophied during ontogeny. Examples: *Plectambonites*, *Strophomena*, *Chonetes*, *Productus* and *Lyttonia*. Ordovician–Permian.

Order Pentamerida
Biconvex valves usually with curved hinge lines; spondylia and cruralia variably developed in ventral and dorsal valves. Examples: *Syntrophia*, *Pentamerus* and *Stricklandia*. Cambrian–Devonian.

Order Rhynchonellida
Commonly biconvex and curved hinge lines; crura variably developed in dorsal valve. Occasional punctate forms. Examples: *Rhynchonella*, *Stenoscisma* and *Rhynchopora*. Ordovician–Recent.

Order Atrypida
Biconvex valves with dorsoventrally directed spiralia, with variably developed jugum.

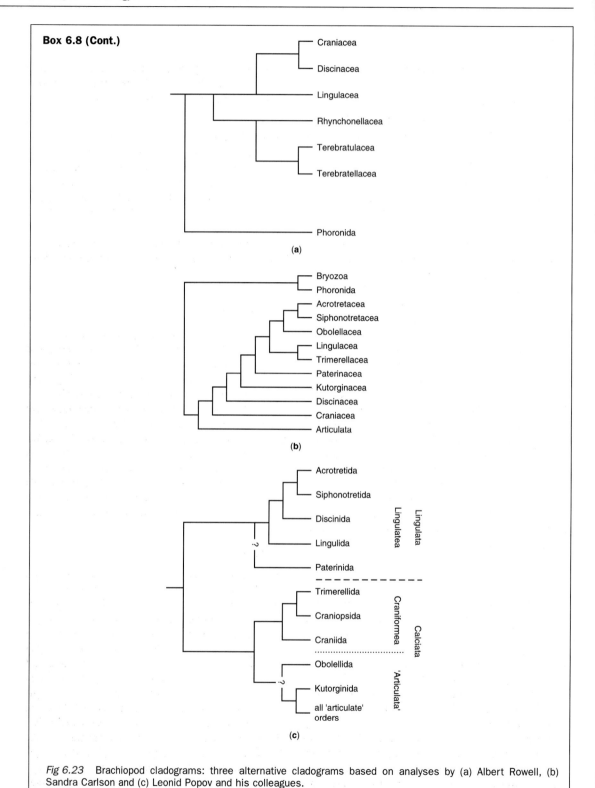

Box 6.8 (Cont.)

Fig 6.23 Brachiopod cladograms: three alternative cladograms based on analyses by (a) Albert Rowell, (b) Sandra Carlson and (c) Leonid Popov and his colleagues.

Box 6.8 (Cont.)

Examples: *Atrypa, Catazyga* and *Pentlandella*. Ordovician–Devonian.

Order Athyrida
Usually biconvex valves, short hinge line, posterolaterally directed spiralia. Examples: *Athyris, Composita* and *Meristella*. Ordovician– Jurassic.

Order Spiriferida
Wide strophic hinges and laterally directed

spiralia; both punctate and impunctate forms. Examples: *Spirifera, Cyrtia, Mucrospirifer, Punctospirifer* and *Spiriferina*. Ordovician– Jurassic.

Order Terebratulida
Biconvex, with short curved hinge lines, an open pedicle foramen and short or long looplike brachidia. Examples: *Dielasma, Digonella, Terebratulina* and *Tichosina*. Devonian–Recent.

---

cular through ovoid to semicircular shapes, some even becoming bilobed. Profiles are variably biconvex to concavoconvex often with geniculation and resupination. Radial ornaments may be dominated by ribs of various strengths and densities; concentric ornament may be merely growth lines or extravagantly accentuated as rugae, lamellae and spines.

Occasionally species may exhibit an unusually wide range of apparent cross-classification morphological variation. For example, Recent specimens of *Terebratalia transversa* from around the San Juan Islands, western USA, show *Spirifer-, Atrypa-* and *Terebratula*-type morphs with increasing hydroenergy conditions (Figure 6.25). Moreover, a number of brachiopods, such as the strophomenides, especially the productoids, may change their shape and life mode markedly during ontogeny (Figure 6.26) from attached pedunculate lifestyles to free-living recumbent habits.

The Phylum Brachiopoda can be described in terms of three main morphotypes: the lingulates, craniformes and articulates (Figure 6.26).

Lingulates

Lingulate brachiopods are composed of chitinophosphate. In the lingulides, the two valves are usually about the same size, and the pedicle emerges between the posterior edges of the ventral and dorsal valves. Although a few authors have excluded these phosphatic forms from the Brachiopoda, the possession of a pedicle and a lophophore support their continued inclusion in the phylum. The opening and closing of the valves is controlled by a complex series of muscles. Some workers have suggested that the withdrawal of the soft parts into the posterior parts of the shell causes a space problem, thus forcing the valves apart; on relaxation, the animal expands forwards within the shells and the valves close. This

hydrostatic mode of operation can be demonstrated for living *Lingula*, and many similar fossil forms may have operated in the same way.

Craniformes

This grouping contains the forms centred around *Crania*, *Craniops* and *Trimerella*. None of these brachiopods possessed a pedicle during adult life. Many of the craniids were cemented to hard substrates by a poorly calcified lower ventral valve. Similarly the craniopsides were cemented apically to a hard substrate or they may have lain unattached on a soft substrate. The trimerellids were a highly specialized group with large, aragonitic shells, exaggerated platforms, umbonal cavities in both valves and often long interareas. With the exception of *Costitrimerella*, the shells are smooth. The trimerellids commonly formed shell banks, in co-supportive clusters, like those of Silurian pentamerides and living oysters.

Articulates

Articulate brachiopods have a pair of calcitic valves with variable convexity, meeting posteriorly at the hinge line · and opening anteriorly along the commissure. The opening and closing of the valves is based on the operation of a muscle system about a fulcrum secured by teeth. There have been many studies on the function and efficiency of the brachiopod hinge. The valves are closed and kept shut by the adductor muscles which stretch across from the floors of the dorsal to ventral valves. The diductor muscles, however, are attached behind the hinge line to the cardinal process of the dorsal valve. The muscle fibres stretch across diagonally to the floor of the ventral valve; open delthyria and notothyria and various deltidial and notothyrial structures allow the passage of the muscles across the hinge

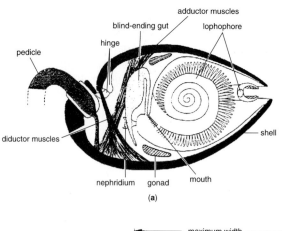

(a)

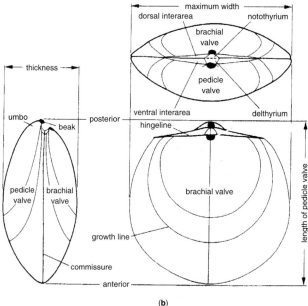

(b)

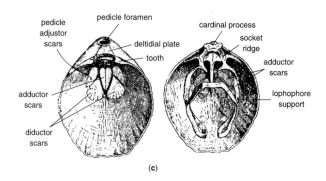

(c)

Fig 6.24 Brachiopod morphologies: (a) internal features of a terebratulide, (b) external terminology of a typical articulate, (c) internal terminology of a terebratulide.

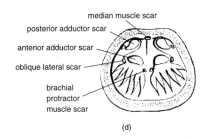

median muscle scar
posterior adductor scar
anterior adductor scar
oblique lateral scar
brachial
protractor
muscle scar

(d)

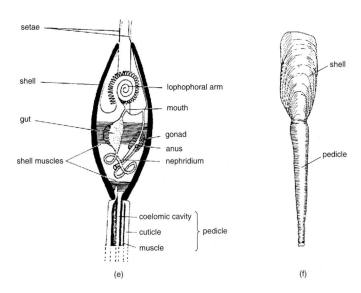

setae

shell

gut

shell muscles

lophophoral arm

mouth

gonad
anus
nephridium

coelomic cavity
cuticle   pedicle
muscle

(e)

shell

pedicle

(f)

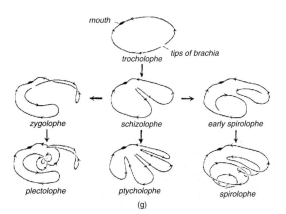

mouth

trocholophe

tips of brachia

zygolophe   schizolophe   early spirolophe

plectolophe   ptycholophe   spirolophe

(g)

Fig 6.24 (Cont.) (d) internal terminology of a craniform calciate, (e) internal features of a lingulate, (f) exterior of a burrowing lingulate and (g) main types of brachiopod lophophore.

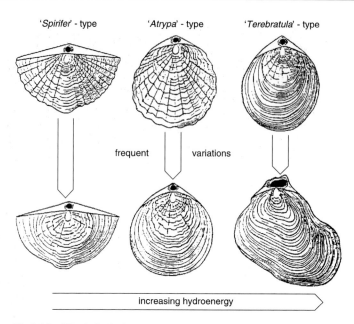

'Spirifer' - type   'Atrypa' - type   'Terebratula' - type

frequent   variations

increasing hydroenergy

Fig 6.25 Morphological variation in *Terebratalia* related to changing hydrodynamic conditions.

areas. When the diductors contract, the valves spring open. Various structures, from the crura of the rhynchonellides to the loops of the terebratulides, support the lophophore.

Most authors recognize eight articulate orders, largely based on the nature of the cardinal process and the form of other internal structures which provide support for muscles and the lophophore. The orthides, with the exception of the triplesiids with forked cardinal processes, had simple linear cardinal processes in contrast to the more complex, mainly bilobed, processes in the strophomenides; both groups also lacked calcified lophophore supports. The pentamerides evolved a variety of platforms such as septalia and cruralia. The rhynchonellides are characterized by crurae which provided lophophore support; however the atrypides, athyrides and spiriferides developed complex spiral brachidia in various orientations, whereas the terebratulides relied on long and short calcareous loops.

More fundamental divisions of the Articulata have been proposed with variable degrees of success. Traditionally the class has been split into two main groups based on the position of the pedicle opening. Protremates like the orthides and strophomenides, have an open delthyrium and notothyrium, often closed by a pseudodeltidium and chilidium, whereas the delthyria of teleotremates, like the rhynchonellides and terebratulides, have a pedicle foramen restricted by deltidial

plates. However, the presence of deltidial plates in otherwise typical protremates questioned the validity of this simple subdivision of the articulates.

Jaanusson (1971) suggested that the articulates could be split into two groups based on the structure of the teeth. Deltidiodonts have simple additive triangular teeth, whereas cyrtomatodont teeth are more complex knob-like structures which grow by resorption (Figure 6.27). The deltidiodont teeth provide an efficient method of articulation during the life of the brachiopod, but after death when the soft parts decay the ventral and dorsal valves are usually separated. However, the functionally advanced cyrtomatodont teeth allow a greater flexibility of relative movement between two valves, while after death the two may remain locked together; consequently most valves of the pentamerides, terebratulides and spiriferides are found articulated or conjoined. Although the deltidiodonts may represent a monophyletic grouping, cyrtomatodont teeth, in for example the pentamerides, spiriferides and terebratulides, may have been derived a number of times from several deltidiodont ancestors, probably within the orthides.

Autecology

Living brachiopods are widespread, and are mainly pedunculate forms, usually attached to hard substrates at a variety of water depths. *Lingula* and its relatives, how-

Fig 6.26 Representatives of the main orders of nonarticulates and articulates. Nonarticulates: (a) *Lingulella* (Ordovician), (b) *Numericoma* (Ordovician), (c) *Dinobolus* (Silurian); articulates: (d), (g) *Schizophoria* (Carboniferous), (e) *Atrypa* (Devonian), (f), (i), (j) *Marginifera* (Permian), (h) *Stenoscisma* (Permian), (k) *Parastrophinella* (Ordovician), (l) *Paeckelmannella* (Permian), (m–o) *Terebratulina* (Pleistocene). (a, c, d, e, g, h, k) approximately ×1; (f, i, j) approximately ×1.5; (l–o) approximately ×2; and (b) approximately ×95.

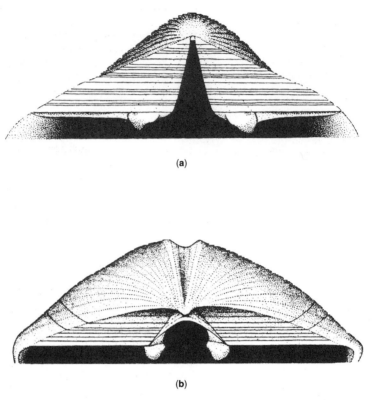

(a)

(b)

Fig 6.27   Teeth of articulates: (a) deltidiodont dentition and (b) cyrtomatodont dentition.

ever, have probably pursued an infaunal life mode, burrowing with the anterior ends of their valves, for nearly 600 million years. In high latitudes, brachiopods range from intertidal depths to over 6000 m; they are most common in fjord-type settings in Canada, Norway and Scotland, and around Antarctica and New Zealand. In the northern hemisphere, the association between the horse mussel, *Modiolus modiolus* and *Terebratulina retusa* is relatively widespread. In the tropics, many species are micromorphic, living in cryptic environments such as crevices within reefs or in the shade of corals and sponges.

Although the life modes of the most common living brachiopods are relatively restricted, the wide variety of fossil brachiopod morphologies suggests that they pursued a wide range of strategies in many ecological niches. Vogel (1986) has identified four main functional types, based on the lingulids, strophomenids, spiriferids and the pygopids. The lingulids had flat valves which suited an indirect opening mechanisms and an infaunal burrowing life mode. The strophomenides developed a mantle with a large surface area to aid food entrainment,

whereas the pygopids required their keyhole outline to partition fresh from waste materials entering and leaving the animal. The complex development of the spiriferid shell and its brachidia is less easy to understand. Brachiopods were most common during the Palaeozoic when they dominated the low-level, filter-feeding benthos. Bassett's (1984) classification of life strategies of Silurian brachiopods (Figure 6.29a) is widely applicable for brachiopods of all ages.

Many bizarre body plans arose during the Permian: one of the most spectacular groups, the richthofeniids, mimicked the corals (see Box 6.10); some Permian spiny strophomenide brachiopods lived pseudoinfaunally (Figure 6.29b), in muddy substrates (Grant, 1966). Brachiopods are less common in Mesozoic and Cenozoic faunas, following major changes in the benthos at the end of the Permian. Despite the post-Permian decline of the phylum, the brachiopods exhibited a remarkable range of adaptations based on a simple body plan and a well-defined role in the fixed, low-level benthos (Figure 6.29c).

Box 6.9 Brachiopod shell structure

Brachiopod shells are multilayered complexes of both organic and mineral materials. Although the microstructure of brachiopods has been investigated for over 100 years, the work by Alwyn Williams (e.g. Williams, 1968) forms the basis for most of our understanding of the secretion of the brachiopod shell (Figure 6.28). Shell structure has proved an important tool in the classification of the phylum. The shells of most articulate brachiopods are composed of three layers. The outer layer, the periostracum, is organic, and underneath are the mineralized primary and secondary layers. These layers are secreted by cells in the generative zone of the mantle. These cells move towards the shell edge shortly after formation. The cells first secrete a gelatinous sheath and then the periostracum followed by fine granular calcite in the thin primary layer. The cells then rotate and become fixed as the shell continues to grow. Thicker secondary shell layers of calcite fibres are then secreted, and in some brachiopods, a third prismatic layer is then secreted.

There are many variations on this basic model. The lingulates have phosphatic material combined into the shell fabric. Many shells are perforated with punctae, in life holding finger-like extensions of the mantle or caeca; punctae have proved useful in the higher classification of the articulates. Some strophomenides have pseudopunctae with fine inclined rods or taleolae embedded in the shell fabric.

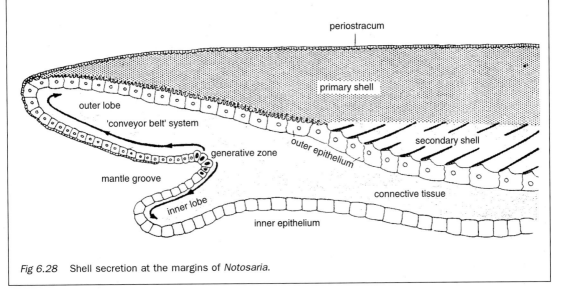

Fig 6.28   Shell secretion at the margins of *Notosaria*.

Synecology

The sheer numerical abundance and diversity of brachiopods in many fossil assemblages has guaranteed the phylum an enormous role in the reconstruction of palaeocommunities and interpretation of palaeoenvironments. Ziegler (1965) established a series of depth-related shelly communities across a spectrum of depths from the intertidal zone to the basin for Lower Silurian animals in the Anglo-Welsh area (Figure 6.31). The *Lingula*, *Eocoelia*, *Pentamerus*, *Stricklandia* and *Clorinda* communities are developed in a range of depths from intertidal conditions to the shelf edge. Boucot (1975) has rationalized these communities and termed them 'benthic assemblage zones' (see also Chapter 2).

Ager (1965) focused on the relationship between brachiopod shell morphology and substrate quality and water depth in Mesozoic ecosystems of the Tethys Ocean (Figure 6.32). Ager defined a range of possible situations occurring throughout the Alpine belt:

1. shallow-water, coarse clastics,
2. shelf sands,
3. peri-reef facies,
4. rocky seafloors,
5. shelf muds,
6. deep-water muds in slope and basinal settings, and
7. epiplanktonic associations.

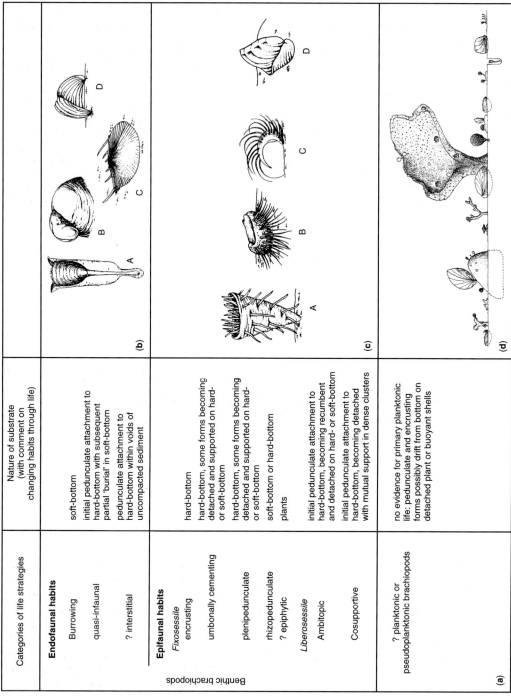

Categories of life strategies	Nature of substrate (with comment on changing habits through life)
Endofaunal habits	
Burrowing	soft-bottom
quasi-infaunal	initial pedunculate attachment to hard-bottom with subsequent partial 'burial' in soft-bottom
? interstitial	pedunculate attachment to hard-bottom within voids of uncompacted sediment
Epifaunal habits	
Fixosessile	
encrusting	hard-bottom
umbonally cementing	hard-bottom, some forms becoming detached and supported on hard- or soft-bottom
plenipedunculate	hard-bottom, some forms becoming detached and supported on hard- or soft-bottom
rhizopedunculate	soft-bottom or hard-bottom
? epiphytic	plants
Liberosessile	
Ambitopic	initial pedunculate attachment to hard-bottom, becoming recumbent and detached on hard- or soft-bottom
Cosupportive	initial pedunculate attachment to hard-bottom, becoming detached with mutual support in dense clusters
? planktonic or pseudoplanktonic brachiopods	no evidence for primary planktonic life; pedunculate and encrusting forms possibly drift from bottom on detached plant or buoyant shells

Benthic brachiopods

Fig 6.29 Composite of life modes through time. (a) Classification of brachiopod lifestyles. (b) Silurian life strategies with A, infaunal; B, cemented and C, D, free-living forms. (c) Late Palaeozoic life strategies: A, coral-like; B, recumbent; C, inverted; D, semi-infaunal. (d) Cretaceous life strategies, mainly dominated by pedun-culate forms attached to large and small patches of substrate, with some infaunal and free-living taxa.

Box 6.10 Brachiopods, functional morphology and paradigms

Martin Rudwick, in an attempt to consider brachiopod functional morphology more objectively, introduced the paradigm approach (Figure 6.30). A number of functions can be assigned any biological structure. The more likely functions of structures can be modelled either mathematically or physically and then tested. A good example is the costation (zig-zag rib pattern) of the anterior commissure of the brachiopod. It can be shown, in numerical terms, that costation increases the length of the commissure and hence the intake area that may be held open without increasing the gape of the two shells. Thus, an increased volume of nutrient-laden fluid may flow into the mantle cavity, while grains of sediment with diameters exceeding the shell gape are excluded.

During the Permian, aberrant productoids, the richthofeniids, mimicked corals and built biological frameworks in the Salt Ranges of Pakistan and the Glass Mountains of Texas. These brachiopods have a cylindrical ventral valve attached to the substrate and a small cap-like dorsal valve. It is difficult to understand how the animal fed. A possible scenario involves the flapping of the upper, dorsal valve to generate currents through the brachiopod's mantle cavity. Rudwick (1961) filmed the flow of water through the cylindrical lower, ventral valve as the upper valve was moved up and down. Fluid did in fact move efficiently through the animal, bringing in nutrients and flushing out waste. However, the paradigm failed the test of field-based evidence. Richard Grant illustrated a specimen of the athyride *Composita* apparently in life position attached to the upper valve. Vigorous flapping of the valve was thus unlikely and it would not have been an ideal attachment site for an epifauna; rather, these aberrant animals possessed lophophores with a ciliary pump action to move currents through the valves.

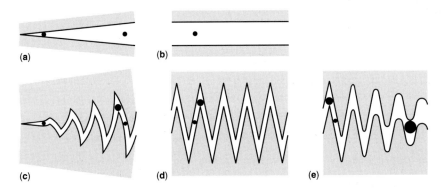

Fig 6.30   The function of the zig-zag commissure: (a) and (b) show an increased gape, allowing entry of more nutrients but also larger clastic particles; (c)–(e) demonstrate that the zig-zag development allows larger areas of entry for nutrients while restricting the diameter of clastic particles.

Origin and evolution

By the early Cambrian the nonarticulate acrotretides, lingulides and obolellids, together with the less easily-classified kutorginides and paterinides and the articulate orthides, were all present. A variety of closely related brachiophorate groups had probably evolved in the late Precambrian prior to skeletalization during the Cambrian radiation. There was a massive radiation of the group during the early Ordovician which may have been associated with increased magmatic and tectonic activity around the oceans and the creation of many islands and archipelagoes, which provided new habitats. The orthides, including the punctate dalmanellaceans and enteletaceans, and the strophomenides, including the plectambonitaceans, and the rhynchonellides, all became well established during the period. With the exception of the more bizarre oyster-like strophomenides of the later Palaeozoic, all the main shell morphologies had evolved. The end-Ordovician, glacially driven, extinction event redirected the evolution of the group. Some 10 million years afterwards the brachiopod fauna had fully recovered but was now dominated by pentamerides and a variety of spire-bearers. The Frasnian extinction, probably associated with global cooling of the tropical oceans,

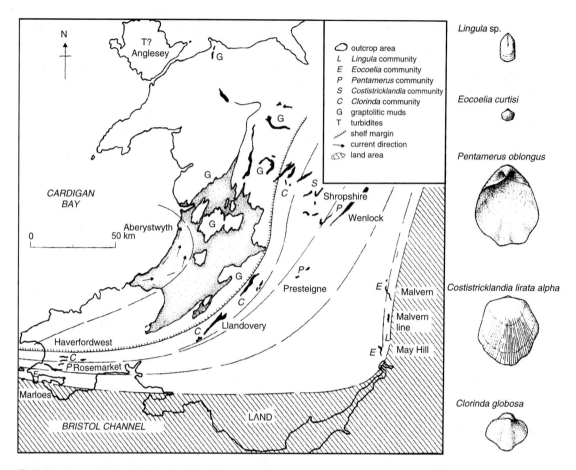

Fig 6.31   Lower Silurian depth-related palaeocommunities developed across the Welsh and Anglo-Welsh region.

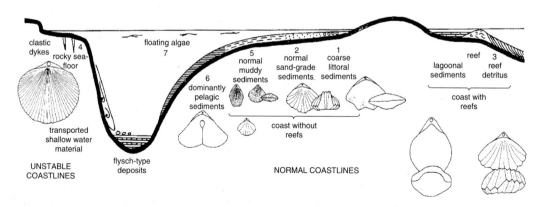

Fig 6.32   Mesozoic palaeocommunities developed across Alpine Europe.

Box 6.11 Commensal relationships

Brachiopod shells have acted as substrates for a variety of small epifaunal organisms. For example, Ager (1961) and his colleagues documented in detail the progressive colonization of the living shells of Devonian spiriferids (Figure 6.33). Each shell supported a sere and, eventually, climax community of epifauna. Each component arrived in a specific order. Many organisms (e.g. reptant tabulate *Aulopora*) grew towards the anterior commissure. It was assumed that the epifauna could take advantage of the nutrient-laden inhalent currents of the brachiopod; waste was expelled through the lateral commissures. It could be argued that many scavengers prefer waste material; nevertheless, the hypothesis of an anterior inhalent current is supported by experimental studies in flumes and the development and operation of the lophophore. The relationship was commensal: the epifauna enjoyed both support and nutrition at no expense to the host.

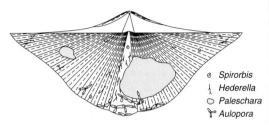

Fig 6.33 Brachiopods and commensalism: a Devonian spiriferid sequentially colonized by *Spirobis*, *Hederella*, *Paleschara* and *Aulopora*.

---

finally wiped out the pentamerides and removed many groups of atrypides in a stepwise manner.

During the Carboniferous and Permian, the productacean strophomenides developed huge diversity. However the end-Permian extinction removed the dominant groups of the Palaeozoic, low-level benthos, the orthides and strophomenides, although the former had been in decline since the Silurian. The end-Permian event probably enhanced the changeover of a benthos dominated by filter-feeding brachiopods to one with deposit-feeding bivalves. The latter are more mobile and can colonize and expand their distributions more rapidly than the fixed brachiopods; moreover the fusion of the mantle lobes and possession of siphons allows shallow burrowing as a routine life mode. Many Triassic brachiopod faunas were highly specialized assemblages of small species. The famous fauna from the St Cassian Beds of the Italian Alps is dominated by small spiriferides and terebratulides perhaps adapted to low oxygen levels.

The end-Cretaceous event saw a significant reduction in diversity of the dominant rhynchonellides and terebratulides of the late Mesozoic. As regards the articulates, only members of those orders together with the aberrant thecidaceans survive today, the latter usually in cryptic habitats within the tropics where their brooded larvae are photonegative.

Biogeography

Although brachiopods are part of the fixed benthos, they have a mobile larval stage, and thus some dispersion is possible. However, it is probable that most articulate brachiopods had a nektobenthonic larva with more limited travel capabilities than the nonarticulates. Brachiopod provincialism was marked during the early Palaeozoic and has been influential in constraining plate tectonic, and more recently terrane models, for the evolution of the Iapetus Ocean and the Appalachian Caledonian mountain belt (see Chapter 2). Provincialism has also been documented from mid Palaeozic faunas, where, particularly during the Devonian, the division between Old and New World faunas has been known for many years. During the early Carboniferous, almost 50% of known brachiopod genera migrated to higher latitudes; global warming, confirmed by data from terrestrial environments, was apparently responsible for the changing biogeographical patterns in the Dinantian.

During the Permian, Zechstein and Boreal brachiopod faunas were usually quite different from the more tropical biofacies of the Tethys area; statistical analysis has now established several biogeographic regions during the early Permian. The biogeography of Mesozoic brachiopod faunas has aided the terrane analysis of the complex tectonostratigraphy of the Western Cordillera. Sandy (1991) documented the relationships between the Jurassic and Cretaceous brachiopod faunas of the Western Cordillera, noting high similarities between the faunas of the western USA and north-west Europe during the late Cretaceous, suggesting that the American terranes had moved to high latitudes by that time.

Bryozoans

There are about 4000 living and 16 000 fossil species of bryozoans, or 'moss animals'. All are colonial coelomates and most are marine. Bryozoans superficially

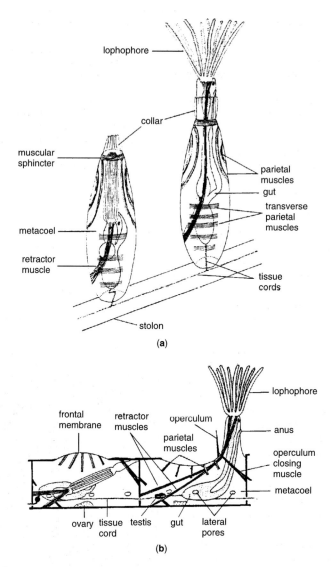

Fig 6.34   Morphology of two living bryozoans: (a) morphology of a stenolaemate and (b) a gymnolaemate.

resemble the corals and hydroids, but they are most similar to minute colonial phoronids with tiny individuals or zooids, commonly less than 1 mm in diameter (Figure 6.34). Each zooid is coelomate with a mouth and anus, together with a circular or horse-shoe-shaped lophophore equipped with tentacles. Individuals are enclosed by a gelatinous, leathery or calcareous exoskeleton usually in the form of slender tubes or box-like chambers, or zooecia. The primary function of most zooids is the capture of food, but a few are specialists in protection, reproduction and waste disposal.

Main fossil bryozoan groups

The first bryozoans probably appeared just prior to the Arenig; nearly ten families of both trepostomes and cryptostomes are recorded from the Arenig while one family of uncertain affinities, the Vinellidae, is known from Lower Ordovician rocks in Estonia. The Stenolaemata dominated Palaeozoic bryozoan faunas. The trepostomes, or stony bryozoans, had stick-like skeletons or zooaria with minute, densely packed zooecia with polygonal apertures. The group diversified dur-

---

Box 6.12 Bryozoan classification

Class Phylactolaemata
Cylindrical zooids with a horseshoe-shaped lophophore. Statoblasts arise as dormant buds. Freshwater, with non-calcified skeleton. Over 12 genera. Possibly early Jurassic–Recent.

Class Stenolaemata
Cylindrical zooids with a calcareous skeleton. A membraneous sac surrounds each polypide; a lophophore protrudes through an opening at the end of the skeletal tube. Marine, with an extensive fossil record; contains the orders Trepostomata, Cystoporata, Cryptostomata, Cyclostomata and the Fenestrata. About 550 genera. Ordovician (Arenig)–Recent.

Class Gymnolaemata
Cylindrical or squat zooids of fixed size with a circular lophophore, usually with calcareous skeleton. Majority are marine but some found in brackish and freshwater environments. Includes the orders Ctenostomata and Cheilostomata. Over 650 genera. Ordovician (Ashgill)–Recent.

---

ing the Ordovician to infiltrate the low-level benthos. Genera such as *Monticulipora*, *Prasopora* and *Tabulipora* are typical of Ordovician assemblages.

The cryptostomes, although originating during the early Ordovician, were more abundant during the mid and late Palaeozoic as the trepostomes declined; in some respects the group forms a path towards the mat-like fenestrates which were particularly common in the Carboniferous (Figure 6.35). *Fenestella*, itself, may be in the form of a mat, cone or funnel. The branches of the colony are connected by dissepiments; rectangular spaces or fenestrules separate the branches which contain the biserially arranged zooids. *Archimedes*, however, is composed of a sequence of nested funnels of fenestellid mats forming a spiral zooarium and perforated with minute apertures. Cowen and Rider (1972) have modelled the feeding strategies of these cone-shaped colonies. Silurian colonies with outward zooecia sucked in water through the sides, expelling it through the top, whereas Carboniferous colonies with inward-facing zooecia probably drew water in through the top of the colony and flushed it out through the sides.

In general, both the cryptostomes and fenestrates outstripped the trepostomes during the late Palaeozoic, being particularly partial to reef environments. Although both groups disappeared at the end of the Permian, they were still conspicuous members of the late Permian benthos; both *Fenestella* and *Synocladia* form large vase-shaped colonies in the communities of the Zechstein reef complex in the north of England and elsewhere. The stony bryozoans lingered on until the mid Triassic.

The cyclostomes are mostly tube-shaped forms with well-integrated, often branching, tree-like growth modes; some forms adopted an encrusting strategy. The first representatives of the order are known in Lower Ordovician rocks, but the group peaked during the mid Cretaceous in spectacular style, with a diversity of over 70 genera. Many genera such as *Stomatopora*, consisting of a series of bifurcating encrusting branches, have very long stratigraphical ranges; moreover, *Stomatopora* may have pursued an opportunist life strategy, establishing pioneer communities, with a rapid, efficient method of growth.

The Gymnolaemata are represented in the fossil record by two orders, the ctenostomes and the cheilostomes. The ctenostomes first appeared in the early Ordovician and many genera have since pursued boring and encrusting life strategies: *Penetrantia* and

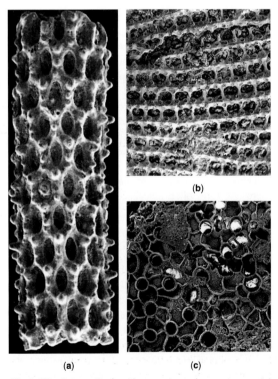

Fig 6.35 Some Carboniferous bryozoan genera: (a) *Rhabdomeson* (×30); (b) *Rectifenestella* (×15); (c) *Fistulipora* (×15).

Box 6.13 Competition and replacement in cyclostome and cheilostome clades?

Perhaps one of the most obvious changes in bryozoan faunas through time involves the decline of the cyclostomes and the diversification of the cheilostomes at and around the Cretaceous–Tertiary boundary. Since both groups occupy similar ecological niches and are comparable morphologically, many workers have assumed that the cyclostomes, originating during the Ordovician and, diversifying in the Cretaceous, were out-competed by the cheilostomes at the end of the Cretaceous. However, Lidgard *et al.* (1993) have analysed this transition in detail and the results are far from conclusive. Both groups continued to participate together in bryozoan communities during the Tertiary and much of the apparent decline in cyclostome numbers may be due to the relative diversification or expansion of the cheilostomes, comprising the grades Cribrimorpha, Anasca and Ascophora, that began to dominate these assemblages in the Cenozoic.

Terebripora are borers, whereas the modern genus *Bowerbankia* has an erect skeleton with semi-spirally arranged zooecia clustered around a central branch. The cheilostomes, however, dominate the class and are the most diverse of all the bryozoan groups. Cheilostomes have polymorphic zooids, adapted for different functions, which are usually linked within the highly integrated colony. This advanced group appeared during the late Jurassic; they are particularly common in shallow-water environments of the Upper Cretaceous and Lower Tertiary of the Baltic and Denmark. *Lunulites*, for example, is free-living and discoidal, with zooids arranged in radiating rows, whereas *Membranipora* is an erect, encrusting form often associated with sea urchins and also common on kelp.

Astogeny and heterochrony

Bryozoans usually form highly integrated colonies. However, as with many colonial animals, the growth of the colony or its astogeny provides a major constraint on the evolution and phylogeny of most bryozoan groups. Moreover, the apparent morphological plasticity in the group is illustrated by evolutionary reversals and widespread homeomorphy. Two main groups are consistently recognized: (i) small, paedomorphic plastic bryozoans occupying shallow-water low-diversity zones and (ii) taller peramorphic bryozoans with muted plasticity, more typical of deeper-water, high-diversity zones. Anstey (1987), in a detailed study of 40 taxa of Palaeozoic bryozoans, suggested that evolutionary change was promoted by paedomorphosis of offshore forms speciating in shallow-water environments.

Ecology and life modes

Virtually all bryozoans are part of the sessile benthos, occurring from mainly the sublittoral zone to the edge of the continental shelf at depths of about 200 m. Nevertheless, a few intertidal forms are known, while some bryozoans have been dredged from depths of over 8 km in oceanic trenches; moreover, over 50 species have been recorded from the hulls of ships. Most species are sensitive to substrate types, turbulence, water depth, temperature and salinity. The shape of colonies can be very plastic, adapting to environmental conditions. Bryozoans are thus typical facies fossils, exhibiting marked ecophenotypic variation.

Bryozoans have successfully pursued several different life modes. Encrusting, erect, unattached or rooted phenotypes all reflect adaptive strategies in response to ambient environmental conditions. Shallow-water colonies, particularly in the subtidal zone, are and were dominated by encrusting, erect, rooted and free-living forms; but deeper-water environments, at over 1 km depth, are characterized by mainly attached and rooted forms. Nevertheless, dense networks of bryozoan colonies have occasionally formed reefs or bryoherms, particularly during the mid Silurian and the Carboniferous.

Further reading

Boardman, R.S. and Cheetham, A.H. (1987) Phylum Bryozoa. In R.S. Boardman, A.H.Cheetham, and A.J. Rowell (eds) *Fossil invertebrates*. Blackwell Scientific Publications, Oxford, pp. 497–549.

Clarkson, E.N.K. (1993) *Invertebrate palaeontology and evolution*, 3rd edition. Chapman and Hall, London.

Cocks, L.R.M. (1985) Brachiopoda. In J.W. Murray (ed.) *Atlas of invertebrate macrofossils*. Longman, London, pp. 53–78.

James, M.A., Ansell, A.D., Collins, M.J., Curry, G.B., Peck, L.S. and Rhodes, M.C. (1992) Biology of living brachiopods. *Advances in Marine Biology*, **28**, 175–387.

McKinney, F.K. and Jackson, J.B.C. (1989) *Bryozoan evolution*. Unwin Hyman, London.

Oliver, W.A. jr and Coates, A.G. (1987) Phylum Cnidaria. In R.S. Boardman, A.H. Cheetham and A.J. Rowell (eds) *Fossil invertebrates*. Blackwell Scientific Publications, Oxford, pp. 140–193.

Rowell, A.J. and Grant, R.E. (1987) Phylum Brachiopoda. In R.S. Boardman, A.H. Cheetham, and A.J. Rowell (eds) *Fossil invertebrates*. Blackwell Scientific Publications, Oxford, pp. 445–496.

Rudwick, M.J.S. (1970) *Living and fossil brachiopods*. Hutchinson, London.

Ryland, J.S. (1970) *Bryozoans*. Hutchinson, London.

Scrutton, C.T. and Rosen, B.R. (1985) Cnidaria. In J.W. Murray (ed.) *Atlas of invertebrate macrofossils*. Longman, London. pp. 11–46.

Taylor, P.D. (1985) Bryozoa. In J.W Murray (ed.) *Atlas of invertebrate macrofossils*. Longman, London, pp. 47–52.

References

Ager, D.V. (1961) The epifauna of a Devonian spiriferid. *Quarterly Journal of the Geological Society of London*, **117**, 1–10.

Ager, D.V. (1965) The adaptation of Mesozoic brachiopods to different environments. *Palaeogeography, Palaeoclimatology* and *Palaeoecology*, **1**, 143–179.

Anstey, R.L. (1987) Astogeny and phylogeny: evolutionary heterochrony in Paleozoic bryozoans. *Paleobiology*, **13**, 20–43.

Bassett, M.G. (1984) Life strategies of Silurian brachiopods. *Special Papers in Palaeontology*, **32**, 237–263.

Boucot, A.J. (1975) *Evolution and extinction rate controls*. Elsevier, Amsterdam.

Coates, A.G. and Jackson, J.B.C. (1987) Clonal growth, algal symbiosis, and reef formation by corals. *Paleobiology*, **13**, 363–378.

Copper, P. (1985) Fossilized polyps in 430-Myr-old *Favosites* corals. *Nature*, **316**, 142–144.

Cowen, R. and Rider, J. (1972) Functional analysis of fenestellid bryozoan colonies. *Lethaia*, **5**, 147–164.

Grant, R.E. (1966) A Permian productoid brachiopod: life history. *Science*, **152**, 660–662.

Jaanusson, V. (1971) Evolution of the brachiopod hinge. *Smithsonian Contributions to Paleobiology*, **3**, 33–46.

Lidgard, S., McKinney, F.K. and Taylor, P.D. (1993) Competition, clade replacement, and a history of cyclostome and cheilostome bryozoan diversity. *Paleobiology*, **19**, 352–371.

Rudwick, M.J.S. (1961) The feeding mechanism of the Permian brachiopod. *Prorichthofenia*. *Palaeontology*, **3**, 450–457.

Sandy, M.R. (1991) Biogeographic affinities of some Jurassic–Cretaceous brachiopod faunas from the Americas and their relation to tectonic and paleooceanographic events. In D.I. MacKinnon, D.E. Lee and J.D. Campbell (eds) *Brachiopods through time*. Balkema, Rotterdam, pp. 415–422.

Scrutton, C.T. (1989) Ontogeny and astogeny in *Aulopora* and its significance illustrated by a new non-encrusting species from the Devonian of southwest England. *Lethaia*, **23**, 61–75.

Scrutton, C.T. and Clarkson, E.N.K. (1990) A new scleractinian-like coral from the Ordovician of the Southern Uplands of Scotland. *Palaeontology*, **34**, 179–194.

Vogel, K. (1986) Origin and diversification of brachiopod shells: viewpoints of constructional morphology. In P.R. Racheboeuf and C.C. Emig (eds) *Les Brachiopodes Fossiles et Actuels*. Biostratigraphie du Paléozoique, **4**, pp. 399–408.

Wells, J. (1963) Coral growth and geochronometry. *Nature*, **197**, 948–950.

Williams, A. (1968) A history of skeletal secretion in brachiopods. *Lethaia*, **1**, 268–287.

Wright, A.D. (1979) Brachiopod radiation. In M.R. House (ed.) *The origin of major invertebrate groups*. Systematics Association, London, pp. 235–252.

Ziegler, A.M. (1965) Silurian marine communities and their environmental significance. *Nature*, **207**, 270–272.

7 Radialians II: echinoderms and hemichordates

Key Points

- Echinoderms have a water-vascular system and tube feet, mesodermal skeletons of calcitic plates with a stereom structure, and pentameral symmetry.
- During the Cambrian–early Ordovician, many bizarre echinoderms evolved.
- Pelmatozoans were mainly fixed echinoderms; they include the blastoids, crinoids and cystoids.
- The echinozoans were mobile benthos; during the Mesozoic irregular echinoids, adapted for burrowing, evolving from regular forms dominant in the Palaeozoic.
- Asterozoans (starfish) were more important in post-Palaeozoic rocks.
- Calcichordates are traditionally classed with the echinoderms; some evidence suggests that they may be ancestral to the chordates.
- Graptolites are hemichordates closely related to the living rhabdopleurids.
- Dendroids have three types of thecae and many stipes, whereas graptoloids have few stipes and monomorphic thecae.
- Graptolites had benthic (dendroids), planktonic (dendroids and graptoloids) and automobile (graptoloids) lifestyles; they evolved rapidly and were widespread.

Introduction

The echinoderms and hemichordates are closely related. Both these radialian groups are deuterostomous with a dipleurula lava and an enterocoelous coelom developed from extensions of the embryonic gut. Whereas the stalked echinoderms dominated the higher-level benthos of the Palaeozoic, mobile forms were common throughout the Mesozoic and Cenozoic. Fossil hemichordates such as the graptolites dominated the water columns of the early Palaeozoic oceans.

Echinoderms

The Phylum Echinodermata is uniquely equipped with a water-vascular system; water is forced around the plumbing by muscular action while tube feet, extending from the system, are often adapted for food processing, locomotion and respiration. The 6000 or so living echino-derm species include the sea lilies, sea urchins, sand dollars, starfish and sea cucumbers (Figure 7.1). Although many species today live in the intertidal or subtidal zones, the group occupied a wide range of marine environments and pursued a variety of life strategies in the geological past. Fossil echinoderms are relatively common, and many Palaeozoic limestones are packed with their distinctive skeletal debris of calcitic plates.

Apart from the water-vascular system, echinoderms have a number of other distinctive features. All members of the phylum have a mesodermal skeleton constructed from porous plates of calcite; each plate is usually a single crystal of calcite and is thus easy to recognize in thin sections. In addition, the plates have a unique ultrastructure of rods linked to form a 3D lattice; this network or stereom was permeated by fingers of soft tissue which occupied the spaces or stroma in the lattice. Finally, a five-rayed or pentameral symmetry, occasionally modified by a secondary bilateral symmetry, is typical of the echinoderms. The phylum is generally split into the mobile, non-stalked eleutherozoans and the mainly fixed, stalked pelmatozoans (see Box 7.1).

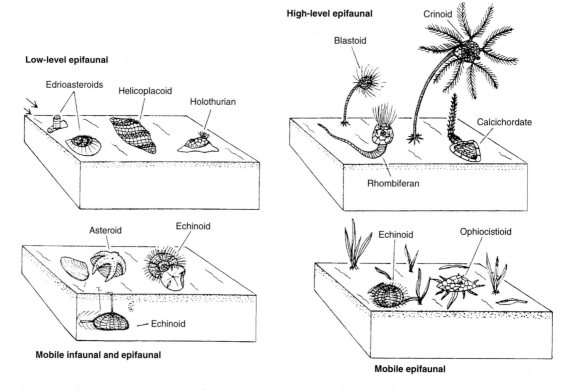

Fig 7.1 Life modes of the main echinoderm body plans.

Box 7.1 Echinoderm classification

In broad terms, the Echinodermata may be divided into two main sister groups: the stalked pelmatozoans and the mobile eleutherozoans. A number of the more bizarre Lower Palaeozoic forms, known from only a few specimens at single localities, are difficult to classify (see Box 7.2).

Subphylum Pelmatozoa

Superclass Cystoidea
Classification is still problematic. Apart from the blastoids, some or all may be paraphyletic.

Class Eocrinoidea
Globular or flat theca with two to five ambulacra bearing brachioles. Early Cambrian –Silurian.

Class Paracrinoidea
Irregularly arranged plates comprise globular to lenticular theca; two to five ambulacra commonly with pinnules. Hydropore adjacent to mouth. Stem attached to three basal plates. Mid to late Ordovician.

Class Blastoidea
Flask-shaped theca with commonly three basal plates; ambulacra with elongate lancet plate and rows of side plates. Mid Ordovician–late Permian.

Class Diploporita
Globular theca with many plates in irregular to regular pattern; three to five food grooves with brachioles. Diplopores perforate thecal plates. Mid Cambrian–mid Devonian.

Class Rhombifera
Globular theca with two to five ambulacra commonly extending from the mouth to the edge of the upper surface. Pore structures cross adjacent thecal plates arranged in rhomboid pattern, and comprise respiratory system. Late Cambrian–late Devonian.

Superclass Crinozoa
Class Crinoidea
Calyx with lower cup and upper tegmen. Examples: sea lilies and feather stars. Early Ordovician–Recent.

Subphylum Eleutherozoa

Class Edrioasteroidea
Disc-shaped theca with straight or curved ambulacra; the mouth is situated centrally and the anus is sited on the interambulacra. Late Precambrian–late Carboniferous.

Class Asteroidea
Between 5 and 25 arms with large tube feet extend from central disc. Examples: starfishes or sea stars. Early Ordovician–Recent.

Class Ophiuroidea
Five long, thin, flexible arms, consisting of vertebrae and with small tube feet, extend from a large, circular central disc. Examples: brittle stars or basket stars. Early Ordovician–Recent.

Class Echinoidea
Test is usually globular with plates differentiated into ambulacral and interambulacral areas. Mouth on the underside; the anus is on the upperside or sited posteriorly. Examples: sea urchins, heart urchins, sand dollars. Late Ordovician–Recent.

Class Holothuroidea
Cucumber-shaped body with leathery skin; muscular mesoderm and spicules. A ring of modified tube feet surround the mouth. Examples sea cucumbers. Early Ordovician –Recent.

Crinozoans

The crinoids are usually sessile animals with pentameral symmetry, rooted by a stalk, for at least part of their life cycle, to 'the seabed. Modern forms may live in dense clusters or 'forests' ranging from the warm waters of the tropics to the icy conditions of polar latitudes. In general, following a short-fixed larval stage, most living species are unattached, pelagic forms; these 'feather stars' prefer the clear-water conditions of the continental shelf. The sea lilies, however, are attached, sessile forms occupying the deeper-water environments of the continental slope. The majority of fossil taxa were almost certainly part of the shallow-water sessile benthos. The success of the crinoids may be measured by the records of over 6000 known fossil species and a range from the early Ordovician to the present day.

Morphology and life modes

The crinoids consist of a segmented stalk composed of columnals or ossicles attached to the seabed by root-like attachment structures or holdfasts. The stalk or stem is

Box 7.2 Origin of the echinoderms and the status of the helicoplacoids

During the major early Cambrian radiation of the echinoderms, some bizarre forms appeared. At least nine genera were present, some having pentameral symmetry; the others were not pentameral. One such group, the helicoplacoids (Figure 7.2), is unique in having only three ambulacral areas wrapped around their spindle-shaped bodies. Moreover, the group lacks appendages and probably lived with their shorter ends anchored in the sediment. Helicoplacoids have plates with the distinctive stereom structure, ambulacra and a mouth sited laterally together with an apical anus. These are clearly primitive echinoderms, which survived by suspension feeding. The echinoderms were already diverse during the early Cambrian. Nevertheless, the generalized morphology of *Helicoplacus* may be close to the stem group of all subsequent Echinodermata. Both the pelmatozoan and eleutherozoan body plans may have been derived from *Helicoplacus*-type animals.

Fig 7.2   *Helicoplacus* from the Lower Cambrian (×10).

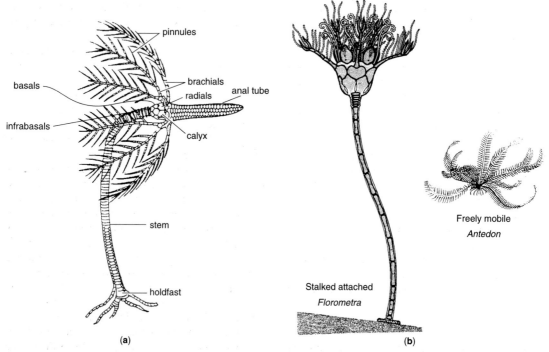

Fig 7.3   (a) Morphology of the Ordovician *Dictenocrinus*. (b) Two main crinoid life strategies, fixed and mobile.

Box 7.3 Columnal classification

The majority of crinoid assemblages are represented by disarticulated ossicles (Figure 7.4). Conventional taxonomy based on descriptions of complete, articulated specimens is thus not possible. Nevertheless, ossicles have many distinctive features, arguably with more well-defined characteristics than many groups of macrofossils. Single stems consist of many ossicles with central canals or a lumen usually carrying nerve fibres. Both the ossicles and lumens have distinctive shapes which are the basis of a form taxonomy of the group. Stems may be either homeomorphic, composed of similarly-shaped ossicles, or heteromorphic with a variety of different-shaped ossicles; xenomorphic columns have different morphologies in different regions of the same individual. Moreover, stems may be subdivided into zones which may be internally homeomorphic or heteromorphic. Columnal taxonomy has proved useful in describing taxa (so-called col. taxa) of pelmatozoan ossicles, particularly crinoid ossicles, of stratigraphical significance.

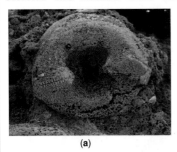
(a)

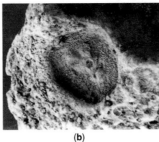
(b)

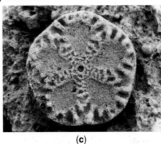
(c)

Fig 7.4 Some crinoid ossicle types. (a) Articular facet of a columnal of the bourguaticrinid *Democrinus* (?) sp., with fulcral ridge of the synarthrial articulation; lumen opens at bottom of '8'-shaped depression, (×15). (b) Cirral scar on a nodal of the isocrinid *Neocrinus* with well-preserved stereom structure microstructure and knob-like synarthrial fulcrum (×18). (c) Articular facet of a columnal of the isocrinid *Neocrinus* with symplectial articulation around the five petal-like areola areas, (×9).

itself attached to the 'case' containing the main functional part of the animal, the calyx, cup or theca. The calyx or aboral cup is built of two rings of calcitic plates: the basals and the overlying radials in a monocyclic configuration. In a number of taxa, comprising the dicyclic forms, a second circle of smaller plates, the infrabasals, interface between the basals and the stem, providing further articulation. The upper, oral surface of the calyx is covered by a flexible membrane or rigid roof, called the tegmen, which houses a number of important structures: the mouth is usually situated centrally at the convergence of five radially arranged feeding grooves; the anus is sited posteriorly with the outlet often modified by an anal tube enhancing the efficiency of waste disposal. Arms or brachials extend upwards from the calyx and together they comprise the crown.

Two main life strategies were pursued by the crinoids (Figure 7.3). The majority of fossil crinoids and about 25 Recent genera are stalked forms. Modern oceans, however, are dominated by mobile comatulids. *Antedon* is one of nearly 100 non-stalked genera which, after a short fixed stage, is free to crawl and swim with the aid of flexible arms and cirri.

Classification and evolution

The first recognizable crinoids with typically constructed cups and columnal-bearing stems (see Box 7.3), such as *Dendrocrinus*, appear some time later during the Arenig. A major expansion in the early Ordovician tropics marked a period of intense morphological experimentation and many adaptive radiations.

Virtually all the Palaeozoic crinoids were stalked, and have traditionally been grouped into three main subclasses, the Inadunata, the Flexibilia and the Camerata. Inadunate crinoids comprise a large and varied group, originating in the early Ordovician and continuing until the Permian. The inadunates have a rigid calyx with either free or loosely attached brachials; the cup bases are monocyclic or dicyclic.

Camerate crinoids are characterized by large cups with both monocyclic and dicyclic plate configurations. The uniserial or biserial brachials, supporting delicate branches called pinnules, are firmly attached to the cup; the tegmen is heavily plated, obscuring the food grooves and mouth, but developed laterally with an anal tube.

There are about 60 genera of flexible crinoids with a dicyclic plate configuration comprising three infrabasals. The brachials are uniserial lacking pinnules, and the tegmen is flexible with a mosaic of small plates. The stems have circular cross-sections and lack cirri.

With the exception of some Triassic inadunates, all post-Palaeozoic crinoids belong to the Articulata, a monophyletic infraclass of the Cladida, recently referred to the 'crown-group articulates'; Simms and Sevastopulo (1993) considered that a few Palaeozoic forms with articulate similarities such as *Ampelocrinus* and *Cymbiocrinus* may be 'stem-group articulates'. Over 250 genera are recognized, with almost two-thirds of known genera extant.

Microcrinoids are a highly specialized crinoid morphotype developed within both Inadunata, during the Palaeozoic, and the Articulata, during the Mesozoic. Microcrinoids are minute, never more than 2 mm in size; they may be paedomorphic forms living together with more typical crinoid communities.

Blastozoans

This subphylum contains three of the more minor, nevertheless important, stalked echinoderm groups: the blastoids, the cystoids and the eocrinoids. The blastozoans are all extinct; these pelmatozoans were usually equipped with a short stem, but often lacked brachia or arms. Blastozoans were probably high-level filter feeders, particularly characterized by pores or brachioles punctuating the thecal plates. The cystoids appeared near the base of the Cambrian and became extinct during the Silurian. The eocrinoids, sometimes classed as cystoids, probably contained ancestors to both the cystoids and the crinoids.

Cystoids

Mid Palaeozoic blastozoans with respiratory pore structures modifying the thecal plates have been traditionally placed within the Cystoidea. This mixed bag includes two classes whose high-level classification is based on the pore structure of the thecae (Figure 7.5). The cystoids were very widespread during the mid Palaeozoic, with their spherical or sac-like theca commonly having over 1000 irregularly arranged plates. Moreover, the group has brachioles lacking pinnules and characteristically the plates are equipped with distinctive pore structures.

Diploporita The diploporites had thecal plates punctuated by pairs of pores either covered with soft tissue (diplopores) or a layer of stereom with the pore pairs joined by a network of minor canals (humatipores). These pores probably held a bulbous respiratory bag and allowed for the efficient entry and exit of coelomic fluid. Both stalked and non-stalked forms are present in this group, suggesting a wide range of strategies from a fixed sessile mode to free-living recumbent styles. The diplo-

porites were very widespread from the early Ordovician to the early Devonian and probably evolved from a late Cambrian blastozoan ancestor.

Rhombifera The rhombiferans appeared during the late Cambrian, equipped with brachioles and distinctive rhombic patterns of respiratory pores crossing thecal plate sutures. Two patterns based on endothecal pore structures with either pectinirhombs or cryptorhombs characterize the Order Dichoporita, whereas mosaics of exothecal humatirhombs typify the Fistulipora. The rhombiferans became common during the early Ordovician and continued with a near-cosmopolitan distribution until the late Devonian. They were probably replaced by the better-adapted blastoids during the Silurian and Devonian.

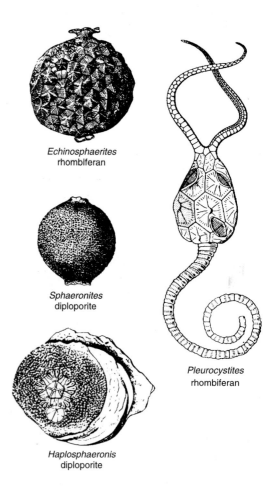

Echinosphaerites rhombiferan

Sphaeronites diploporite

Haplosphaeronis diploporite

Pleurocystites rhombiferan

Fig 7.5 Some Ordovician cystoid genera: *Echinosphaerites* and *Sphaeronites*, ×0.75; *Haplosphaeronis* and *Pleurocystites*, ×1.5.

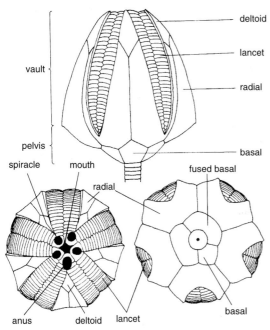

Fig 7.6 Basic blastoid morphology.

Blastoids

The blastoids – small, extinct, pentamerally symmetrical animals with short stems and hydrospires adapted for respiration (Figure 7.6) – are represented by over 80 genera in rocks of Silurian to Permian age. The blastoid cup or theca is usually globular, composed of a ring of three

basal plates, surmounted by a circle of five larger radial plates; the deltoids are small interambulacral plates interfacing with the ambulacral areas with their own lancet plates. The mouth is surrounded by five large openings or spiracles.

Although blastoids are relatively rare, a few horizons are packed with specimens, particularly in the early Carboniferous; for example, levels associated with Viséan reefal facies in northern England have abundant blastoids, as do the Permian limestones on the island of Timor where, for example, *Timoroblastus* and *Schizoblastus* occur (Figure 7.7).

The blastoids first appeared during the mid Ordovician, probably from an Ordovician ancestor with brachioles and a reduced number of plates, and initially competed, ecologically, with the rhombiferan cystoids. The evolutionary history of the group was marked by changes in the shape of the theca and variations in the length of the ambulacra. Two main groups are recognized: the more primitive Fissiculata are characterized by hydrospire folds and the Spiraculata with, as the name suggests, well-developed spiracles.

Eocrinoids

The eocrinoids were the earliest of the brachiole-bearing echinoderms, with a huge range of thecal shapes with primitive holdfasts and an irregular to regular arrangement of plates (Figure 7.8). Sutural pores rather than thecal pores, located along the joins between the plates, are characteristic of the earliest eocrinoids; in others there is a total lack of respiratory structures. Moreover, eocrinoids differ from the crinoid groups in having biserial brachial appendages; they are therefore blastozoans. Over 30 genera have been described from rocks of early Cambrian to late Silurian age. The origins of the other blastozoan classes are probably to be found within this heterogenous group; for example, the aberrant late

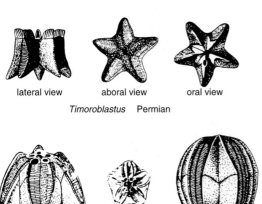

lateral view   aboral view   oral view

Timoroblastus   Permian

lateral view   oral view   lateral view

Pentremites   *Schizoblastus*
Carboniferous   Carboniferous – Permian

Fig 7.7 Some blastoid genera (×0.6).

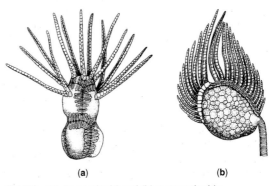

(a)   (b)

Fig 7.8 (a) An eocrinoid and (b) a paracrinoid.

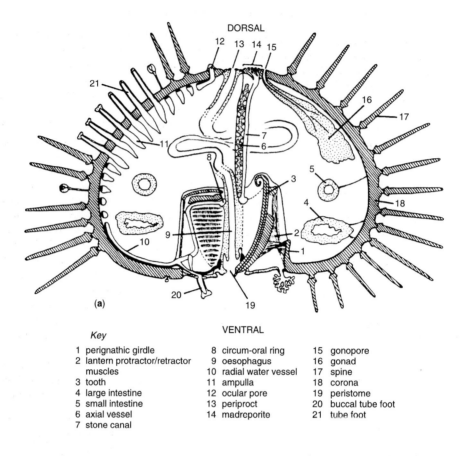

(a)

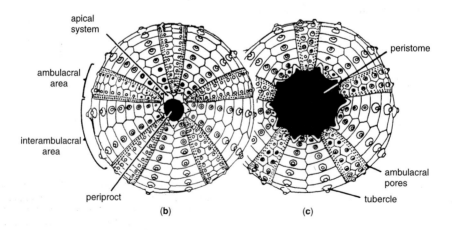

(b)   (c)

Fig 7.9 Echinoid morphology. (a) Internal anatomy in cross-section; (b) dorsal and (c) ventral views of *Echinus*.

Cambrian eocrinoid *Cambrocrinus* has been cited as an ancestor for the rhombiferan cystoids. Whereas many eocrinoids were high-level suspension feeders with the first columnal-constructed stems, some lay reclined or recumbent on the seabed. Two genera illustrate some of the variation in the group. The early Ordovician form *Macrocystella* possesses a spindle-shaped theca with regularly arranged plates and a tapering stem, whereas the Ordovician *Cryptocrinus* has a globular theca with a more irregular arrangement of plates.

Echinozoans

Echinoids have robust, rigid endoskeletons or tests composed of plates of calcite coated by an outer skin covered by spines. The tests are usually either globular or discoidal to heart-shaped. Echinoids are most common in shallow-water marine environments where they congregate in groups as part of the nektobenthos.

There were two sudden divergences, however, from the regular echinozoan morphology which generated irregular burrowing echinoids in the Jurassic, and the quasi-infaunal sand dollars during the Palaeocene; both events were probably rapid and permitted major adaptive radiations of parts of the group into new ecological niches (see Box 7.4).

Basic morphology

The test of most regular echinoids (e.g. the common sea urchin, *Echinus esculentus*) is hemispherical (Figure 7.9). The lower, adapical or oral surface is perforated by the mouth, whereas the upper, apical or aboral surface has the anal opening. The sea urchin is part of the active mobile benthos.

The test is built of a network of many hundreds of calcite plates organized into 10 segments, composed of pairs of interlocking calcitic plates, radiating from the oral surface and converging on the aboral surface. The five narrower segments or ambulacral areas (ambs) interface with the ocular plates and carry the animal's tube feet. These alternate with the wider interambulacral areas (interambs) which abut against the genital plates and are armed with large spines. Together, all 10 columns comprise the corona, which forms the majority of the test.

The central part of the aboral surface has a ring of five genital plates, each perforated by a hole to allow the release of gametes; the madreporite is commonly larger than the other genital plates and has numerous minute pores that connect the water-vascular system to the external sea-water. These alternate with the ocular plates, terminating the ambulacral areas, and each houses further outlets for the water vascular system. This part of the apical system surrounds the periproct or anal opening

---

Box 7.4 Echinoid classification

The traditional split of the class into regular and irregular forms is no longer considered to reflect the true phylogeny of the echinoids. Whereas the irregular echinoids are probably monophyletic, arising only once, the regular echinoids do not form a monophyletic clade. The group is currently subdivided into three subclasses (Figure 7.10): the first contains the more primitive forms; the second and third, the more advanced forms.

Subclass Perischoechinoidea
  Regulars with ambulacra in 2–20 columns, interambulacra with many columns. Lantern with simple grooved teeth. Late Ordovician–Permian.

Subclass Cidaroidea
  Regulars with two columns in ambulacra. Interambularcal plates have a large tubercle. Devonian–Recent.

Subclass Euechinoidea
  Post-Palaeozoic taxa, both regular and irregular. Both ambulacra and interambulacra with twin columns. Late Triassic–Recent.

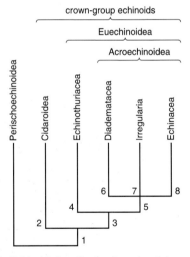

Fig 7.10 Echinoid classification based mainly on cladistic analysis. 1, Ten ambulacral and ten interambulacral areas. 2, Upright lantern without foramen magnum. 3, Distinctive perignathic girdle. 4, Distinctive ambulacral areas. 5, Upright lantern with deep foramen magnum. 6, Grooved teeth. 7, Stout teeth. 8, Keeled teeth.

which is partially covered by a number of smaller plates attached to a membrane. On the underside of the test, the peristome (containing the mouth parts) is also covered by a membrane coated with small plates. The mouth holds a relatively sophisticated jaw apparatus, called Aristotle's lantern, comprising five individual jaws each based on a single curved sabre-like tooth; operating like a mechanical grab, the lantern grasps at detritus, forcing particles into the animal's digestive system.

The animal's various organs seem to be suspended within the echinoid test supported by fluid. The water-vascular system copes with a number of functions. The stone canal rises vertically from the central ring around the oesophagus, to unite with the madreporite. Five radial water vessels depart from the central ring to service the ambulacral areas; smaller vessels are attached to each tube foot and its ampulla, where variations in water pressure drive the animal's locomotory system.

The digestive system lacks a stomach and operates through the oesophagus together with a large and small intestine; waste material is expelled through the rectum into the anus and externally by way of the periproct. An unsophisticated nervous system comprising a nerve ring and five radial nerves connects with the ambulacral pores where the nerve ends divide externally to form a sensory net.

The regular to irregular transition The irregular echinoid morphotype evolved rapidly and apparently involved some large architectural changes to adapt the animal to a buried strategy. *Plesiechinus hawkinsi* is one of the first irregular echinoids, appearing during the early Jurassic with an asymmetrical test, numerous short spines, large adapical pores and a posteriorly-placed periproct together with presumed keeled teeth. Ten million years later, the full toolkit of adaptations had evolved for a burrowing life mode. A secondary bilateral symmetry is superimposed on an existing pentameral symmetry to form a heart-shaped or flattened ellipsoidal test. The periproct migrates from a position on the apical surface to the posterior side of the test to eject waste laterally. One of the ambulacral areas is modified to form a food groove, and a series of tube feet are extendable with flattened ends to assist respiration.

One of the earliest sand dollars, the clypeasteroid *Togocyamus*, appeared during the Palaeocene, and some 20 million years later in the Eocene more typical sand dollars had evolved to command a cosmopolitan distribution. The flattened test was adapted for burrowing whereas the accessory tube feet could encourage food along the food grooves and drape the test with sand. The highly accentuated petals helped respiration with an increased surface area for the tube feet, and the development of a low lantern with horizontal teeth signalled changes in feeding patterns.

Modes of life

The regular *Echinus* and the irregular *Echinocardium* probably mark the ends of a spectrum of life modes from epifaunal mobile behaviour to a number of infaunal burrowing strategies (Figure 7.11). Mobile regular forms such as *Echinus* graze on both hard and soft substrates and in caves and crevices on the sea-floor; these sea urchins may be omnivores, carnivores or herbivores. Irregular forms display a range of adaptations appropriate to an infaunal mode of life where burrows are carefully constructed in low-energy environments. Extreme morphologies are developed in the sand dollars (Clypeasteroidea) permitting rapid burial just below the sediment–water interface in shifting sands.

Evolution

The relatively sparse early record of the group may be partly the result of the absence of a rigid skeleton; nevertheless, the echinoids were not a common element of the Palaeozoic benthos. The enigmatic *Bothriocidaris* has been described from the Ordovician of Estonia and the Upper Ordovician of the Girvan district, south-west Scotland. Although variously classified as an echinoid, cystoid or holothurian, some authorities considered that *Bothriocidaris* and *Eothuria* may be non-aligned echinoids, arising during the frantic Ordovician radiation of the group. *Aulechinus*, from the Upper Ordovician Lady Burn Starfish Bed, Girvan, is one of the most primitive and first echinoids, with only two plate columns in the ambulacral areas. During the Palaeozoic there was generally an increase in the number and size of ambulacral areas and in the sophistication of Aristotle's Lantern, although most genera remained relatively small.

There was a significant decline in echinoid diversity during the late Carboniferous, and by the Permian only about half a dozen species are known. Large proterocidarids were highly specialized detritus feeders and the small omnivorous *Miocidaris* and *Xenechinus* were opportunists. *Miocidaris*, ancestral to the cidaroids, together with a precursor of the euechinoids survived the end-Permian extinction event (Smith and Hollingworth, 1990) to radiate extravagantly during the early Mesozoic and ensure the survival of the echinoids. Following the end-Permian extinctions, the regular echinoids diversified during the late Triassic and early Jurassic with more

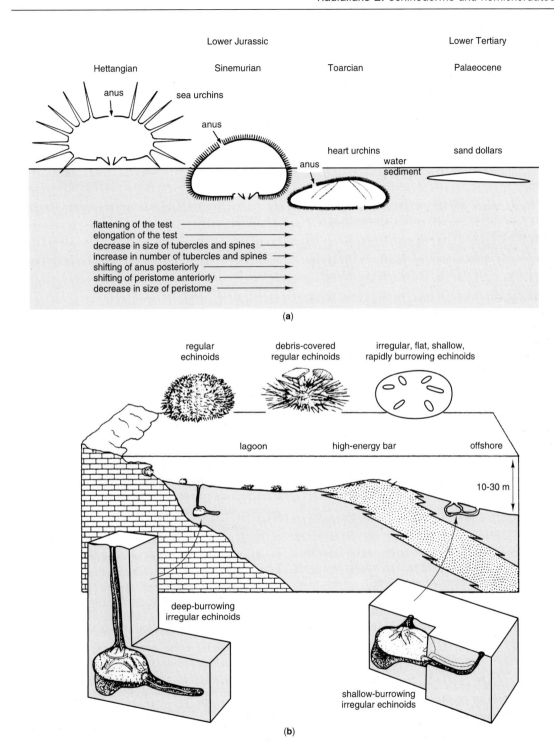

Fig 7.11   Echinoid life modes: (a) transition from the sea urchins through the heart urchins to the sand dollars; (b) habitats and modes of life of echinoids.

Box 7.5 Microevolution of *Micraster*

Serial morphological changes, associated with adaptive biological improvement for an infaunal life mode, in the lineages of *Micraster* have been known and debated for nearly a century. Moreover, microevolution in the *Micraster* plexus has provided possibilities to test phyletic gradualist and punctuated equilibria models. In the best known lineage, *M. leskei*–*M. decipiens*–*M. coranguinum*, the following morphological changes occur (Figure 7.12):

1. development of a higher, broader (heart-shaped) form associated with an increase in size and thickness of the test;
2. the peristome (mouth) moves anteriorly and the posteriorly situated periproct (anus) has a lower position on the side of the test, with a broader subanal fasciole;
3. the madreporite increases in size at the expense of the adjacent specialized plates;
4. a more tuberculate and deeper anterior ambulacra is evolved;
5. more granulated periplastronal areas develop.

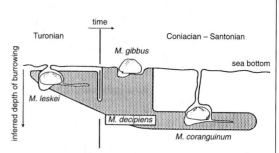

Fig 7.12 Evolution of the Late Cretaceous heart urchin, *Micraster*.

These changes have been related to life in progressively deeper burrows, but a few studies have noted the lack of matching trace fossils in the Chalk. Nevertheless, there is no doubt that *Micraster* was a burrower, probably to escape predation; the adaptations may have been geared to a greater efficiency of burrowing in shallow depths where such traces are easily destroyed by the reworking of sediments by bottom currents.

advanced regulars dominating the early Mesozoic record. The irregulars appeared during the early Jurassic, and substantially increased in numbers during the period (Figure 7.13).

Asterozoans

Asterozoans are common in modern seas and oceans, having appeared during the early Ordovician. The subphylum contains two main groups: the asteroids (or starfish) and the ophiuroids (or brittle-stars). These animals have a star-shaped outline with usually five arms radiating outwards from the central body or disc. The water-vascular system is open. The mouth is situated centrally on the underside of the animal (i.e. on the oral or dorsal surface) whereas the anus, if present, opens ventrally on the adoral surface. The asterozoans are characterized by a mobile lifestyle within the benthos, where many are carnivores. However, since asterozoan skeletons disintegrate rapidly after death, because of feeble cohesion between skeletal plates, recognizable fossils are relatively rare.

Distribution and ecology of the main groups

Three classes of asterozoans – the primitive Somasteroidea, the Asteroidea, and the Ophiuroidea –

have been recognized. The Somasteroidea include some of the earliest starfish-like animals, described from the Lower Ordovician of Gondwana. These echinoderms have pentagonal-shaped bodies with the arms initially differentiated from parts of the oral surface. In some respects this short-lived group, which probably disappeared during the mid Ordovician, displays primitive starfish characters intermediate between a pelmatozoan ancestor and a typical asterozoan descendant.

Typical asteroids have five arms, extending in a digitate configuration from the disc, which is coated by loosely fitting plates permitting considerable flexibility of movement (Figure 7.14). Additional respiratory structures, papulae, project from the coelom through the plates of the upper surface; this back-up system would aid the high metabolic rates of these active starfish. The first true starfish were probably derived from the somasteroids during the early Ordovician and were relatively immobile sediment shovellers. Some of the first starfish (e.g. *Hudsonaster*, from the middle Ordovician) have similar plate configurations to the young growth stages of living forms such as *Asterias*. Although relatively uncommon in Palaeozoic rocks, the group was important during the Mesozoic and Cenozoic and is now one of the most common echinoderm classes (see Box 7.6).

The ophiuroids appeared during the early Ordovician and the group is probably paraphyletic. Classification is

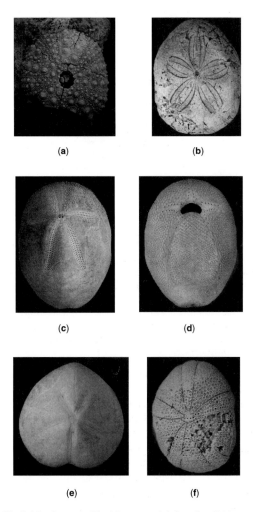

Fig 7.13 Some echinoid genera. (a) Regular *Echinometra* (Pleistocene–Recent); (b) sand dollar *Clypeaster* (Tertiary); (c), (d) apical and oral views of the spatangoid *Brissus* (Recent); (e) heart urchin *Micraster* (Cretaceous); (f) helectypoid *Echinoneous* (Pleistocene). (a, c, d, e) (×0.75); (b) (×0.4); (f) (×1.4).

based on arm structure and disc plating. The ophiuroid body plan is distinctive, with a subcircular central disc and five long thin, flexible arms. The mouth is situated centrally on the lower surface of the disc. Most of the disc is filled by the stomach and, in the absence of an anus, waste products are regurgitated through the mouth. The arms consist of highly specialized ossicles or vertebrae. Ophiuroids are common in modern seas and oceans, preferring deeper-water environments below 500 m; their basic architecture differs little from some of the first members of the group, e.g. *Taeniaster* from the middle Ordovician of the United States.

Box 7.6 Asterozoan predation

Modern starfish are vicious and voracious predators, enjoying a diet of shellfish. Asteroids, for example, can prise apart the shells of bivalves with their sucker-armoured tube feet far enough to evert their stomachs through their mouths and into the mantle cavity of the animal, where digestion of the soft parts takes place. Donovan and Gale (1990) suggested that this predatory life mode might have significantly influenced the evolution of the brachiopods. The radiation of the Neoasteroidea might have inhibited the post-Permian diversification of some brachiopod groups. The strophomenides were arguably the most diverse Permian order. Many members pursued a reclined, quasi-infaunal life strategy and may have presented an easy kill for the predatory asteroids.

Homalozoans

This subphylum contains some of the most bizarre and controversial fossil animals ever described. Variably described as carpoids, homalozoans or calcichordates, most authorities consider the group different from the radiate Echinodermata, and strictly speaking it should be excluded from that phylum. Detailed study and analysis of the relatively few fossils in this group has produced some possible evidence of their close chordate affinities. In fact, Jefferies (1986) has proposed that the superphylum Dexiothetica should contain both the Chordata and the Echinodermata.

The calcichordates were marine animals ranging in age from mid Cambrian to possibly late Carboniferous, with a calcitic, echinoderm-type skeleton lacking radial symmetry (Figure 7.15). A chordate-like reconstruction suggests the body consists of a head and a tail used for locomotion. Moreover, structures indicating a fish-like brain, cranial nerves, gill slits and a filter-feeding pharynx similar to that in tunicates have been described for a number of well-preserved specimens. Two main types of calcichordate are recognized: the cornutes were often boot-shaped and appear to have a series of gill slits on the left side of the roof of the head, whereas the mitrates, derived from a cornute ancestor, were more bilaterally symmetrical with covered gill slits on both sides. Jefferies considered the cornutes to be the common stem group for all the extant chordate groups, and he classed the mitrates as primitive vertebrates.

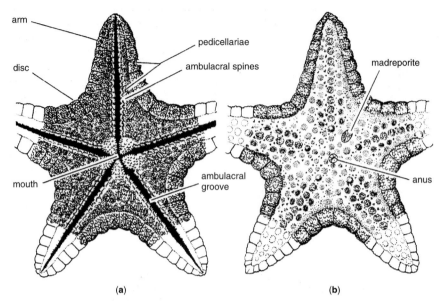

Fig 7.14   Morphology of the asterozoans: ventral (a) and dorsal (b) surfaces.

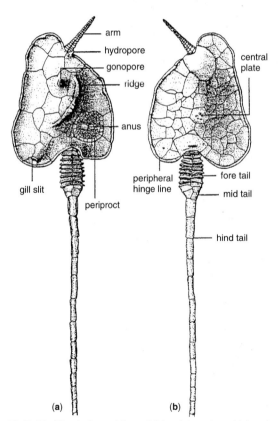

Fig 7.15   Morphology of the calcichordates: dorsal (a) and ventral (b) surfaces.

Hemichordates

The Phylum Hemichordata contains about 100 living species characterized by a notochord for at least half of their life cycle. They are small, soft-bodied animals with bilateral symmetry and a lack of segmentation. The phylum contains two very different classes: firstly, the tiny, mainly colonial, pterobranchs which pursue a life in the sessile benthos; and secondly, the larger infaunal acorn or tongue worms, which burrow mainly in subtidal environments, and comprise the enteropneusts. The hemichordates have a mixture of characters suggesting links with the lophophorates, the echinoderms and the invertebrate and vertebrate chordates. Although the longitudinal rod-like structure, the notochord, is now known to be unrelated to a true backbone, the hemichordates have gill slits in the throat and a nerve cord, together associating the group with the chordates.

Graptolites

The graptolites or Graptolithina are stick-like fossils, often common in Lower Palaeozoic black shales, where the group is of critical importance in biostratigraphical correlation. The majority of Ordovician and Silurian biozones are based on graptolite species or assemblages.

Box 7.7 Echinoderm Lagerstätten and the Lady Burn Starfish Bed

The multiplated echinoderm skeleton disintegrates rapidly after death; although individual plates or ossicles have a high preservation potential, the complete skeletons do not. Nevertheless, occasionally rapid burial or transportation into anoxic conditions may result in the exceptional preservation of the echinoderm skeleton. Starfish beds, usually characterized by accumulations of complete echinoderms, occur sporadically throughout the fossil record. The Leintwardine Starfish Bed of the Anglo-Welsh area contains late Silurian echinoderms within fine-grained turbidites, whereas the Lower Jurassic Starfish Bed of south Dorset is dominated by ophiuroids suddenly buried by a thick layer of sandstone. However, one of the most remarkable echinoderm Lagerstätten occurs within the Upper Ordovician succession of the Craighead inlier, north of the Girvan valley, south-west Scotland. The Lady Burn Starfish Bed is one of several sandstone units within a deep-water mudstone sequence. Entire crinoids, cystoids, echinoids and calcichordates were carried downslope and rapidly buried on the unstable slopes of a submarine fan system.

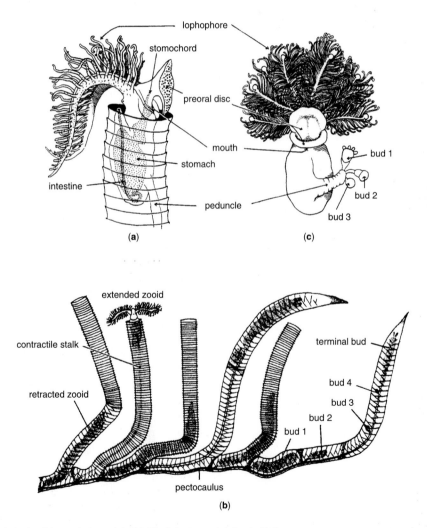

Fig 7.16 Rhabdopleurid morphology: (a), (b) *Rhabdopleura*, (c) *Cephalodiscus*.

Graptolites (from the Greek 'stone writing') usually occur as flattened carbonized films resembling hieroglyphics in black shales, often reworked by currents, although fortunately complete, unflattened specimens have been extracted from cherts and limestone by acid-etching techniques. Although the group is extinct, graptolites are probably related to the pterobranch chordates: both have similarly organized colonies and skeletons.

Modern hemichordate analogues

The pterobranchs superficially resemble the bryozoans: both are colonial organisms and individual zooids feed with tentaculate, ciliated arms (Figure 7.16). The group has a long geological history, with early records such as *Rhabdotubus* from the Middle Cambrian and *Graptovermis* from the Lower Ordovician. The living genera *Cephalodiscus* and *Rhabdopleura* have been used as analogues for many aspects of graptolite morphology, ontogeny and palaeoecology. Both groups have a periderm with fusellar tissue while the dendroid stolon may be analogous to the pterobranch pectocaulus. *Rhabdopleura* is known first from the Silurian and occurs in oceans today mainly at depths of a few thousand metres. The genus is minute with a creeping colony hosting a series of exoskeletal tubes each containing a zooid with its own lophophore-like feeding organ comprising a pair of arms. The zooids are budded from a stolon and interconnected by a contractile stalk, the pectocaulus.

Cephalodiscus, however, is rather different, being constructed from clusters of stalked tubes budded from a basal disc. In further contrast to *Rhabdopleura*, species of *Cephalodiscus* usually have five pairs of ciliated feeding arms. Significantly, individual zooids in the *Cephalodiscus* colony can in fact crawl outside the colony and often farther afield onto adjacent objects. Moreover living *Cephalodiscus* can construct external spines.

Graptolite morphology

The basic graptolite template consists of an organic (scleroproteinaceous) skeleton characterized by a growth pattern of half rings of periderm interfaced by zigzag sutures, similar to the construction of the pterobranchs. Each colony or rhabdosome grew from a small cone, the sicula, as a series of branches or stipes (Figure 7.17). The stipes may be isolate or linked together by lateral struts to resemble a reticulate lattice. A series of variably cylindrical tubes is developed along the stipes; these thecae house the individual zooids of the colony. Aggregates of rhabdosomes, synrhabdosomes, have been documented for a variety of species. Such skeletal complexes have been ascribed to asexual budding or common attachment to a single float or patch of suitable substrate; a more recent taphonomic explanation involves the entrainment of groups of rhabdosomes by marine snow, a bonding material composed of organic debris and mucus. Nevertheless, there is evidence for both natural and taphonomic synrhabdosomes. Although about eight

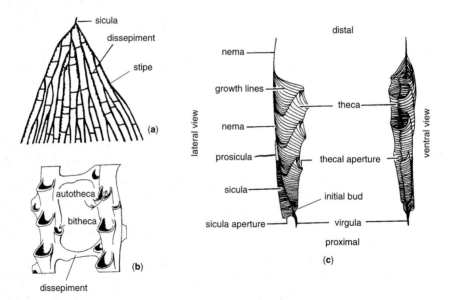

Fig 7.17 Graptolite morphology: (a) dendroid morphology with (b) detail of the thecae; (c) graptoloid morphology.

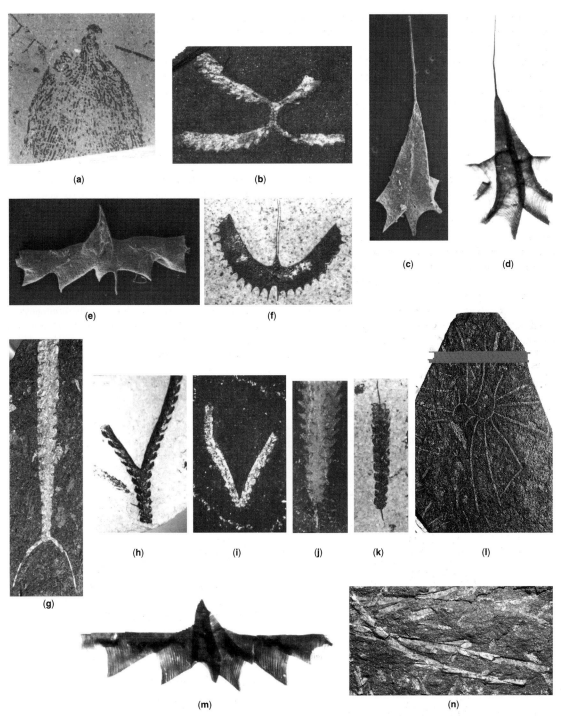

Fig 7.18 Some graptolite genera: (a) *Rhabdinopora* (×2), (b) *Tetragraptus* (×2), (c) *Tetragraptus*, proximal end (×20), (d) *Isograptus*, proximal end (×20), (e) *Xiphograptus* (×20), (f) *Isograptus* (×10), (g) *Appendispinograptus* (×2), (h) *Dicranograptus* (×2), (i) *Dicellograptus* (×2), (j) *Orthograptus* (×2), (k) *Undulograptus* (×2), (l) *Nemagraptus* (×2), (m) *Didymograptus* (*Expansograptus*) (×20), (n) *Atavograptus* (×2). (a) is an early Ordovician dendroid; (b)–(f), (k), (m), are early Ordovician graptoloids; (g)–(j), (l) are late Ordovician graptoloids and (n) is a Silurian monograptid.

orders are now recognized in the Class Graptolithina, only two, the Dendroidea with many branches and polymorphic thecae, and especially the Graptoloidea, with fewer branches and monomorphic thecae, have important geological records (Figure 7.18; see Box 7.8).

Dendroidea

The Dendroidea is the older of the two main groups, possibly first appearing in the early Cambrian and disappearing during the later Carboniferous. The dendroid rhabdosome was multibranched, like a bush, with its many stipes connected laterally by dissepiments. Three types of theca, of different sizes – the autotheca, bitheca and stolotheca – were developed along the stipes. The earlier genera were benthic, attached to the seafloor by a short stalk and basal disc. Probably during the latest Cambrian a few genera, including *Rhabdinopora*, detached themselves to pursue a floating life strategy attached to floating objects in the plankton; others were probably free floating. The latter, together with minute brachiopods and the occasional trilobite, probably formed a major part of the early Palaeozoic plankton.

Dendroid taxa   The dendroids were characterized by multistiped rhabdosomes often supported by dissepiments and hosting three types of thecae (Figure 7.17).

Dendrograptus was a benthic genus, bush-like, erect and attached to the sea-floor by a holdfast. *Dictyonema* had both planktonic and benthic species and ranged in age from the late Cambrian to the late Carboniferous. The rhabdosome was conical to cylindrical in shape. The benthonic forms possessed holdfasts together with incompletely developed or no sicula, whereas a variety of vane-like floats and a nema for attachment were developed in planktonic species. The planktonic nematophorous members are now assigned to *Rhabdinopora*.

The following anisograptid genera are in some ways intermediate between the typical dendroids and graptoloids and may be classified with either group. Here they are included with the dendroids although some include them with the graptoloids. *Radiograptus*, for example, developed large spreading colonies. Both *Kiaerograptus* and some early species of *Bryograptus* had both auto- and bithecae; the latter had triradiate rhabdosomes with initially three primary stipes. *Clonograptus* had a horizontal, biserially symmetrical rhabdosome with stipes generated by dichotomous branching from an initial biradiate configuration.

Graptoloidea

Compared with the dendroids, the graptoloid rhabdosome is superficially simpler and consists of an initial

---

Box 7.8 Graptolite classification

The stolonoids, an encrusting or sessile group, restricted to Poland, may be Pterobranchia.

Class Graptolithina

Order Dendroidea
   Multibranched colonies; stipes, commonly supported by dissepiments, having autothecae, bithecae and a stolotheca. Anisograptids intermediate between the dendroids and graptoloids are retained here. Early Cambrian–mid Carboniferous.

Order Tuboidea
   Similar to dendroids but characterized by irregular branching and reduced stolothecae. Autothecae and bithecae commonly form clusters. Early Ordovician–mid Silurian.

Order Camaroidea
   Encrusting life mode; endemic to Poland. Autothecae with expanded, sack-like bases. Bithecae small and irregularly spaced.

Stolotheca black and hard. Early Ordovician.

Order Crustoidea
   Encrusting life mode; endemic to Poland. Autothecae with complex apertures. Early –mid Ordovician.

Order Graptoloidea
   The jury is still out on the detailed classification of this group. The position of the anisograptids (included here with the dendroids) is uncertain, as is the status of the dichograptids and the retiolitids. Colonies with few stipes (between one and eight), a nema and sicula and a single type of theca. Early Ordovician–early Devonian.

Suborder Dichograptina
   Primitive graptoloids lacking both bithecae and virgellae. Early–mid Ordovician.

Suborder Virgellina
   Virgella always present. Early Ordovician–early Devonian.

sicula, divided into an upper prosicula and a lower meta-sicula, with at its apex, distally, a long thin needle or nema (Figure 7.17). The metasicula, like the rest of the rhabdosome, is composed of fusellar tissue; the virgella projects under the aperture, proximally, at the base of the sicula, and is characteristic of the Suborder Virgellina. Initially the thecae grow outwards from the sicula in sequence as the rhabdosome develops. On the basis of the classification followed here, most genera possess only autothecae, although a few more primitive forms display bithecae.

Graptoloid taxa   The architecture of the graptoloid skeleton was governed by three main aspects of the animal's morphology: the number of stipes or branches, their mutual attitudes and the shape of the thecae. Morphology in this order is based on permutations of these parameters; the following genera illustrate this variation (Figure 7.18).

Tetragraptus, common during the Arenig, typically had four stipes arranged in horizontal, pendent or reclined attitudes with simple, overlapping thecae. *Didymograptus* was twin-stiped, commonly with the branches in horizontal, pendent or reclined orientations; thecae were simple. *Isograptus*, however, had two relatively wide stipes, reclined with a long thread-like sicula. *Nemagraptus* had a very distinctive rhabdosome consisting of two sigmoidal stipes initially diverging from the sicula at about 180°, with additional stipes, themselves curved, arising at intervals along the main branches. Thecae are long and thin, and diverge at small angles from the stipes. *Dicellograptus* had a pair of stipes adopting reclined attitudes, but often the branches were curved or even coiled; the thecae were characterized by extravagant sigmoidal shapes and incurved apertures. *Monograptus* was a uniserial scandent form with a straight or curved rhabdosome and a nema embedded in the dorsal wall but projected distally. *Rastrites* possessed a curved rhabdosome with long, straight, widely-separated thecae with hooked ends. *Cyrtograptus* had a spirally coiled rhabdosome with secondary branches or cladia orientated like the arms of a spiral galaxy.

Retiolitids   The retiolitids are a spectacular group of apparently scandent biserials with a reduced, minimalist periderm consisting of a network of bars or lists probably surrounded by a net rather like parts of a spider's web (Figure 7.19). The group appeared in the Llandeilo and continued successfully till the latest Silurian and as currently understood is polyphyletic. The rhabdosome of various groups of retiolitids may

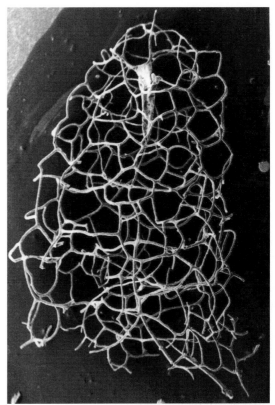

Fig 7.19   Retiolitid *Phorograptus* (Middle Ordovician) (×30).

have functioned like a sponge, drawing in fluid and nutrients through the perforated periderm and expelling waste upwards.

Growth and ultrastructure of the graptolites

Detailed studies on the ultrastructure of graptolites using both scanning and transmission electron microscopes have established two types of skeletal tissue: fusellar tissue, consisting of bundles of short, branching fibrils occurs together with cortical tissue in the form of longer, parallel fibres. Fusellar material was secreted as a series of half rings, with the cortical tissue overlapping the fusellar layer both inside and outside the rhabdosome (Figure 7.20). Secretion may have been by mobile zooids, free to patrol the exterior of the colony while still attached by a flexible cord to the rest of the colony. Zooids may even have detached and reattached themselves entirely from the colony; this may explain the secretion of spines over 15 mm by minute zooids less than 1 mm in length.

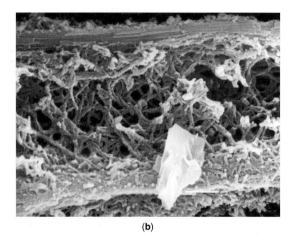

(b)

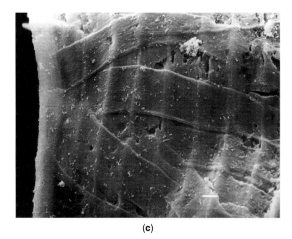

(c)

(a)

Fig 7.20 Graptolite ultrastructure: (a) collage of *Geniculograptus* rhabdosome showing banded fusellar tissue (×50); (b) detailed section through part of rhabdosome showing relatively thin parallel sheet fabric (top) and criss-cross fusellar fabric (below) (×1000); (c) detail of aperture exterior of *Geniculograptus* showing development of bandages (×500).

Modes of life and feeding strategy

There is little doubt that the earliest bush-like dendroids with stout nemae were attached to the seabed and functioned as part of the sessile benthos. Detachment in various benthic genera occurred in the late Cambrian, with some such as *Rhabdinopora* pursuing an attached nematophorous strategy in the surface waters while others merely floated, unaided, in the plankton. More controversial is the mode of life of the various graptoloid groups. Conventionally the graptoloids were considered part of the plankton by Bulman (1964) and others. Passive drifting may have been aided by fat and gas bubbles in graptolite tissue or even by vane-like extensions to the nema.

Kirk (1969) proposed that, far from being passive members of the plankton, the graptolites were automobile, moving up and down in the water column. During periods of intense feeding, fluid and nutrients would be drawn through the theca with a reactive upward movement of the colony in the water column. During feeding cycles the colony could transit vertically into the nutrient-rich photic zone and, when replete, the rhabdosomes would sink to positions in the water column where the specific gravity of the colony matched that of the surrounding sea-water. Clearly to function in this way, thecae must have faced upwards, thus inverting the conventional orientation of many graptolite taxa.

Many aspects of the functional morphology and pro-

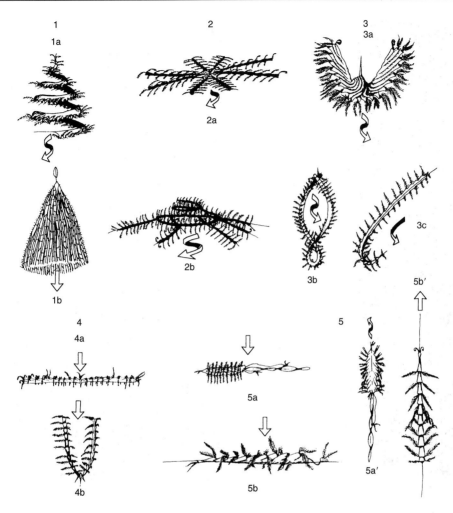

Fig 7.21 Graptolite life modes: 1, conical forms with spiral motion; 2, flat or slightly conical forms with slow, slightly spiral velocities; 3, mono- or biramous forms with spiral movement due to asymmetry; 4, forms with high angles between stipes having linear movement; 5, straight forms with mainly linear descent.

posed life mode of the graptolites have been investigated by computer and physical models, emphasizing the importance of harvesting strategies to the success of the colony (Rigby, 1993). Underwood (1994) has synthesized much of the this information in his constructions of graptolite feeding strategies and life modes (Figure 7.21).

Graptolite evolution

Graptolite evolution has been described in terms of four main stages: first, the transition from sessile to planktonic strategies in the dendroids during the late Cambrian and early Ordovician; second, at the end of the Tremadoc

the appearance of the monomorphic thecae of the graptoloids; third, the development of the biserial rhabdosome late in the Arenig; and finally the origin of the uniserial monograptids (Figure 7.22). The small, stick-like benthic organisms reported from Middle Cambrian rocks on the Siberian Platform and ascribed to the graptolites may be better assigned to the Pterobranchia. The first undoubted graptolites include the dendroids *Callograptus*, *Dendrograptus* and *Dictyonema* occurring in middle Cambrian rocks of North America. However, by the late Cambrian the diversity of the dendroid fauna had peaked; the fauna included genera such as *Aspidograptus* and *Dictyonema* which resembled small bushes and were attached to the substrate by holdfasts or

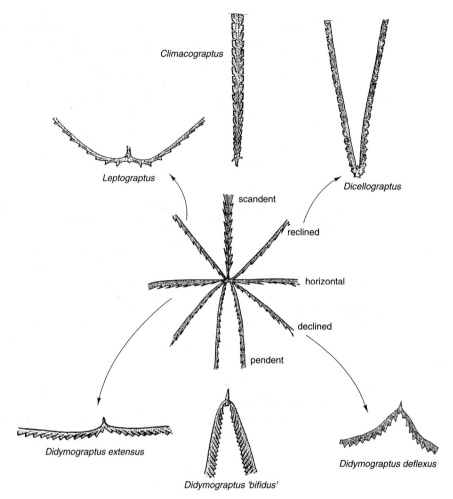

Fig 7.22 Evolution of stipes.

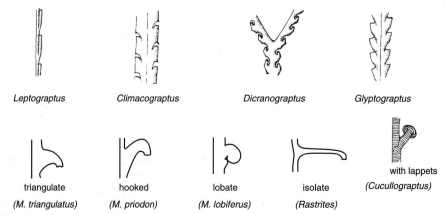

Fig 7.23 Evolution of thecae. *M, Monograptus.*

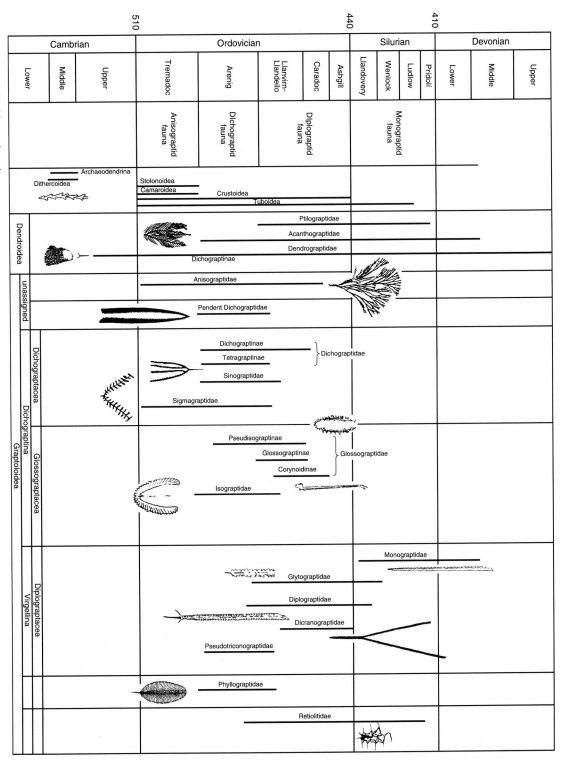

Fig 7.24 Graptolite biostratigraphy.

more complex root-like structures. During the late Cambrian and early Ordovician, some dendroids made the transition from the sessile benthos to the plankton; attachment discs continuous with the nema suggest these genera hung suspended in the surface waters and pursued an epiplanktonic life strategy. Both *Radiograptus* and *Dictyonema* have been cited as possible ancestors for the planktonic graptolites, and perhaps *Staurograptus* was in fact the first graptolite floater. The Tremadoc seas witnessed the radiation of the anisograptids.

The explosion of dichograptid genera during the Arenig introduced a variety of symmetrical graptolites with from eight to two stipes orientated in declined, pendent and scandent attitudes. A twin-stiped dichograptid probably was ancestral to the next wave of graptolites, the diplograptids, which radiated at the base of the Llanvirn.

The single-stiped monograptids dominated Silurian graptolite faunas and despite their apparent simplicity the group developed a huge variety of forms (Figure 7.23). The last graptolites, species of *Monograptus*, disappeared during the early Devonian in China, Eurasia and North America. Nevertheless, this uniserial morphology had survived for over 30 million years.

Graptolite biostratigraphy

Traditionally four sequential graptolite 'faunas' have been recognized (Bulman, 1958) through the Tremadoc to Silurian interval (Figure 7.24). The anisograptid fauna, with *Rhabdinopora* and allied genera, characterizes the Tremadoc; although Upper Tremadoc graptolite faunas are rare, the genera *Bryograptus*, *Kiaerograptus* and *Aorograptus* are important and have been described in detail from western Newfoundland by Williams and Stevens (1991).

The appearance of the Arenig dichograptid fauna is signalled by *Tetragraptus*, associated with didymograptids and some relict anisograptids. The younger diplograptid fauna contains four smaller units: the *Glyptograptus–Amplexograptus* (Llanvirn–Llandeilo), *Nemagraptus–Dicellograptus* (Lower Caradoc), *Orthograptus–Dicellograptus* (Upper Caradoc–Ashgill) and *Orthograptus–Climacograptus* (Lowest Silurian) subfaunas. The monograptid fauna contains a variety of evolving single-stiped forms. The last graptoloids disappear in the lower Devonian.

Graptolite biogeography

Since the majority of graptolites lived either in the water column or within the plankton, quite different factors influenced their distribution in contrast to say the controls on the coeval benthos. Provinciality was most marked in the early Ordovician (late Arenig–Llanvirn) when two main provinces, the Atlantic and Pacific, were recognized. The Atlantic province, including the then high-latitude regions of Avalonia and Gondwana, was characterized by pendent *Didymograptus* species; the Pacific province, including low-latitude regions such as the Laurentian margins, had more diverse faunas with isograptids, cardiograptids and oncograptids (Cooper *et al.*, 1991).

Further reading

Berry, W.B.N. (1987) Phylum Hemichordata (Including Graptolithina). In R.S. Boardman, A.H. Cheetham, and A.J. Rowell (eds) *Fossil invertebrates*. Blackwell Scientific Publications, Oxford, pp. 612–635.

Clarkson, E.N.K. (1993) *Invertebrate palaeontology and evolution*, 3rd edition. Chapman and Hall, London.

Palmer, D. and Rickards, R.B. (1991) *Graptolites–writing in the rocks*. Boydell Press, Woodbridge, UK.

Rickards, R.B. (1985) Graptolithina. In J.W. Murray (ed.) *Atlas of invertebrate macrofossils*. Longman, London, pp. 191–198.

Smith, A.B and J.W. Murray (1985) Echinodermata. In J.W. Murray (ed.) *Atlas of invertebrate macrofossils*. Longman, London, pp. 153–190.

Sprinkle, J. and Kier, P.M. (1987) Phylum Echinodermata. In R.S. Boardman, A.H. Cheetham and A.J. Rowell (eds) *Fossil invertebrates*. Blackwell Scientific Publications, Oxford, pp. 550–661.

References

Bulman, O.M.B. (1958) The sequence of graptolite faunas. *Palaeontology*, **1**, 159–173.

Bulman, O.M.B. (1964) Lower Palaeozoic plankton. *Quarterly Journal of the Geological Society of London*, **119**, 401–418.

Cooper, R.A., Fortey, R.A. and Lindholm, K. (1991) Latitudinal and depth zonation of early Ordovician graptolites. *Lethaia*, **24**, 199–218.

Donovan, S.K. and Gale, A.S. (1990) Predatory asteroids and the decline of the articulate brachiopod. *Lethaia*, **23**, 77–86.

Jefferies, R.P.S. (1986) *The ancestry of the vertebrates*. British Museum (Natural History), London.

Kirk, N. (1969) Some thoughts on the ecology, mode of life, and evolution of the Graptolithina. *Proceedings of the Geological Society of London*, **1659**, 273–293.

Rigby, S. (1993) Graptolite functional morphology: a discussion and critique. *Modern Geology*, **17**, 271–287.

Simms, M.J. and Sevastopulo, G.D. (1993) The origin of the articulate crinoids. *Palaeontology*, **36**, 91–109.

Smith, A.B. and Hollingworth, N.T.J. (1990) Tooth structure and phylogeny of the Upper Permian echinoid *Miocidaris keyserlingi*. *Proceedings of the Yorkshire Geological Society*, **48**, 47–60.

Underwood, C.J. (1994) The position of graptolites within Lower Palaeozoic planktic ecosystems. *Lethaia*, **26**, 198–202.

Williams, S.H. and Stevens, R.K. (1991) Late Tremadoc graptolites from western Newfoundland. *Palaeontology*, **34**, 1–47.

8 Spiralians: arthropods and molluscs

Key Points

- The arthropods are legged invertebrates, with a segmented body plan and jointed appendages; Cambrian arthropods were no more diverse than living faunas.
- Trilobites appeared in the early Cambrian, and during the Palaeozoic they evolved advanced visual systems and enrollment structures.
- Insects first appeared during the early Devonian and rapidly diversified; there are probably at least 10 million species of living insects.
- The crustaceans include many familiar groups such as the crabs, lobsters and shrimps, together with the barnacles and ostracodes.
- The largest arthropods were chelicerates and included the giant eurypterids which patrolled marine marginal environments during the Silurian and Devonian.
- Early molluscs were short-lived, unusual forms but with the molluscan features of a mantle, mineralized shell and a radula.
- Bivalves have a bewildering variety of shell shapes, dentitions and muscle scars, adapted for a wide range of marine life strategies.
- Gastropods undergo torsion in early life; they have a single shell, commonly coiled, and occupy a wide range of environments.
- Cephalopods are the most advanced molluscs, with a well-developed head, senses and a nervous system; they include nautiloids, ammonoids and the coleoids.
- In the Mesozoic, molluscs developed protective strategies, such as robust armour, deep infaunal life modes or variation in shape and colour to confuse predators.

Introduction

The arthropod–mollusc lineage, together with the annelids, have similar developmental characteristics. All three groups are included within the spiralians, having a spiral cleavage pattern during cell division in the embryo. This group is also schizocoel, with the coelom developing from a split within the mesoderm and most members have trochophore-type larvae. Arthropods and molluscs achieved a spectacular diversity of forms, habitats and lifestyles, with both groups dominating post-Palaeozoic invertebrate faunas and participating in the Mesozoic arms race.

Arthropods

Arthropods are a group of common, legged invertebrates, accounting for about three-quarters of all species living on the planet today, largely because of the phenomenal success of the insects. The basic body plan, conspicuously segmented, with jointed appendages adapted for feeding, locomotion and respiration, together with a tough exoskeleton, first appeared during the early Cambrian and has since been exploited in the wide variety of extant and fossil arthropod groups pursuing many lifestyles (Figure 8.1). All members of this phylum have both segmented bodies and appendages; moreover,

the animal is differentiated into a head, thorax and abdomen, with often the head and thorax fused to form the prosoma. The possession of modified appendages around the mouth allowed many arthropods to process a wide variety of foods.

The arthropod exoskeleton is constructed mainly from the organic substance chitin; however, it is often hardened or sclerotized by calcium carbonate or calcium phosphate, so that the potential for preservation is increased. The endoskeleton acts as a base for the attachment of locomotory muscles, permitting rapid movement, and is not usually mineralized. Although many arthropods undergo metamorphosis, virtually all the main groups grow by moulting or ecdysis: first the endoskeleton is dissolved and then the old exoskeleton is detached along sutures while the new exoskeleton is generated. Spent exuviae are all that remain of the previous skeleton or cuticle of the animal.

During a geological history of at least 600 million years, arthropods have adapted to life in marine, freshwater and terrestrial environments. Controversy and lively debate, however, currently surround the origins of the phylum. Some authorities contend that the phylum arose from a single common ancestor, while others consider that arthropodization may have occurred separately in at least three related groups. Whether the arthropods are monophyletic or polyphyletic has yet to be settled; nevertheless, they almost certainly shared a common ancestor with annelid worms, probably during the late Proterozoic.

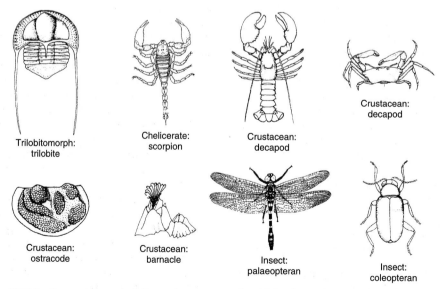

Trilobitomorph:
trilobite

Chelicerate:
scorpion

Crustacean:
decapod

Crustacean:
decapod

Crustacean:
ostracode

Crustacean:
barnacle

Insect:
palaeopteran

Insect:
coleopteran

Fig 8.1 Some of the main arthropod groups: a variety of forms based on a simple body plan of a tough exoskeleton and jointed limbs.

Early arthropod faunas

A large variety of apparently different arthropod types formed much of the basis for the Cambrian radiation. Over 20 different groups of arthropods have been described from the mid Cambrian Burgess Shale and related deposits; many have been accorded phylum status, emphasizing the explosive nature of the radiation. Gould (1989) has argued that morphological disparity during the Cambrian was greater than at any time subsequently during the rest of the Phanerozoic. Nevertheless, cladistic and phenetic analyses of both morphological and taxonomic criteria by Briggs *et al.* (1993) suggest otherwise. Rather, the disparity amongst the Cambrian arthropods is not markedly different from that seen across living taxa.

Subphylum Trilobitomorpha

The Trilobitomorpha contain highly derived arthropods lacking specialized mouth parts with an exoskeleton comprising the cephalon, thorax and pygidium, together with trilobitomorph appendages having lateral branches developed from the walking limbs. The trilobitomorphs are mainly represented by the trilobites.

Trilobite morphology

Trilobites were a unique and successful arthropod group, common throughout the Palaeozoic until their extinction at the end of the Permian. Some of the earliest arthropods were trilobites, and marine Cambrian strata (as well as many later deposits) are usually correlated on the basis of trilobite assemblages. The group formed an important part of the mobile benthos, although a few groups were adapted to pelagic life modes. The trilobite exoskeleton (Figure 8.2), as the name suggests, is divided longitudinally into three lobes; the axial lobe protects the digestive system, whereas the two pleural lobes cover appendages. In virtually all trilobites a well-defined cephalon, thorax and pygidium are developed. The trilobite exoskeleton is composed almost entirely of calcite.

The cephalon has a raised axial area, the glabella, with a series of glabellar furrows. Eyes are commonly developed laterally (see Box 8.2), with the facial or cephalic suture separating the inner fixed and the outer free cheeks. Although there are many blind trilobites, this may be a secondary condition; despite loss of vision, the facial sutures remained. There are four main types of suture (Figure 8.3).

On the ventral surface, underneath the cephalon, two, possibly three, plates were associated with anterior soft parts including the mouth. The rostral plate was situated at the anterior margin. The hypostoma, on the ventral surface, was a plate of variable shape and size, thought to have supported the mouth region; in some trilobites the hypostoma was attached to the back of the rostral plate at the anterior margin, while in others it was detached and supported by the ventral integument. Recent research has emphasized the importance of the hypostoma in the classification of the Trilobita. The small metasoma is doubtfully known from only a few taxa and apparently lay behind the mouth. The dorsal margin was protected by a flange or doublure.

With the exception of the agnostids and eodiscids, which have two and two to three thoracic segments respectively, trilobites are polymeric, usually having up to 40 thoracic segments. The trilobite pygidium is usually a plate of between one and 30 fused segments. Most Cambrian trilobites have small micropygous pygidia, whereas later forms are either heteropygous, where the pygidium is smaller than the cephalon; or macropygous, where the pygidium is larger.

Like virtually all arthropods, the trilobites grew by

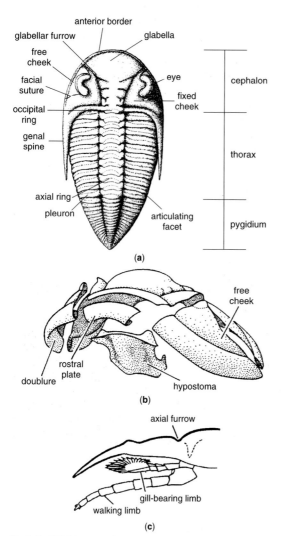

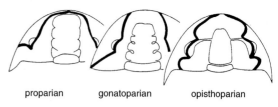

Fig 8.3 Facial sutures: the tracks of the proparian, gonatoparian and opisthoparian sutures; the lateral suture (not illustrated) follows the lateral margin of the cephalon.

ecdysis or moulting (Figure 8.4). Ontogeny involved the periodic discarding of spent exoskeletons or exuviae. Initial moult stages are quite different from the adults. After a nauplius larval stage, the protaspid stage is a minute disc with a segmented median lobe destined to become the glabella. The next, meraspid, stage has a discrete transitory pygidium, where thoracic segments form at its anterior margin and are released at successive moults to form the thorax. The holaspid stage has a full complement of thoracic segments but growth continues through further moults, and maturity may not have been reached until well after the holaspid stage was initially achieved. Clearly, in many trilobite-dominated faunas, counts of skeletal remains will significantly over-represent the numbers of living animals in the community. Many researchers divide the number of exuviae by between six and eight to obtain a more realistic census of the trilobite population in a typical community.

During the Palaeozoic, a number of groups, including asaphids, calymenids, phacopids and trinucleids, evolved a variety of sophisticated structures to enhance enrolment. During times of stress, to avoid unpleasant environmental conditions or perhaps an attentive predator, trilobites could roll up like a carpet. Spheroidal enrollment involved articulation of all the thoracic segments to form a ball, whereas in the less common discoidal mode of enrollment the thorax and pygidium were merely folded over the cephalon. Cambrian trilobites could certainly enroll, but it was not until the Ordovician

Fig 8.2 Trilobite morphology: (a) external morphology of the Ordovician trilobite *Hemiarges*; (b) generalized view of the anterior of the Silurian trilobite *Calymene* revealing details of the underside of the exoskeleton; (c) details of the limb pair associated with a segment of the exoskele-

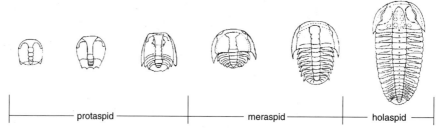

Fig 8.4 Moult phases of the Bohemian trilobite *Sao hirsuta* Barrande; protaspid stages about ×9, meraspid stages about ×7.5 and the holaspid about ×0.5.

Box 8.2 Vision in trilobites

Trilobites have the oldest visual system. The eyes are compound, analogous to those of the crustaceans and insects. There are two main types of lens arrangement variably developed across the group (Figure 8.5). The trilobite eye (Clarkson, 1979) generally consists of many lenses of calcite with the c-axis normal to the surface of the eye. The more primitive and widespread holochroal eye has many close-packed lenses, all about the same size, covered by a single membrane. The more advanced and complex schizochroal condition has no modern analogue and has larger discretely arranged lenses in rows or files. It is uncer-tain how this system operated in detail; presum-ably it offered higher quality images than those of the holochroal system. Moreover, both mature holochroal and schizochroal configurations appar-ently developed from immature schizochroal con-ditions. Thus early growth stages of holochroal eyes in quite different groups such as those of the Cambrian eodiscid *Shizhudiscus*, with the oldest visual system in the world, and *Paladin* from the Carboniferous have broadly similar schizochroal arrangements, suggesting that the latter system developed by paedomorphosis. Eye reduction has been documented in some lineages in response to endobenthonic habits.

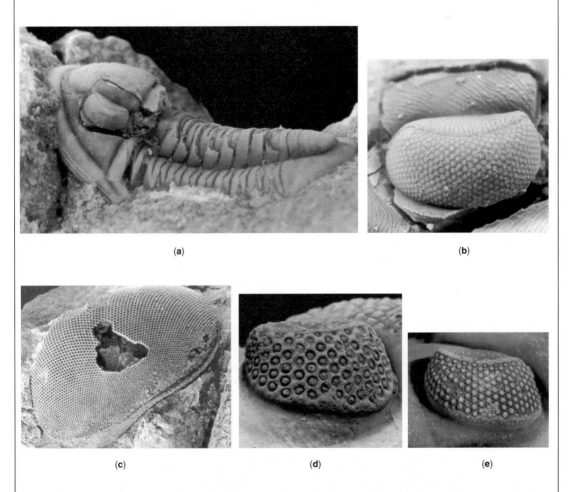

(a)

(b)

(c)

(d)

(e)

Fig 8.5 Vision in trilobites: (a) lateral view of a complete specimen of *Cornuproetus*, Silurian, Bohemia (×4); (b) detail of the compound eye of *Cornuproetus* (×20); (c) holochroal compound eye of *Pricyclopyge*, Ordovician, Bohemia (×6); (d) schizochroal compound eye of *Phacops*, Devonian, Ohio (×4); (e) schizochroal compound eye of *Reedops*, Devonian, Bohemia (×5).

that true coaptative structures locking parts of the skeleton against each other, first appeared. For example, in the phacopids, tooth and socket pairs were developed on the cephalic and pygidial doublure respectively; these opposing structures clicked together to hold the trilobite in a tight ball, presenting only the exoskeleton to the world outside.

Main trilobite groups and lifestyles

Although some workers have split the group into two orders, the Agnostida and the Polymerida, most currently recognize about nine orders of trilobite, based on a spectrum of characters, including the analysis of ontogenetic stages and more recently the location and morphology of the hypostome. In the most primitive conterminant condition the hypostome is similar in shape to the glabella and is attached to the anterior part of the doublure; natant hypostomes were unattached, whereas the impendent hypostome was attached to the doublure, but its shape bore no resemblance to that of the glabella above.

The redlichiids include *Olenellus*, with 18–44 spiny, thoracic segments and are typical of the Atlantic province; *Redlichia*, more typical of the Pacific province; and the large spiny, micropygous *Paradoxides*, common in high-latitude middle Cambrian faunas (Figure 8.6). Some authorities have excluded the distinctive agnostids from the Trilobita. The order is characterized by small to minute, usually blind forms with subequal cephala and pygidia and only two thoracic segments; the agnostids were probably planktonic which may account for their very wide distribution.

The naraoids, including *Naraoia* and *Tegopelte*, were not calcified and lacked thoracic segments. The group was restricted to the mid Cambrian. *Naraoia* was first described from the Burgess Shale as a branchiopod crustacean, and it was only recently identified as a soft-bodied trilobite. Corynexochid trilobites were a mixed bag of taxa; the order includes genera with conterminant hypostomes such as *Olenoides* and large smooth forms such as *Bumastus* and *Illaenus*, having impendent hypostomes.

The lichids contain mainly spiny forms with conterminant hypostomes. Apart from *Lichas*, the order also contains the spiny odontopleurids such as *Leonaspis*. Phacopids were mainly proparian trilobites with schizochroal eyes and lacking rostral plates, ranging from the Lower Ordovician to Upper Devonian. The order includes the large tuberculate *Cheirurus*, *Calymene* with a marked gonatoparian suture, and *Dalmanites* with long genal spines, kidney-shaped eyes, spinose thoracic seg-

ments and the pygidium extended as a telson-like spine.

The ptychopariids are all characterized by natant hypostomes and include some specialized groups; for example, *Triarthrus* possibly modified for burrowing, the blind *Conocoryphe* and *Harpes* with a marked sensory fringe. Asaphids had either conterminant or impendent hypostomes and include *Asaphus*, *Ceratopyge* and pelagic forms such as *Cyclopyge* and *Remopleurides*, and the stratigraphically important trinucleids such as *Onnia*, *Cryptolithus* and *Tretaspis*.

The proetids were isopygous forms with large glabellae, long hypostomes, genal spines and large holochroal eyes. The group ranged from the Lower Ordovician to the Upper Permian. *Proetus* was a small form with a relatively large inflated and often granular glabella, continuing from the Ordovician to the Devonian. *Phillipsia*, one of the youngest members of the order, was a small isopygous genus with large crescent-shaped eyes and an opisthoparian suture.

Trilobites display a range of morphologies presumably adapted to a spectrum of lifestyles within the Palaeozoic marine realm (Figure 8.7). Most trilobites were almost certainly nektobenthonic, leaving a variety of tracks and trails in the marine sediments of Palaeozoic seas (see Chapter 12). With the exception of the phacopids that may have hunted, the majority of trilobites probably fed on detritus gripped, shred and brought to the mouth with the aid of spines at the bases of the legs. Many trilobites developed spinose exoskeletons. The spines reduce the weight : area ratio of the organism and initial studies suggested floating, planktonic life strategies for these spiny trilobites, occasionally with inflated glabellae. In addition, spines increase the body size of the trilobite, make it less attractive to predators and provide a larger buffer between vital soft parts and the edge of the animal. More recently, a support function on a soft muddy substrate has been argued. Downward-directed spines probably held the thorax and pygidium well above the sediment–water interface. Moreover, spines, during flexure of the skeleton, probably aided shallow burrowing. Spines are most extravagantly developed in the odontopleurids.

Some, such as *Cybeloides* and *Encrinurus*, evolved eyes on stalks, while others (e.g. *Trinucleus*) lost them altogether in favour of possible sensory setae; these specialized forms may have periodically concealed themselves in the sediment as endobenthos. *Trimerus* had a cephalon and pygidium fashioned in the shape of a shovel which may have helped it plough through the sediment. The cyclopygid *Opipeuter*, from the Lower Ordovician of Spitzbergen, Ireland and Utah, was per-

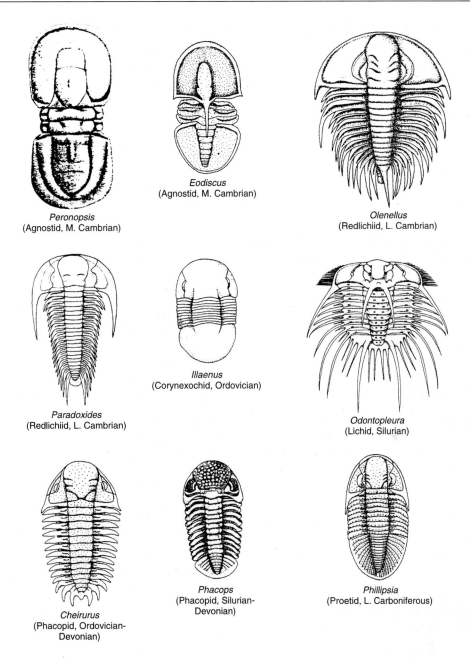

Peronopsis
(Agnostid, M. Cambrian)

Eodiscus
(Agnostid, M. Cambrian)

Olenellus
(Redlichiid, L. Cambrian)

Paradoxides
(Redlichiid, L. Cambrian)

Illaenus
(Corynexochid, Ordovician)

Odontopleura
(Lichid, Silurian)

Cheirurus
(Phacopid, Ordovician-
Devonian)

Phacops
(Phacopid, Silurian-
Devonian)

Phillipsia
(Proetid, L. Carboniferous)

Fig 8.6 Some common trilobite taxa: *Peronopis* (×9), *Eodiscus* (×6), *Olenellus* (×1.5), *Paradoxides* (×0.5), *Illaenus* (×0.75), *Odontopleura* (×0.75), *Cheirurus* (×0.75), *Phacops* (×0.5) and *Phillipsia* (×1.5).

haps analogous to a giant amphipod. The flexibility of a long slender exoskeleton and large eyes like those of living planktonic isopods, together with its widespread distribution, suggests that *Opipeuter* pursued a pelagic lifestyle.

Nevertheless, certain broad morphotypes appear repeatedly in different lineages, perhaps as a response to convergent life strategies. There are eight ecomorphic groups ranging from the turberculate, mobile phacomorphs to the smooth infaunal illaenimorphs.

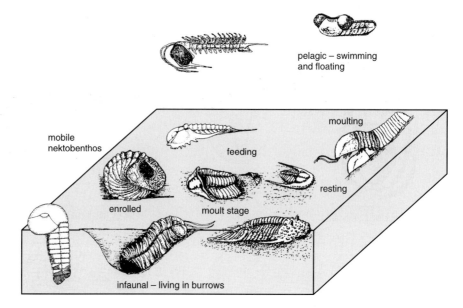

Fig 8.7 Lifestyles of the trilobites: a mosaic of selected Lower Palaeozoic trilobites in various life attitudes.

Trilobite distribution and evolution

Trilobite faunas have formed the basis for many palaeo-geographical reconstructions of the Cambrian and Ordovician world. During the Cambrian, biogeographical patterns were complex, but some provinces have been defined, such as the high-latitude Atlantic (with redlichiides) and the low-latitude Pacific (with olenellids) regions. Statistical analysis of Ordovician trilobite faunas established a low-latitude bathyurid province (Laurentia), an intermediate- to high-latitude asaphid province (Baltica) and a high-latitude *Selenopeltis* province (Gondwana). Despite a number of modifications, this basic pattern remains unchanged (see also Chapter 2).

A number of distinctive trilobite palaeocommunities have been recognized during the early Palaeozoic. The broad onshore–offshore spectrum from shallow-water illaenid–cheirurid associations to deep-water olenid communities established for the early Ordovician (Fortey, 1975) has been modified by a number of authors for other Ordovician and Silurian assemblages (Figure 8.8). In general terms, the shallow-water, pure carbonate illaenid–cheirurid communities had the greatest longevity.

Not surprisingly, the Trilobita have been a major source of evolutionary data. Models for both punctuated equilibria and gradualism have been developed for trilo-bite microevolution in Devonian and Ordovician taxa respectively (see also Chapter 3). Clarkson's (1988) survey of microevolutionary change in Upper Cambrian olenid trilobites from the Alum shales of Sweden provided further evidence of gradualistic change. Macroevolutionary change, however, was apparently effected by heterochrony. Paedomorphosis during ontogeny of the animal as a whole or applied to particular organs such as the eyes generated new species and new biological structures.

Trilobites (such as *Choubertella* and *Schmidtiellus*) first appeared in the early Cambrian and the group survived until the end of the Permian, when the last genera, such as *Pseudophillipsia*, disappeared. In a history of 350 million years, the basic body plan was essentially unchanged, but many modifications promoted trilobite abundance and diversity (Figure 8.10).

A number of evolutionary trends have been described through the Palaeozoic history of the trilobites. With the evolution of coaptative structures, there is a trend towards better and more efficient articulatory and enrollment mechanisms. There was a reduction in the size of the rostral plate and in some groups there was an increase in spinosity and a trend from micropygy to isopygy. Schizochroal visual systems appeared, by paedomorphosis, during the early Ordovician in the phacopids.

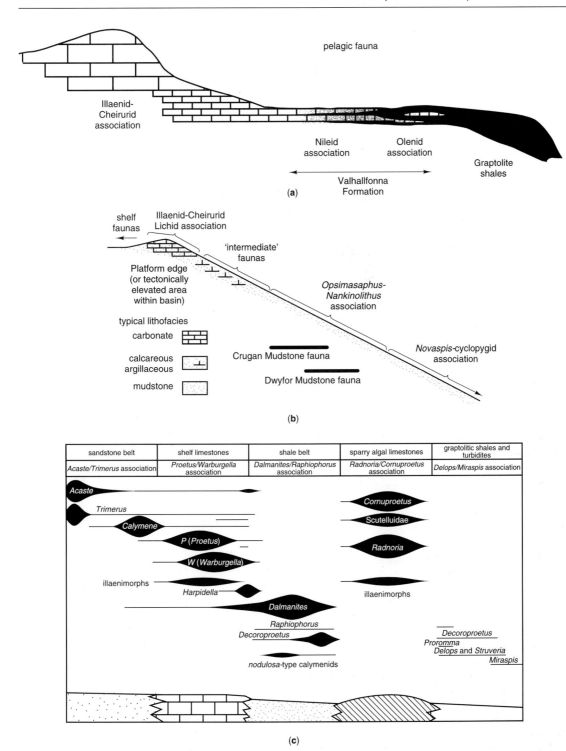

Fig 8.8 Trilobite communities: overview of (a) early Ordovician (Arenig), (b) late Ordovician (Ashgill) and (c) mid Silurian (Wenlock) trilobite associations in relation to water depth and sedimentary facies.

Box 8.3 Trilobite abnormalities and injuries

Trilobites have left a rich record of abnormalities and injuries, some evidence that they faced problems during ecdysis and that they were the objects of predation (Figure 8.9). There were three main types (Owen, 1985):

1. injuries sustained during moulting,
2. pathological conditions resulting from disease and parasitic infestations,

3. teratological effects arising through some embryological or genetic malfunctions.

Predation scars occur asymmetrically: they are three times as likely to be present on the right-hand side of the exoskeleton. If predators preferred to attack the right then perhaps there was already a lateralization of their nervous system and other organs. It can be argued, however, that these specimens were the survivors and that the preferential areas for fatal attack were, in fact, on the left-hand side of the trilo-

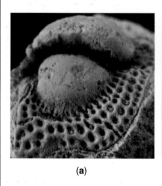 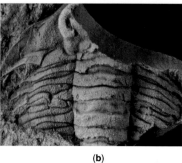 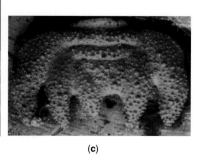

(a) (b) (c)

Fig 8.9 Pathological trilobites. (a) *Onnia superba* – the fringe in the lower part of the photograph has an indentation and a smooth area, probably regeneration following an injury during moulting (×4). (b) *Autoloxolichas* – the deformed segments on the left-hand side may be either genetic or the result of repair following injury (×3). (c) *Sphaerexochus* – only two ribs are developed on the right-hand side, probably a genetic abnormality (×25).

Subphylum Uniramia

The Uniramia contain a variety of forms, including the millipedes, centipedes, symphylans and pauropods, together comprising the myriapods, and also, for example, the springtails, dragonflies, cockroaches and locusts comprising the insects. The last group alone may prove to have as many as 10 million living species when the rich faunas of the tropics have been completely described. The subphylum also includes the onychophorans, with flexible segmented bodies, and unjointed limbs propelled by changes in blood pressure analogous to the water-vascular system of the sea urchins. The uniramians have unbranched or uniramous appendages, a simple gut, a single pair of antennae and a pair of mandibles, and with a toughened head capsule. Myriapods are essentially insects without wings, and with many more legs.

Myriapods first appeared during the late Silurian, when *Kampecaris*-like forms were responsible for a variety of terrestrial trails. Some of the largest forms (e.g. the giant *Arthropleura*) forged their way through the lush, green vegetation of the late Carboniferous forests.

Insects or hexapods have six limbs. The oldest insect is probably the springtail *Rhyniella praecursor* from the Lower Devonian Rhynie Chert of the Orcadian Basin of north-east Scotland. There was an early diversification of insects; early and mid Devonian faunas are now well known from Rhynie, Gaspé, Québec and Gilboa, New York State, and these probably coincided with the diversification of the land plants (Labandeira and Sepkoski, 1993). Nevertheless by the late Carboniferous a diverse insect fauna had evolved, with forms such as the dragonflies and mayflies capable of powered flight. Moreover, by the end of the Permian, most of the familiar insect orders had appeared. During the later Mesozoic and Cenozoic, significant coevolutionary relationships were established between plants and insects, particularly between flowering plants and insect pollinators (see Chapter 10).

Subphylum Crustacea

As the name suggests, the crustaceans have a hard crusty carapace. The group is aquatic, mainly marine, with

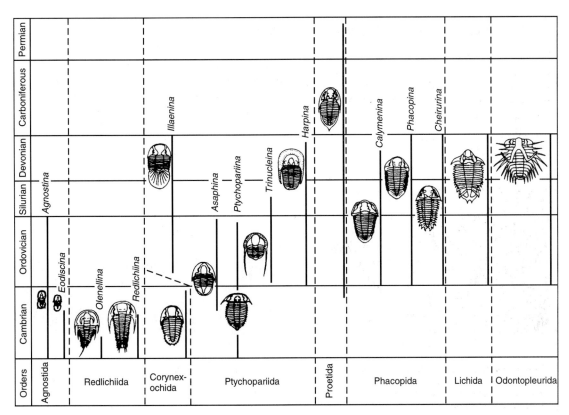

Fig 8.10   Stratigraphical distribution of the main trilobite groups.

gills, mandibles, two pairs of antennae and stalked compound eyes. The heavily armoured crabs and lobsters typify this diverse subphylum, which first appeared in the Cambrian; however, the barnacles and ostracodes are also crustaceans with a notable geological record.

There are at least eight main classes of crustacean. With the exception of the ostracodes, which are usually considered part of the microfauna and are dealt with elsewhere (Chapter 11), only two groups, the Cirripedia and the Malacostraca, have significant geological records. The cirripedes or barnacles have shells, or capitula, consisting of several plates and these animals are adapted to an encrusting lifestyle. Two groups, the acorn barnacles and goose barnacles, have contrasting life strategies. The acorn barnacles, such as *Balanus*, have capitula consisting of overlapping plates and they are attached to rocks and other shells; the group rapidly diversified from an origin during the late Cretaceous and are locally common. The goose barnacles are pseudoplankton, living attached to floating debris, and have a relatively poor fossil record.

The malacostracans include two subclasses, the phyl-

locarids and the eumalacostracans. The phyllocarides have large bivalved carapaces, seven abdominal somites and a telson with a pair of furcae, extending posteriorly. *Canadaspis* from the Burgess Shale may be one of the first crustaceans. Living phyllocarids are usually minute in contrast to their larger Palaeozoic ancestors. The eumalacostracans include the decapods, shrimps, lobsters and crabs, together with the less common branchiopods. Some of the most spectacular Carboniferous eumalacostracans have been described from the Granton Shrimp Bed (Figure 8.11).

Many heavily armoured mollusc-eating groups, such as the crabs and lobsters, diversified during the so-called Mesozoic Marine Revolution and became established components of the modern fauna.

Subphylum Chelicerata

The chelicerates are a diverse and heterogenous group including the mites, scorpions and spiders. The familiar horseshoe crab, *Limulus*, together with the extinct sea

Fig 8.11 Carboniferous shrimps: (a) *Tealliocaris woodwardi* from the Gullane Shrimp Bed, near Edinburgh (×4); (b) *Waterstonella grantonensis* from the Granton Shrimp Bed, near Edinburgh (×2); (c) *Crangopsis socialis* and *Waterstonella grantonensis* from the Granton Shrimp Bed (magnification ×2).

scorpions, the eurypterids, are also members of the subphylum, defined in terms of a prosoma (head and thorax) with six segments, an opisthosoma (abdomen) with at most 12 segments and a pair of chelicerae (pincers) attached to the first segment of the prosoma (Figure 8.12).

There are two main classes of chelicerates: the Merostomata contains *Limulus* and the eurypterids, whereas the Arachnida comprises mainly the terrestrial spiders and scorpions. This traditional split of the subphylum into marine merostomes and nonmarine arachnids may be incorrect since both groups probably had both marine and nonmarine representatives. The bizarre *Sanctacaris* from the Middle Cambrian Burgess Shale may be a plesiomorphic sister to the entire group of Chelicerata.

The xiphosures have a relatively large, convex prosoma, approximately equal in length to the opisthosoma, which usually contains fewer than 10 segments; the telson is commonly long and spiny, and the ophthalmic ridges together with the cardiac lobe are usually well preserved. Although some xiphosure-like taxa, such as *Paleomerus*, have been described from the Lower Cambrian, the early taxa, such as species of *Chasmataspis* from Tennessee, are probably of early Ordovician age. A trend towards larger size and a shorter fused abdomen is seen in most groups. Carboniferous taxa, for example *Belinurus* and *Euproops*, have well-developed cardiac lobes and ophthalmic ridges together with fused abdomens. *Mesolimulus* from the Upper Jurassic Solnhofen Limestone, however, is smaller than living taxa although it too was marine and left clear evidence of its appendages in a trackway in the lagoonal muds of Bavaria.

The eurypterids were some of the largest known arthropods, approaching 2 m in length, and ranging in

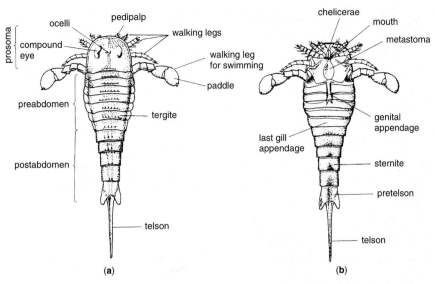

Fig 8.12 Chelicerate morphology, displaying features of the (a) dorsal and (b) ventral surfaces.

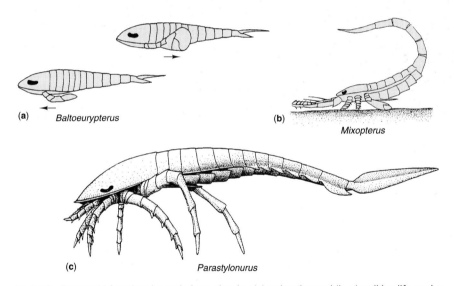

Fig 8.13 Eurypterid functional morphology, showing (a) swimming and (b, c) walking life modes.

age from Ordovician to Permian (Figure 8.13). They occupied a variety of environments from marine to fresh water and some may have been amphibious. Much of our knowledge of eurypterid morphology has been derived from superbly preserved specimens etched from Silurian dolomites cropping out on the Estonian island of Saaremaa. The exoskeleton is long and relatively narrow. The subrectangular prosoma bears a variety of appendages; the first pair of appendages were chelicerae adapted for grasping, while others were modified for movement, with the last pair of large paddle-like appendages probably adapted for swimming. The opisthosoma, comprising the pre- and post-abdomen, consists of 12 segments. The telson, variably developed as a long spine or a flattened paddle, emerged posteriorly.

With the exception of generalist feeders such as *Baltoeurypterus*, most eurypterids were predators attacking fishes and other eurypterids. In the Silurian rocks of the Anglo-Welsh area, *Pterygotus* and its allies are asso-

Box 8.4 Exceptional arthropod-dominated faunas

Arthropods are common in a number of Lagerstätten deposits, suggesting that they probably attained a much greater diversity in the past than the rest of the fossil record suggests. Nearly 40% of the animals described from the mid Cambrian Burgess Shale are arthropods. Apart from typical trilobites such as *Olenoides*, there are also soft-bodied taxa, for example *Naraoia* and the larger *Tegopelte*. However, the commonest and first-discovered Burgess arthropod is the elegant trilobitomorph *Marrella*. The fauna contains many other arthropods such as *Canadaspis*, probably the first phyllocariid crustacean. There are many unique or nonaligned arthropods in the fauna: *Anomalocaris*, *Emeraldella*, *Leanchoilia*, *Odaraia*, *Sidneyia* and *Yohoia* are not easily classified with established groups. The small and bizarre *Hallucigenia* was probably an onychophoran, while *Sanctacaris* was a chelicerate.

The Upper Cambrian calcareous concretions or orsten from mainland Sweden and the Swedish island of Öland have yielded a phosphatized fauna dominated by crustaceans and ostracodes, together with agnostid trilobites. Many of these diverse forms were minute, living in microhabitats within or above the muds of Cambrian seas. These faunas are quite distinct from the earlier Burgess Shale-type faunas and provide a window on a habitat occupied by a wide range of body plans on a microscopic scale, possibly adapted to life below the sediment–water interface. The early Devonian faunas of the Hunsrückschiefer of the German Rhineland contain beautifully preserved phyllocariid crustaceans such as *Nahecaris*, together with a number of other arthropods apparently lacking living counterparts; *Cheloniellon* is large and ovoid with a pair of antennae, nine segments and a conical telson, whereas *Mimetaster* is similar to *Marrella* from the Burgess Shale.

The late Carboniferous Mazon Creek fauna of Illinois occurs across two facies associated with a deltaic system. The marine Essex fauna developed on the delta front and is dominated by fishes including coelacanths and some of the earliest lampreys. However, huge crustaceans are associated with the weird *Tullimonstrum* which may, like many members of the Burgess fauna, represent a unique phylum although it could be a heteropod gastropod. The nonmarine Braidwood assemblage has a formidable array of arthropods, including 140 species of insect together with centipedes, millipedes, scorpions and spiders. The fauna, together with a flora of over 300 species of land plant, occupied a lowland swamp milieu between the sea and coal forests. Shrimps and ostracodes apparently inhabited ponds within the swamps.

Terrestrial assemblages, such as the Montsech fauna from the Lower Cretaceous of north-east Spain, have yielded new information on the evolution of spiders. Three web-weaving species equipped to attack an abundant insect life inhabited settings around coastal lagoons (Selden *et al.* 1991). Cretaceous and Tertiary amber (see Chapter 1) also preserves some remarkable arthropod faunas.

ciated with normal marine faunas, whereas *Eurypterus* itself preferred inshore environments; *Hughmilleria* and related forms, however, dominated brackish to freshwater communities.

Over 25 genera of eurypterids have been described. Although some arthropods with eurypteroid features have been noted from rocks of early Cambrian age, the first definite eurypterids are known from the Lower Ordovician rocks of Deepkill, New York State, where probable *Pterygotus* and *Waeringopteris* occur together with *Eocarcinosoma*. The group was most abundant during the Silurian and Devonian, but only one family, the Hibbertopteroidea, survived till the end of the Permian.

The arachnids are a huge group of terrestrial carnivores containing over 100 000 known species of mites, scorpions, spiders and ticks. The prosoma consists of six segments, with a pair of chelicerae, a pair of sensory or feeding pedipalps and four pairs of walking legs. The opisthosoma has 12 segments. Arachnids breathe mainly through so-called lung books, or tracheae, or both.

Molluscs

The Mollusca is the second largest animal phylum after the Arthropoda, with over 120 000 living species. The phylum includes the slugs, snails, squids, cuttlefish and octopuses, in addition to all manner of shellfish such as the clams, mussels, snails, squids and oysters. Although some molluscs are the size of sand grains, the giant squid, *Architeuthis,* can grow to over 20 m in length, and is the largest of all the invertebrates. Molluscs today occupy a wide range of habitats, from the abyssal depths of the oceans, across the continental shelves and intertidal mudflats to lakes, rivers and dry land. Molluscs are unsegmented soft-bodied animals, although some show

---

Box 8.5 Classification of the molluscs

Following Peel (1991), the term 'monoplacophoran' is now reserved for a grade of organization.

Class Aplacophora
  Worm-like spiculate molluscs. Possibly Carboniferous (or older)–Recent.

Class Polyplacophora
  Multiple shell usually with eight plates, large muscular foot and a series of gill pairs. Cambrian–Recent.

Class Tergomya
  Exogastrically coiled, univalved, bilaterally symmetrical, often planispirally coiled or cap-shaped. Mid Cambrian–Recent.

Class Helcionelloida
  Endogastrically coiled, univalved, untorted molluscs. Early Cambrian–early Devonian.

Class Gastropoda
  Univalved, shell usually coiled, having head with eyes and other sense organs, muscular foot for locomotion. Internal organs rotated through up to 180° during torsion early in ontogeny. Late Cambrian–Recent.

Class Bivalvia
  Twin-valved, joined along dorsal hinge line commonly with teeth and ligament; lacking head but with well-developed muscular foot and often elaborate gill systems. Early Cambrian–Recent.

Class Rostroconchia
  Superficially similar to bivalves, but with single bilobed shell having a dorsal bridge connecting the two lobes. Early Cambrian–Late Permian.

Class Scaphopoda
  Long, cylindrical shell, open at both ends. Mid Ordovician–Recent.

Class Cephalopoda
  Most advanced molluscs with head and well-developed sensory organs together with tentacles. Late Cambrian–Recent.

---

serial repetition of internal organs, usually with an external or secondary internal skeleton of calcium carbonate. The mantle, an extended outfolding of the body wall, lines the shell and secretes the skeletal material. Apart from a mineralized shell, many molluscs are also characterized by a rasping feeding organ, the radula, composed of chitin and containing a few or many hundreds of teeth, primarily used for scraping food.

Molluscan shells are secreted as calcium carbonate, mainly aragonite, from the edge of a variably developed mantle. The bivalves evolved a range of shell fabrics from simple prismatic structures through nacreous and foliated phases to cross-lamellar fabrics. Shell structure can be helpful in the higher classification of the group (see Box 8.5), but there is also some convergence among derived taxa.

The archimollusc and molluscan radiation

Traditionally, molluscan evolution has been seen as a radiation from a hypothetical ancestor, or archimollusc, with a minimal molluscan morphology; this approach has been modified and merged with a recent cladogram for the phylum (Figure 8.14). The archimollusc is close to the monoplacophoran grade, but probably shared common ancestors with both the aplacophorans and polyplacophorans. Both the bivalves and scaphopods were probably derived from the rostroconchs (Runnegar, 1983).

Early molluscs

The early Cambrian was a time of experimentation, with a variety of short-lived, often bizarre, molluscan groups, such as the helcionelloids, dominating many faunas (Figure 8.15). Most workers now agree that the first molluscs were descended from forms like living flatworms. They were probably spiculate animals with the gills situated posteriorly, and similar to modern soft-bodied aplacophorans. The aplacophorans and the shelled molluscs shared a common ancestor, probably during the late Precambrian. Significantly, the articulated remains of a halkieriid mollusc (see Chaper 5) from the Lower Cambrian rocks of North Greenland have promoted new discussion on the identity of the earliest molluscs (Peel, 1991). The halkieriid not only displays an armour of a series of sclerites, commonly described in the past as discrete elements, but also two large mollusc-like shells at the front and back of the worm-like body.

The hyoliths, long conical calcareous shells with an operculum-covered aperture, are often associated with the molluscs; the group ranged from the Cambrian to Permian with some of the 40 known genera reaching

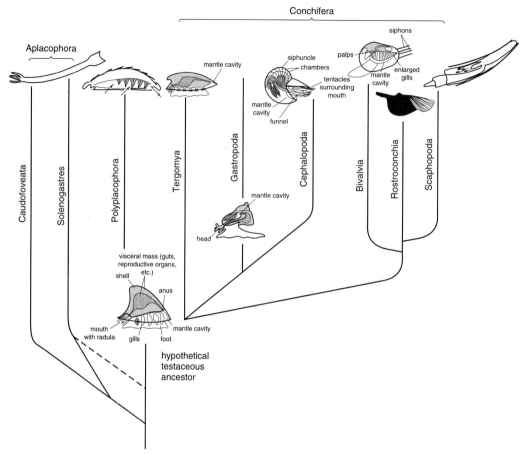

Fig 8.14 Mollusc evolution from the theoretical archimollusc integrated with a cladistic-type framework.

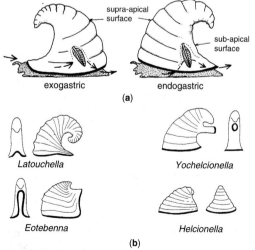

Fig 8.15 Early molluscs: range of morphologies in the helcionellids: exo- and endogastric conditions (a), and some genera (b).

lengths of 200 mm. Current studies assign the group to its own phylum, distantly related to the molluscs.

Class Bivalvia

Bivalves are some of the commonest shelly components of beach sands. Many taxa are farmed and harvested for human consumption, and pearls are a valuable by-product of bivalve growth. Although bivalves may be the least intellectually equipped of the molluscs, the class has developed a spectacular variety of shell shapes and life strategies, during a history spanning the entire Phanerozoic, nearly all based on a simple bilaterally symmetrical exoskeleton. The first bivalves were marine, shallow burrowers; epifaunal, deep-burrowing and boring strategies, together with migrations to freshwater habitats, were secondary innovations. There are over 4500 genera of bivalves, with less than half described from the fossil record. In view of the wide range of life strategies and their relationships to particular sediments, the bivalves are good facies fossils. Nonmarine bivalves have been used extensively, in the absence of other groups, to zone parts of the Upper Carboniferous, and Charles Lyell counted marine bivalves to plot his Lyellian curve for the Tertiary (the increasing proportion of living forms in fossil faunas through the Tertiary was used to subdivide the system). But despite these examples, and the use of inoceramids and rudists in Cretaceous carbonates, their biostratigraphical precision is limited (see Chapter 2).

Basic morphology

Bivalves are twin-valved molluscs superficially resembling the brachiopods, and they are common in modern seas and oceans (Figure 8.17). In contrast to the Brachiopoda, bivalve shells are always composed of calcium carbonate, usually aragonite, and each has a plane of symmetry parallel to the commissure separating the left and right valves from each other, i.e. the two valves are mirror images of each other. Bivalves have sometimes been termed lamellibranchs or pelecypods but the group was first named the Bivalvia by Linnaeus in 1758.

The bivalve exoskeleton has two lateral valves, left and right, united dorsally along the hinge line by a tough ligament; here, teeth and sockets interlock and prevent shearing between the two shells. The valves open ventrally. The valves are secreted by mantle lobes; the attachment of the mantle to the shell is marked by the pallial line which may be indented posteriorly into a pallial sinus marking the extension of the siphons. The earliest-formed parts of each shell, the beaks or umbones, may be separated by the cardinal area supporting the dorsal ligament. The valves are closed by the contraction of a pair of adductor muscles, situated anteriorly and posteriorly. While the shells are closed, the hinge ligament is constrained between the dorsal parts of the shells; when the adductors relax, the ligament expands and the shells spring open. The scars of these shell-closing muscles may usually be seen as clear shiny and depressed areas inside both valves.

In many cases, the umbones of the valves generally point or face obliquely anteriorly, the ligament is usually more extended or is grown only posterior to the umbones, the pallial sinus is situated posteriorly, and the anterior adductor is often the larger of the two scars. When the valves are held with the join between the two valves vertically, and the anterior end pointing away from the observer, then the right and left valves are in the correct positions.

There are eight patterns of teeth and sockets in the bivalves (Fig. 8.18):

1. taxodont: teeth numerous and arranged in a radial or subparallel pattern, fanning out upwards;
2. actinodont: similar to taxodont arrangement but fanning out downwards;
3. dysodont: small, simple teeth;
4. isodont: very large teeth on either side of ligament pit;
5. schizodont: large, sometimes grooved teeth;
6. heterodont: a mixture of cardinal and lateral teeth;
7. pachyodont: exceptionally large, blunt teeth; and
8. desmodont: teeth reduced or absent.

Box 8.6 Computer-simulated growth of molluscan shells

Most valves of any incrementally grown shelled organism can be modelled as an expanding cone, either straight or coiled, and in fact the ontogeny of living *Nautilus* was known to approximate to a logarithmic spiral in the 18th century. Raup (1966) defined and computer-simulated the ontogeny of shells on the basis of four parameters:

1. the shape of the generating curve or axial ratio of the shell's cross-sectional ellipse;
2. the rate of whorl expansion per revolution (W);
3, the position of the generating curve with respect to the axis (D); and
4. the whorl translation rate (T).

Shells are generated by rotating the generating curve around a fixed axis, with or without translation (Figure 8.16). For example, when $T = 0$, shells lacking a vertical component, such as some bivalves, planispiral ammonites and brachiopods, are simulated, whereas those with a large value of T are typical of high-spired gastropods. Only a small selection of the theoretically possible spectrum of shell shapes occurs in nature. More recent work has applied more complex techniques to simulate ammonite heteromorphs. Nevertheless, only a relatively small percentage of the theoretically available morphospace has actually been exploited by fossil and living shelly organisms. Clearly some fields map out functionally and mechanically improbable or disadvantageous morphologies; other fields have yet to be tested.

Box 8.6 (Cont.)

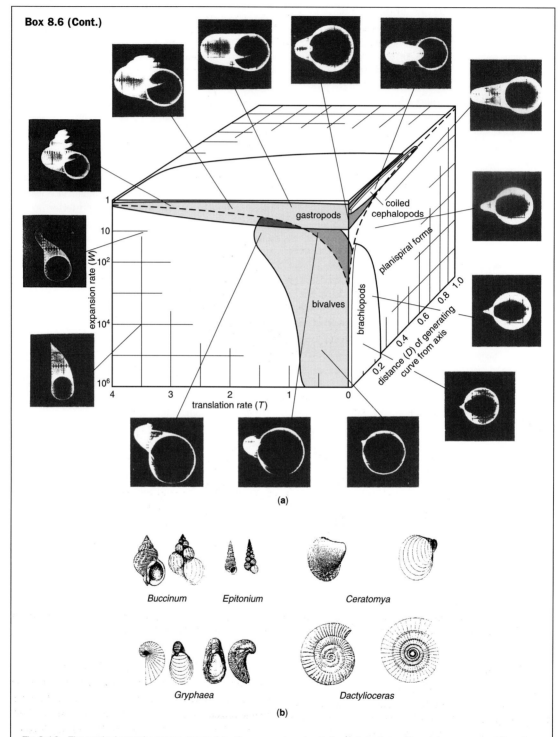

(a)

Buccinum   *Epitonium*   *Ceratomya*

Gryphaea   *Dactylioceras*

(b)

Fig 8.16 Theoretical morphospace created by the computer simulation of shell growth – (a) some computer simulations matched with reality (b).

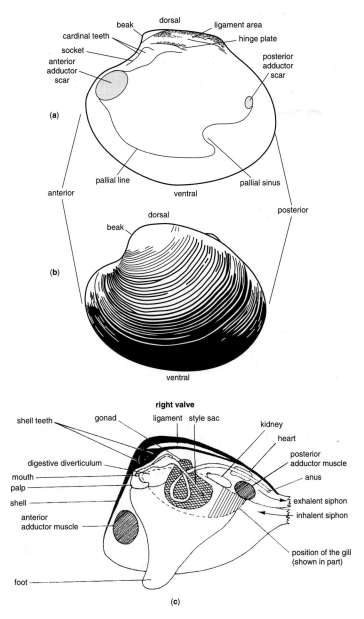

Fig 8.17 Bivalve morphology based on a living *Cerastoderma*: (a) Internal features of the right valve. (b) External features of the left valve. (c) Reconstruction of the internal structures attached to the right valve.

Main bivalve groups

The Bivalvia are subdivided, by zoologists, mainly on the basis of soft-part morphology, such as features of the alimentary canal and the gills together with features of the shell; palaeontologists have usually attempted to use details of the hinge structures (see Figures 8.18 and 8.19).

Five subclasses are recognized: Palaeotaxodonta, Isofilibranchia, Pteriomorpha, Heteroconchia and Anomalodesmata. The palaeotaxodonts are, in some respects, the oldest and most primitive of the subclasses, characterized by taxodont dentition, equivalved shells and protobranch gills. Most are detritus-feeding infaunal marine animals, such as *Nucula*, most abundant today in deeper-water environments. *Ctenodonta* has a typical

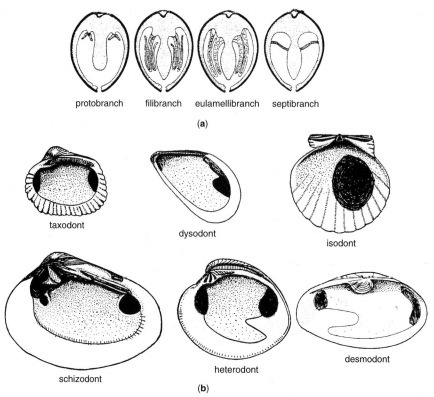

protobranch   filibranch   eulamellibranch   septibranch

(a)

taxodont

dysodont

isodont

schizodont

heterodont

desmodont

(b)

Fig 8.18   (a) Main gill types in the bivalves. (b) Main types of bivalve dentition.

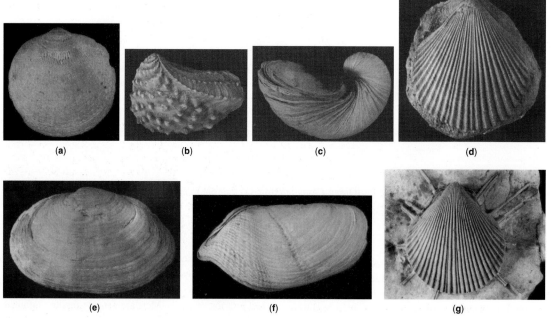

(a)   **(b)**   **(c)**   **(d)**

(e)   **(f)**   **(g)**

Fig 8.19   Some bivalve genera: (a) *Glycimeris* (Miocene), (b) *Trigonia* (Jurassic), (c) *Gryphaea* (Jurassic), (d) *Chlamys* (Jurassic), (e) *Mya* (Recent), (f) *Pholas* (Recent), and (g) *Spondylus* (Cretaceous). (All ×0.75).

taxodont dentition, an elliptical shell and an external ligament; it ranged in age from the Ordovician to the Carboniferous.

The isofilibranchs have inequilateral shells, filibranch gills and either lack or have one or two cardinal teeth with sometimes tiny, ridge-like teeth developed anteriorly or posteriorly or both. Most are fixed epibenthic forms attached by a byssus or nestled into cavities. The group includes the mussels *Modiolus* and *Mytilus,* together with elongate highly ornamented *Orthonota*.

The pteriomorphs are mainly marine fixed benthos, attached by a byssus secreted by a gland on the rear of the foot, or they may be cemented, though some such as the glycimerids and the scallops developed, secondarily, a free-living life mode. The group includes the ark shells *Arca* and *Anadara*, the scallops *Chlamys* and *Pecten*, and the oysters *Crassostrea* and *Ostrea*.

The heteroconchs are a mixed bag of mainly suspension feeders, subordinate during the Palaeozoic, but radiating during the Mesozoic when mantle fusion and the development of long siphons promoted a deep-infaunal life mode. This group includes the typical clams such as the horse-hoof clam *Hippopus* and the wedge-clam *Donax*, as well as the giant clam *Tridacna* together with the razor shells *Ensis* and *Tagelus*, the ship-worm *Teredo* and the cockle *Cerastoderma*. The anomalodesmatans are predominantly suspension-feeding marine forms with modified shells, such as *Pholadomya*.

Lifestyles and morphology

Studies of the diversity of bivalve form and its relation to a spectrum of life strategies suggest a number of recurrent adaptive morphologies (Stanley, 1970). The seven main ecomorphic groups include the following: infaunal shallow burrowing, infaunal deep burrowing, epifaunal attached by byssus, epifaunal cemented, free lying, swimming, and boring and cavity dwelling. Specific assemblages of morphological features are associated with each life mode (Figure 8.20). These studies have been adapted by a number of authors for similar bivalve-dominated communities throughout the Phanerozoic.

Bivalve evolution

The earliest known bivalves have been reported from the basal Cambrian. The praenuculid *Pojetaia* has been described from rocks of Tommotian age in Australia and China, and *Fordilla* is known from Lower Cambrian rocks in Denmark, North America and Siberia. Both genera have two valves separated by a working hinge with a ligament, together with muscle scars and teeth. These

Infaunal shallow burrowers
Glycimeris
equivalved, adductor muscles of equal sizes and commonly with strong external ornament.

Infaunal deep burrowers
Mya
elongated valves, often lacking teeth and with permanent gape and a marked pallial sinus.

Epifaunal with byssus
Mytilus

elongate valves with flat ventral surface and reduction of both the anterior part of the valve and the anterior muscle scar. Attached by thread-like byssus.

Epifaunal with cementation
Ostrea

markedly differently shaped valves, sometimes with crenulated commissures; large single adductor muscle.

Unattached recumbents
Gryphaea

markedly differently shaped valves sometimes with spines for anchorage or to prevent submergence in soft sediment.

Swimmers
Pecten
valves dissimilar in shape and size with very large, single adductor muscle and commonly with hinge line extended as ears.

Borers and cavity dwellers
Teredo
elongate, cylindrical shells with strong, sharp external ornament; cavity dwellers commonly grow in dimly lit conditions following the contours of the cavity.

Fig 8.20 Morphology and adaptations of the main ecological groups of bivalve mollusc.

genera probably postdate the oldest rostroconch, *Heraultipegma*, by a few million years and thus an origin for the bivalves within the rostroconch plexus is probable on both morphological and stratigraphical grounds. The rapid early Ordovician radiation of the class provided a complete set of all the bivalve subclasses by the mid Ordovician. Not only were dysodont, heterodont and taxodont dentitions established, but a variety of trophic types had also developed following the frantic early Ordovician radiation.

After this major diversification, the group stabilized during the remaining part of the Palaeozoic, although some groups evolved extensive siphons which aided deep-burrowing life modes. This adaptation, together with the mobility provided by the bivalve foot, were important advantages over most brachiopods which simultaneously pursued a fixed epifaunal existence. Moreover, the gills probably crossed a functional threshold during the late Ordovician–early Silurian, operating as nutrient filters as well as respiratory organs. The early Mesozoic radiation of the group featured siphonate forms with desmodont and heterodont dentitions, equipped to handle life deep in the sediments of nearshore and intertidal zones where they diversified.

Box 8.7 Rudists

The rudists were aberrant bivalves, ranging in age from the late Jurassic to late Cretaceous, and occupying the Tethyan region as well as shallow seamounts in the Pacific. During a relatively short interval they evolved a bizarre range of morphologies (Figure 8.21), and although many groups apparently mimicked corals, the rudists were probably not reef-building organisms. The rudists were inequivalved with a large attached valve, usually the right valve of conventional terminology, and a small cap-like free valve, though in some taxa both valves were extended. Virtually all rudists had a single tooth flanked by two sockets in the attached valve, and two corresponding teeth and a socket in the free valve. The valves functioned with an external ligament and a pair of adductors attached to internal plates or myophores. In advanced forms, the ligament became invaginated and lost its function; moreover, the gills were apparently functionally dispossessed, with feeding carried out by the mantle margins.

There were three rudist growth strategies (Ross and Skelton, 1993). Elevators had tall conical shells with a commissure raised above the sediment–water interface to free the animal from the risk of ingesting sediment. The elevators were thus similar to solitary corals. Clingers or encrusters were flat bun-shaped forms which usually adhered to hard substrates or overgrew firm sediment surfaces. The recumbents had large shells, extending laterally and extravagantly over the sea-floor like large calcified croissants.

The rudists occupied carbonate shelves throughout the Tethys region, with their larvae island hopping around the tropics, often growing together in a gregarious habit; clusters or clumps probably trapped mud in molluscan-rich structures. As noted above, it now seems likely that the rudists were essentially sediment-dwelling organisms, and were never true reef-building organisms.

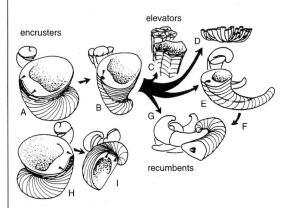

Fig 8.21   Rudist growth strategies: encrusters (A, B, H, and I), elevators (C, D, E) and recumbents (F, G).

Much of the post-Palaeozoic radiation of the class has been related to the effects of increased predation pressure associated with the Mesozoic Marine Revolution (Harper and Skelton, 1993).

Class Gastropoda

The gastropods, the 'belly-footed' molluscs, are the most varied and abundant of the molluscan classes. The group includes the snails and slugs, forms both with and without a calcareous shell. During a history spanning the entire Phanerozoic, gastropods evolved creeping, floating and swimming strategies together with grazing, suspension-feeding, predatory and parasitic trophic styles.

Gastropods are characterized by torsion. In many forms, the mantle cavity containing the animal's respiratory organs is situated above the head, and is considered to have rotated during ontogeny from an originally posterior (non-gastropodan) position. The advantages of this arrangement are unclear, although it has been suggested that this arrangement allows withdrawal of the head into the shell prior to the foot, the upper surface of which carries the protective operculum (Figure 8.22). During development, the head and foot remain fixed relative to each other, but all the visceral mass, the mantle and the larval shell are, in effect, rotated through 180°. The process of torsion is characteristic of the Gastropoda, although in some groups there may be secondary reversal. The coiling of the gastropod shell is unrelated to the rotation of the soft parts. Following torsion, the mantle cavity and anus were open anteriorly and the shell coiled posteriorly in an endogastric position, in contrast to the exogastric style of the 'monoplacophoran' grade shell.

The gastropod shell is commonly aragonitic or calcitic, usually conical, with closure at the pointed apex, and open ventrally at the aperture (Figure 8.22). Each revolution of the shell or whorl meets adjacent whorls along a suture, and the earlier whorls together comprise the spire. Tight coiling about the vertical axis generates a central pillar or columella, but an axial cavity, the umbilicus, is present in many forms. The aperture is commonly oval or subcircular and circumscribed by an outer and inner lip. The head emerges at the anterior mar-

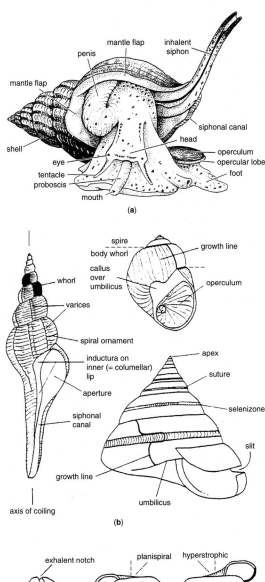

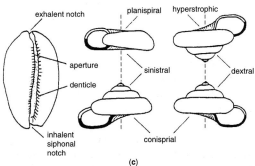

Fig 8.22 Gastropod morphology: (a) annotated reconstruction of a living gastropod; (b) annotated shell morphology of three gastropod shell morphotypes; (c) main types of gastropod coiling strategy.

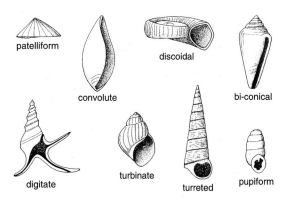

Fig 8.23 Gastropod shell shapes.

gin of the aperture, where the aperture may be notched or extended as a siphonal canal supporting inhalent flow through the siphon. Material is ejected through the exhalent slit in the outer lip in some primitive forms. During ontogeny, the inactive track of this slit is successively infilled with shell material to form the selenizone.

The gastropod shell is conventionally orientated, for the purposes of illustration, with the aperture facing forward and the apex facing upwards. If the aperture is on the right-hand side, the shell is coiled clockwise in dextral mode; sinistral shells have the opposite sense of coiling. The shell surface is commonly modified by strong growth lines, ribs, tubercles and projections. Many gastropods have an operculum which partly closes the aperture when the animal withdraws into its shell. Gastropods developed a huge variety of shell shapes, ranging from the simple patelliform to the complex digitate shell (Figure 8.23).

Main gastropod groups and their ecology

Three subclasses of gastropods are traditionally defined on the basis of respiratory and nervous systems: the Prosobranchia are fully torted with one or two gills, an anterior mantle cavity and cap-shaped or conispiral shells; the Opisthobranchia are largely detorted, with the shell reduced or absent, and mantle cavity posterior or absent; and the Pulmonata also show detorsion with the mantle cavity modified as a lung, and the shells are usually conispiral. Fossil taxa (Figure 8.24) are normally assigned to these categories on the basis of similarities in shell morphology.

The prosobranchs are mainly part of the marine benthos with a few freshwater and terrestrial taxa. The primitive archaeogastropods are marine, mainly grazing, herbivores with cap-shaped or low-spired forms. The Order Archaeogastropoda includes several suborders. The

Fig 8.24 Some gastropod genera: (a) *Murchisonia* (Devonian) (×1.25); (b) *Euomphalus* (Carboniferous) (×0.5); (c) *Lophospira* (Silurian) (×0.5); (d) *Patella* (Recent) (×1); (e) *Platyceras* (Silurian) (×1); (f) *Neptunea* (Plio-Pleistocene) (×0.6); (g) *Viviparus* (Oligocene) (×0.8); (h) *Turritella* (Oligocene ×1).

bellerophontines had planispirally coiled shells with a well-developed slit, ranging in age from the Cambrian to the Triassic. The long-ranging *Bellerophon* was very common in the early Carboniferous. Macluritines have large thick shells lacking a slit-band; for example, *Maclurites* is planispirally coiled and hyperstrophic, with a robust operculum common during the Ordovician.

The pleurotomariines have variably shaped shells, usually conispiral. They dominated shallow-water Palaeozoic environments, although today the group is restricted to deeper-water settings. *Pleurotomaria* had a trochiform shell with a broad selenizone; the older Ordovician–Silurian *Lophospira* had a turbinate shell. The trochines are typical of rocky coasts, grazing on algae; some Palaeozoic taxa (e.g. the Ordovician–Silurian *Cyclonema*) were probably scavengers, whereas some (such as the Silurian–Devonian *Platyceras*) are commonly attached to the anal tubes of crinoids, although this coprophagous life mode has been challenged. The murchisoniines range from the Ordovician to the Triassic, possessing high-spired shells with a siphonal notch. *Murchisonia* is a long-ranging genus

(Silurian–Permian). The euomphalines were mainly discoidal gastropods, such as *Euomphalus* which ranged from the Silurian to the Permian. The patellines (such as the limpet *Patella*) have cap-like shells which allow them to clamp on to rocks in the intertidal zone; when moving over the substrate they graze on algae.

The Order Mesogastropoda consists of prosobranchs that have lost the right gill and usually have conispiral shells, often with siphonal notches. These taxa have diversified into marine, freshwater and terrestrial environments. *Turritella* is a high-spired multiwhorled shell with strong spiral ribs and a simple aperture, whereas *Cypraea* is involute with the earlier whorls completely enclosed by the final whorl. The Order Neogastropoda contains conispiral, commonly fusiform shells with a siphonal notch or canal; most members of the order are carnivorous and they have been important predators in marine environments from the Tertiary onwards. *Neptunea* has a large body whorl and a short siphonal canal whereas *Conus* is biconical with a narrow aperture and a siphonal notch.

The Subclass Opisthobranchia includes marine gastropods with reversed torsion and commonly lacking shells. Typical opisthobranchs include the pteropods and nudibranchs (sea slugs); the latter is a diverse and colourful group.

The Subclass Pulmonata contains detorted gastropods, such as slugs and snails, with the mantle cavity modified to form an air-breathing lung. The group probably ranges in age from the Jurassic to the present, and is characteristic of terrestrial environments. *Planorbis* has a smooth planispiral shell with a wide umbilicus, whereas *Helix* is smooth and conispiral, and *Pupilla* has a smooth pupiform shell.

The gastropods show a considerable diversity of form, and it is difficult to relate given morphotypes to particular life modes. In general terms, however, some gastropods occupying high-energy environments have thick shells, commonly cap-shaped or low-spired, whereas some shells with marked siphonal canals are adapted to creeping slowly across soft substrates. Many infaunal forms occupying high-energy environments are high-spired. Carnivores are often siphonal, whereas herbivores have complete apertural margins and commonly grazed on hard substrates. Thin-shelled taxa are typical of freshwater and terrestrial environments.

Gastropod evolution

The group was derived from a monoplacophoran-grade ancestor by torsion and development of an endogastric condition, where the shell is coiled away from the ani-

mal's head. Some workers have suggested an origin from among coiled forms in the small shelly fauna, such as *Pelagiella*, which may link the monoplacophoran grade through *Aldanella* to the gastropods.

The monophyly of the gastropods has, however, been questioned. It is possible that the main groups – the archaeogastropods, mesogastropods, opisthobranchs and the pulmonates – may be grades of gastropod organization, forming a series of parallel-evolving clades. The neogastropods, however, are a clade derived from either advanced archaeogastropods or primitive mesogastropods during the late Mesozoic.

Most Palaeozoic gastropods were probably herbivores or detritus feeders. The class became more important during the late Palaeozoic and the Mesozoic when many more predatory groups evolved. However, gastropods reached their acme during the Cenozoic, with the neogastropods, in particular, dominating molluscan nektobenthos.

Gastropods are not particularly good zone fossils and few evolving lineages are known in detail. Nevertheless, microevolutionary sequences in the genus *Poecilizontes* from the Pleistocene of Bermuda suggest that new subspecies evolving by allopatric speciation arose suddenly by paedomorphosis. These rapid speciation events, separated by intervals of stasis, are strong supportive evidence of the punctuated equilibrium model of microevolutionary change. Moreover, in a classic study of late Tertiary snails from Lake Turkana, Kenya, Peter Williamson proposed punctuated changes in 14 lineages (see Chapter 3).

Class Cephalopoda

The cephalopods are the most advanced of the molluscs. The close association of a well-defined head with the foot is the source of their name, meaning 'head-footed'. High metabolic and mobility rates, a well-developed nervous system, sharp eyesight and an advanced brain are ideal adaptations for a carnivorous predatory life mode. The molluscan foot is modified to a hyponome and probably also the tentacles; the hyponome squirts out water under high pressure, providing the animal with an efficient form of jet propulsion that permits speeds of up to 70 km/h.

Modern cephalopods belong to three groups. Firstly, living *Nautilus* has an external coiled shell with a thin internal mantle and nearly a hundred tentacles; only five species of this genus are known, although it is traditionally used as an analogue for the behaviour of all extinct, externally shelled cephalopods such as the ammonoids.

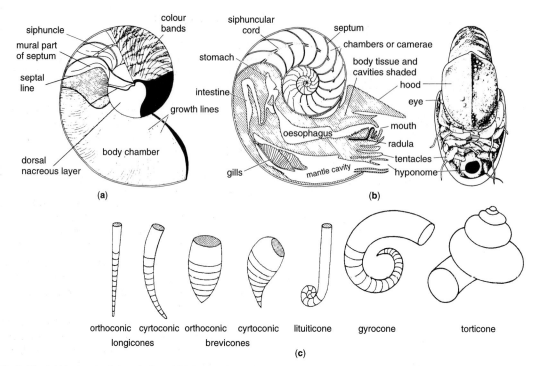

Fig 8.25   (a) Features of the shell and (b) internal morphology of living *Nautilus*. (c) Shell shapes of the nautiloids.

Secondly, the squids and cuttlefish have internal shells, thick external mantles and ten tentacles; this group includes *Spirula*, with an internal planispiral shell. Finally, the paper nautilus, *Argonauta*, and the octopus belong to a group based on a thick external mantle, but usually lacking a skeleton. These living forms are most common in shallow-water belts around the ocean margins.

The traditional split of the class into firstly the Dibranchiata, with one gill pair and including the coleoids, and secondly the Tetrabranchiata, with two gill pairs, has been superseded by a tripartite division into the three subclasses: the Nautiloidea, with straight or coiled external shells with simple sutures (late Cambrian–Recent); the Ammonoidea, with coiled, commonly ribbed external shells with complex sutures (early Devonian–Recent), and the Coleoidea, with straight or coiled internal skeletons (Carboniferous – Recent). This classification avoids decisions regarding the gill configurations of extinct groups such as the ammonoids.

The origin of the cephalopods remains controversial. Peel (1991) has suggested that the group may be derived from within the Class Helcionelloida; both groups are characterized by endogastric coiling and, moreover, the

helcionelloids predate the appearance of the cephalopods by some 10 million years. An alternative view derives the cephalopods from a tergomyan-like *Knightoconus*.

Nautiloidea

Nautilus occurs mainly in the south-west Pacific, normally at depths of 5–550 m. It pursues a nocturnal, nektobenthic life mode as both a carnivore and scavenger; it is prey to animals with powerful jaws such as the perch, marine turtles and sperm whale.

Nautilus has its head, tentacles, foot and hyponome concentrated near the aperture of the body chamber; the visceral mass containing other vital organs is situated to the rear of the body chamber (Figure 8.25). The surrounding mantle extends posteriorly as the siphuncular cord connecting all the previous, now empty, chambers. Each chamber is partitioned from those adjacent by a sheet of calcareous material, the septum. Where the septum fuses with the outer shell, a suture is formed. The form of the suture, or the suture pattern, is fundamental in the classification of the externally shelled cephalopods. The conch is usually orientated as follows: anterior at the aperture, posterior at the apex, venter on

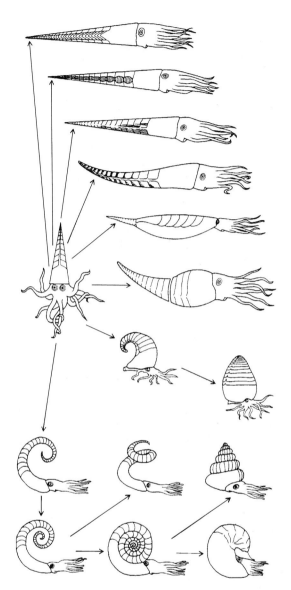

Fig 8.26 Life attitudes and external morphologies of the nautiloids.

the side with the hyponome, usually the outside, and the dorsum opposite. Many different shell shapes have evolved, reflecting a moderate diversity of lifestyles and attitudes (Figure 8.26).

Ammonoidea

Ammonites usually had a planispirally coiled shell comprising the protoconch, phragmocone and body chamber

(Figure 8.27). The protoconch or larval shell records the earliest ontogeny of the animal; the phragmocone is chambered, with each chamber marking successive occupation by the animal, and sealed off from previous chambers by a complex folded septum, like the edge of a sheet of corrugated iron. Where the septum is welded to the shell, a suture is developed, commonly with a complex pattern of frilled lobes and saddles.

There are five main sutural types (Figure 8.28). The orthoceratitic pattern, with broad undulations or rounded lobes and saddles, characterizes mainly nautiloids ranging in age from late Cambrian–Recent. Agoniatitic patterns have a narrow mid-ventral lobe and a broad lateral lobe with additional lobes and saddles, and are typical of early to mid Devonian forms. Goniatitic sutures are characterized by sharp lobes and rounded saddles, and are found in late Devonian–Permian ammonoids. Ceratitic sutures possess frilled lobes and undivided saddles, and ammonitic patterns have both the lobes and saddles fluted and frilled. Based on these sutural patterns, three groups within the ammonoids can be recognized: the goniatites are typical of the Devonian–Permian, the ceratites are typical of the Triassic, and the ammonites dominated the Jurassic and Cretaceous. Nevertheless, these sutural patterns may be cross-stratigraphic, with Cretaceous taxa having both goniatitic and ceratitic grades of suture in homeomorphs of more typical Devonian and Triassic forms.

A long perforated hosepipe-like tube, or siphuncle, connects the outer living chamber with all the previous, empty chambers. Septal necks act like washers, guiding the passage of the siphuncle through each septum. Excepting the clymeniids, the siphuncle is situated along the outer ventral margin of the shell. Sea-water may be pumped in or out of the chambers through the siphuncle

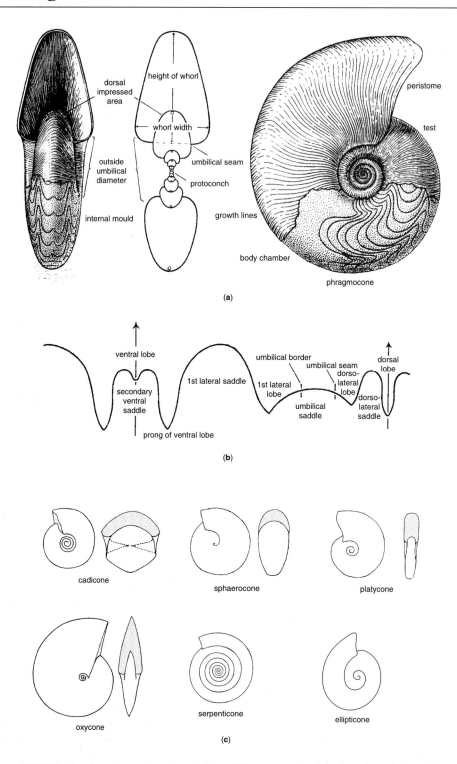

Fig 8.27 Morphology and shape terminology of the ammonoids: (a) external morphology; (b) suture pattern; (c) shell shapes.

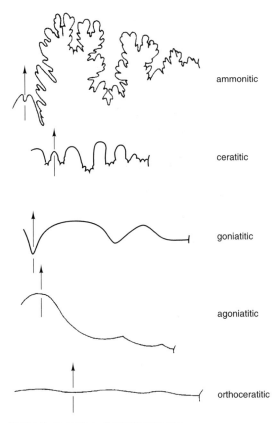

ammonitic

ceratitic

goniatitic

agoniatitic

orthoceratitic

Fig 8.28 Evolution of suture patterns.

in order to alter the buoyancy of the ammonite, similar to mechanisms in the nautiloids and in submarines. The body chamber contains the soft parts of the ammonite. The aperture may be modified laterally with lappets and ventrally with the rostrum. In many taxa, aptychi sealed the aperture externally, although these plates may also have been part of the jaw apparatus.

Main ammonoid groups The subclass is currently split into six orders. The first three, the Anarcestida, the Clymeniida and the Goniatitida, have goniatitic sutures. The anarcestides characterize early to mid Devonian faunas, when forms such as *Anarcestes* and *Prolobites* displayed tightly coiled shells together with a ventral siphuncle. The clymeniids were the only ammonoids with a dorsal siphuncle; they radiated in late Devonian faunas in Europe and North Africa where the group is important for biostratigraphical correlation. Clymeniids developed a variety of shell shapes: *Progonioclymenia* is evolute with simple ribs, *Soliclymenia* evolved triangular whorls, and *Parawocklumeria* is a globular involute form with a trilobed appearance. The goniatitides ranged

in age from the mid Devonian to late Permian, with typical goniatitic sutures consisting of eight lobes and ventral siphuncles. *Goniatites*, for example, was a spherical inflated form with spiral striations, whereas *Gastrioceras* was a depressed, tuberculate form.

The prolecanitidines (early Carboniferous–late Permian) had large, smooth shells with wide umbilici, and sutures grading from goniatitic to ceratitic. *Prolecanites*, for example, was evolute, with a wide umbilicus. The Ceratitida includes most of the Triassic ammonoids, with ceratitic suture patterns and commonly elaborate ornamented shells. Nevertheless, some taxa developed ammonitic-grade sutures, and a number of lineages evolved heteromorphs (see Box 8.9).

The ammonites comprise three orders: the Phylloceratida, the Lytoceratida and the Ammonitida (Figure 8.29). The ammonites appeared first in the early Triassic, with ammonitic sutures, commonly ornamented shells and ventral siphuncles. The first members of the Order Phylloceratida, such as *Leiophyllites*, appear in Lower Triassic faunas and, according to some, this stem group gave rise to the entire ammonite fauna of the Jurassic and Cretaceous. The conservative *Phylloceras* survived from the early Jurassic to near the end of the Cretaceous with virtually no change, after having generated many of the major post-Triassic lineages. The phylloceratides were smooth, involute, compressed forms; the suture had a marked leaf-like or phylloid saddle and a crook-shaped or lituid internal lobe. Although the group had a near-cosmopolitan distribution, its members were most common in the Boreal province, in both shallow and deeper-water environments. The lytoceratides originated at the base of the Jurassic, with evolute loosely coiled shells, as seen in *Lytoceras* itself, which had a near-cosmopolitan distribution, particularly during high stands of sea level. Like the phylloceratides, the order remained conservative; however, it too generated many other groups of Jurassic and Cretaceous ammonites.

The ammonites are thus a rather catch-all polyphyletic net of the lineages branching from the phylloceratides and lytoceratides. Many of the taxa result from iterative evolution, as morphotypes are repeated successively during the Jurassic and Cretaceous; most of the groups are short-lived legs of a palaeontological relay.

Ammonoid ecology and evolution Tests with models have confirmed the probable life attitudes for even the most bizarre heteromorph forms (Figure 8.31). Theoretically at least, virtually all ammonoids could favourably adjust their attitude and buoyancy in the water column. Most ammonoids were probably part of the mobile benthos, although after death, their gas-filled

Fig 8.29 Ammonite genera: (a) *Ludwigia murchisonae* (macroconch) from the Jurassic of Skye; (b) Cluster of *Ludwigia murchisonae* (microconchs) from the Jurassic of Skye; (c) *Quenstedtoceras henrici* from the Jurassic of Wiltshire; (d) *Quenstedtoceras henrici* (showing characteristic suture pattern) from the Jurassic of Wiltshire; (e) *Peltomorphites subtense* from the Jurassic of Wiltshire. (all ×1).

shells could be transported over vast distances by oceanic currents. Many of these benthic ammonoids were endemic, and the shovel-like jaws of some groups were most efficient at the sediment–water interface.

Cretaceous ammonites had shell types related to life modes and environments (Batt, 1993). For example, evolute heavily ornamented forms were probably nektobenthonic, as were spiny cadicones and sphaerocones, nodose sphaerocones and platycones, together with broad cadicones. Evolute planulates and serpenticones, together with small planulates, were probably pelagic in the upper parts of the water column. However, most oxycones were restricted to shallow-shelf depths. Some heteromorphs were nektobenthonic, whereas a few floated in the surface waters.

In many ammonite faunas the consistent co-occurrence of large and small conchs at specific horizons suggests that the macroconch and microconch may be related sexual dimorphs (see Figure 3.12). The macroconch was probably the female with a capacity to hold large quantities of eggs. Traditionally, both conchs were assigned to different genera since they may look disimilar.

The ammonoids probably originated from the bactritid orthocone nautiloids, with protoconchs and large living chambers, during the early Devonian. The anarcestide goniatites, with simple sutural patterns, were relatively scarce during the mid Devonian; however, by the late Devonian other groups such as the clymeniids, with a dorsally situ-

Box 8.9 Heteromorph ammonites

One of the more spectacular aspects of ammonite evolution was the appearance of bizarre heteromorphic ('different shape') shells in many different lineages at a number of different times (Figure 8.30). Heteromorphs first appeared during the Devonian, but were particularly significant in late Triassic and late Cretaceous faunas. Some such as *Choristoceras*, *Leptoceras* and *Spiroceras* appeared merely to uncoil; *Hamites*, *Macroscaphites* and *Scaphites* partly uncoiled and developed U-bends, whereas *Noestlingoceras*, *Notoceras* and *Turrilites* mimicked gastropods, and *Nipponites* adopted shapes almost beyond belief. Initially the heteromorph was considered as a decadent degenerate animal anticipating the extinction of a lineage. Some heteromorphs, however, apparently gave rise to more normally coiled descendants and their association only with extinction events is far from true. Additionally, functional modelling suggests many were perfectly adapted to both nektobenthonic and pelagic life modes.

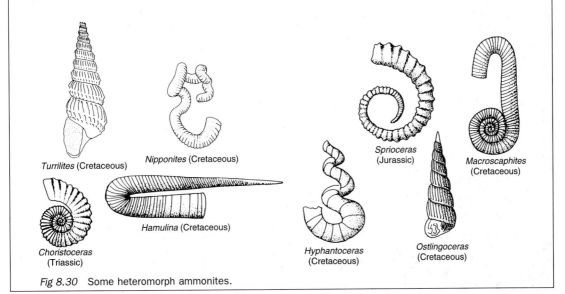

Turrilites (Cretaceous)

Nipponites (Cretaceous)

Hamulina (Cretaceous)

Choristoceras (Triassic)

Sprioceras (Jurassic)

Macroscaphites (Cretaceous)

Hyphantoceras (Cretaceous)

Ostlingoceras (Cretaceous)

Fig 8.30 Some heteromorph ammonites.

ated siphuncle, were common. The goniatitides expanded during the Carboniferous, together with the prolecanitides, from which all the subsequent ammonoids probably originated. During the Triassic, the ceratitides diversified, peaking in the late Triassic; but by the Jurassic the smooth involute phylloceratides, the lytoceratides and the polyphyletic ammonitides were all well established. Complex septa and sutures may have increased the strength of the ammonoid phragmocone, protecting the shell against possible implosion at deeper levels in the water column; more intricate septa also provided a larger surface area for the attachment of the soft parts of the living animal, perhaps aiding more vigorous movement of the animal and its shell.

Coleoidea

The Subclass Coleoidea contains cuttlefish, octopuses and squids, together with the paper nautilus, *Argonauta*.

Coleoids show the dibranchiate condition, with a single pair of gills within the mantle cavity. Although argonauts can be traced back to the mid Tertiary, the living coleoid orders generally have a poor fossil record. In contrast, the skeletons of extinct belemnites are locally abundant in Jurassic and Cretaceous rocks. Belemnites had an internal skeleton, contrasting with the exoskeletons of the shelled cephalopods such as the nautiloids and ammonoids. The belemnite skeleton is relatively simple, consisting of three main parts: the bullet-shaped guard is solid and composed of radially arranged needles of calcite with, at its anterior end, a conical depression or alveolus; the conical phragmocone, equipped with concave septa and a siphuncle, fits tightly within the alveolus, and the spatulate pro-ostracum extends anteriorly (Figure 8.32). This assemblage, situated on the dorsal side of the animal, is analogous to the chambered shells of nautiloids and ammonoids. Soft parts of belemnites, including the contents of ink sacs and tentacle hooks, are occa-

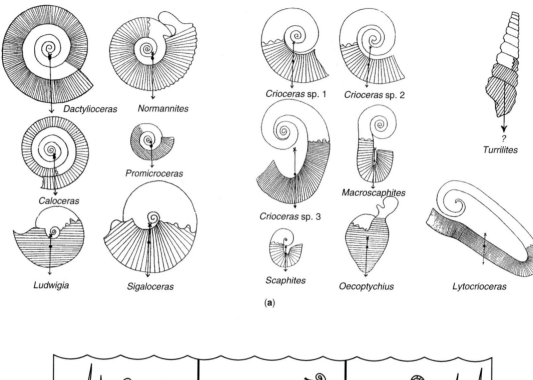

Crioceras sp. 1   Crioceras sp. 2

Dactylioceras   Normannites

Promicroceras

Caloceras

Crioceras sp. 3

Ludwigia   Sigaloceras

Scaphites

?
Turrilites

Macroscaphites

Oecoptychius   Lytocrioceras

(a)

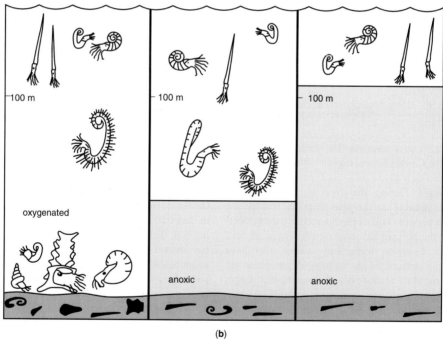

(b)

Fig 8.31 Life attitudes and buoyancy of the ammonites: (a) supposed life orientations of a selection of ammonite genera, with the centre of gravity marked (x); (b) relationship of some ammonite morphotypes to water depth and the development of anoxia.

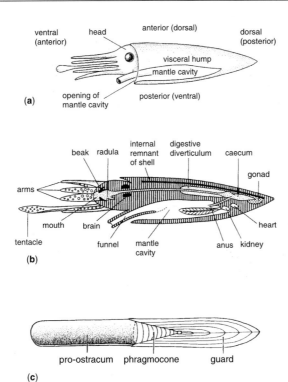

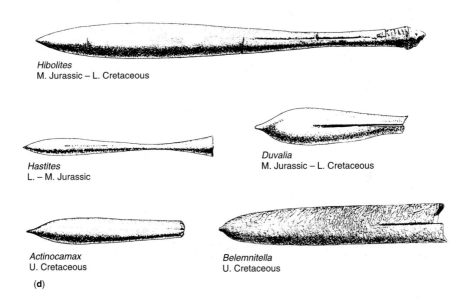

Fig 8.32 Coleoid morphology: (a) reconstruction of a living belemnite; (b) soft-part morphology of the belemnites; (c) internal skeleton of the belemnites; (d) some belemnite genera.

Box 8.10 Belemnite battlefields

Spectacular mass accumulations of belemnite rostra are relatively common in Mesozoic sediments
and, although some authors have used these
assemblages in palaeocurrent studies, few have
addressed their mode of formation. Dense accumulations of bullet-shaped belemnite rostra have promoted the term belemnite battlefields for such distinctive shell beds (Figure 8.33). Doyle and
MacDonald (1993) have classified these accumulations into five genetic types:

1. post-spawning mortalities;
2. catastrophic mass mortalities;
3. predation concentrates, either *in situ* or
   regurgitated;
4. condensation deposits perhaps aided by
   winnowing and sediment by-pass; and
5. resedimented deposits derived from usually
   condensed accumulations.

Clearly some battlefields were generated by several agencies, but post-spawning mortality and winnowing are perhaps the commonest causes.

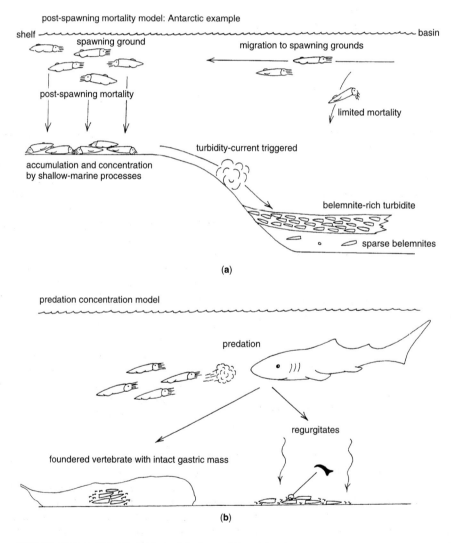

Fig 8.33 Belemnite battlefields and their possible origins.

sionally preserved.

By analogy with modern squids, the belemnites were probably rapid-moving predators living in shoals, with their buoyancy regulated by the chambered phragmocone. The animal probably maintained a horizontal attitude within the water column, preferring the open ocean. However, data from the stomach contents of ichthyosaurs confirm that these molluscs formed part of their diet. Some of the oldest records of belemnites (e.g. *Jeletzkya* from the mid Carboniferous of Illinois) are tentative. The first unequivocal belemnites are from the Middle Triassic rocks of Sichuan Province, China where several species of *Sinobelemnites* occur. Belemnites became extinct at the end of the Cretaceous.

Some of the first supposed belemnites, like the Carboniferous *Paleoconus*, are relatively short stubby forms. Of the early Jurassic belemnites, *Megateuthis* is large, *Cuspiteuthis* has a long slender form, whereas *Dactyloteuthis* is laterally flattened; the later Jurassic *Hibolites* is spear-shaped. The Cretaceous *Belemnitella* has a large bullet-shaped guard, whereas that of *Duvalia* has a flattened spatulate shape. However, despite differences in the detailed morphology of the endoskeltons across genera, most of the Mesozoic belemnites probably looked very similar.

The compact calcareous guards of belemnites have proved ideal for the analysis of oxygen isotope ratios ($^{16}O:^{18}O$) relating to palaeotemperature conditions in Jurassic and Cretaceous seas. These data have indicated warm peaks during the Albian and the Coniacian–Santonian with a gradual cooling from the Campanian onwards. And as with many other Mesozoic groups, belemnite distributions show separate low-latitude Tethyan and high-latitude Boreal assemblages.

Class Scaphopoda

The scaphopods, or elephant-tusk shells, have a single, slightly-curved high conical shell, open at both ends (Figure 8.34). They lack gills and eyes, but have a mouth equipped with a radula and surrounded by tentacles; they also possess a foot, similar to that of the bivalves, adapted for burrowing. Scaphopods are mainly carnivorous, feeding on small organisms such as forams, and spending much of their life in quasi-infaunal positions within soft sediment in deeper-water environments. The first scaphopods, (e.g. *Rhytiodentalium*) appeared during the mid Ordovician and apparently had similar life styles to living forms such as *Dentalium*.

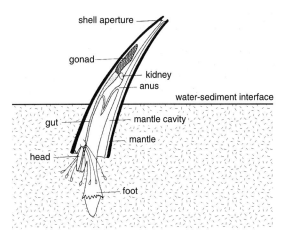

Fig 8.34   Scaphopod morphology.

Class Rostroconcha

Relatively recently, a small class of molluscs, superficially resembling bivalves, but lacking a functional hinge, has been documented; over 35 genera have been described, most from the Lower Palaeozoic. The rostroconchs probably had a foot which emerged through the anterior gape between the shells. However, the two shells are in fact fused along the hinge line, and posteriorly the shells are extended as a platform or rostrum. Ontogeny occurs from an initial dissoconch, with the bilobed form developing by disproportionate growth of shell from the lateral lobes of the mantle. The group appeared first during the early Cambrian when, for example, *Heraultipegma* characterized rostroconch faunas of the Tommotian. Rostroconchs diversified during the Ordovician to reach an acme in the middle Ordovician when all seven families were represented; they probably occupied similar ecological niches to those of the bivalves. However, there followed a decline in abundance and diversity until final extinction at the end of the Permian when only conocardiodes such as *Arceodomus* were still extant.

The rostroconchs developed from within the monoplacophoran plexus with a loss of segmentation; the rostroconchs themselves then generated both the bivalves and the scaphopods (Runnegar and Pojeta, 1974), whereas the gastropods and cephalopods were probably derived independently from a separate monoplacophoran ancestor. In some respects, the rostroconchs may represent a missing link between the univalved and bivalved molluscan lineages.

Evolutionary trends within the Mollusca

A spectacular variety of morphotypes and life modes evolved during the Phanerozoic, from the simple body plan of the archimollusc. Despite the diversity of early molluscs in the Cambrian, the phylum was not notably conspicuous in the tiered suspension-feeding benthos of the Palaeozoic, although many more localized, often nearshore, assemblages were dominated by molluscs.

During the Palaeozoic, bivalves were common in nearshore environments, often associated with lingulide brachiopods, although the class also inhabited a range of deeper-water clastic environments. By the late Palaeozoic, bivalves had invaded a variety of carbonate environments. However, at the end of the Palaeozoic, the appearance of more typical bivalves in shallow-water belts may have displaced the Palaeozoic associations seaward. During the late Mesozoic and Cenozoic, the significant radiation of infaunal taxa may have been a response to increased predation.

The majority of Palaeozoic gastropods were Archaeogastropoda, which commonly dominated shallow-water marine environments and some carbonate reef settings. The Mesozoic was dominated by the Mesogastropoda. The Cenozoic, however, marks the acme of the group, with the radiation of the siphonal carnivorous neogastropods, and with a further diversification of mesogastropods.

The cephalopods evolved through the development of a chambered shell with a siphuncle which gave them considerable control over attitude and buoyancy; this system was refined in the nautiloid groups. The evolution of complex folded sutures in the ammonoids, the exploitation of a pelagic larval stage and a marginal position for the siphuncle apparently set the agenda for the further radiation of the group during the Mesozoic.

Throughout the Phanerozoic, the fleshy molluscs provided a source of nutrition for many groups of predators. Evolution of the phylum was probably in part influenced by the development of predator–prey relationships and minimization of predator success. Thick armoured shells were developed in some groups, while the evolution of deep infaunal life modes was also part of a defensive strategy (Vermeij, 1987).

Further reading

Briggs, D.E.G., Thomas, A.T. and Fortey, R.A. (1985) Arthropoda. In J.W. Murray, (ed.) *Atlas of invertebrate macrofossils*. Longman, London, pp. 199–229.

Box 8.11 Predation and search images

Predation and the development of avoidance strategies, together with a so-called arms race, had an important influence on molluscan evolution. Predators develop a particular search image when seeking their favoured prey. Living terrestrial snails show a wide range of colour patterns and the purpose of this variability may be to confuse predators like the song thrush by presenting a wide range of images. If the predator targetted as prey one particular variant in the population, then other variants would be free to recover until a switch in images was produced. Although such relationships are documented for some Mesozoic and Cenozoic faunas, data are sparse for the Palaeozoic. Hollingworth and Barker (1991) demonstrated that the increase in complexity of colour patterns within neritid gastropods, from the Jurassic onwards, was directly related to the increased attention of predators such as brachyuran crabs. Clarkson *et al.* (1995) have described remarkable intraspecific variation in a bellerophontiform mollusc, *Pterotheca*, from the Silurian rocks of the Pentland Hills. All the specimens of this mollusc are asymmetrical; yet it is difficult to see a functional advantage for this polymorphic variation. Most likely the variation provided predators such as nautiloid cephalopods with a confusion of search images, hence protecting the size and stability of the gastropod population.

Clarkson, E.N.K. (1993) *Invertebrate palaeontology and evolution,* 3rd edition. Chapman and Hall, London.

Gould, S.J. (1989) *Wonderful life. The Burgess Shale and the nature of history.* Hutchinson Radius, London.

Lehmann, U. (1981) *The ammonites–their life and their world.* Cambridge University Press, Cambridge.

Morton, J.E. (1967) *Molluscs.* Hutchinson, London.

Peel, J.S., Skelton, P.W. and House, M.R. (1985) Mollusca. In J.W. Murray. (ed.) *Atlas of invertebrate macrofossils.* Longman, London, pp. 79–152.

Pojeta, J. Jr, Runnegar, B., Peel, J.S. and Gordon, M. Jr (1987) Phylum Mollusca. In R.S. Boardman, A.H. Cheetham and A.J. Rowell (eds) *Fossil invertebrates.* Blackwell Scientific Publications, Oxford, pp. 270–435.

Robison, R.A. and Kaesler, R.L. (1987) Phylum Arthropoda. In R.S. Boardman, A.H. Cheetham and A.J. Rowell (eds) *Fossil invertebrates.* Blackwell Scientific Publications, Oxford, pp. 205–269.

Whittington, H.B. (1985) *The Burgess Shale.* Yale University Press, New Haven.

References

Batt, R. (1993) Ammonite morphotypes as indicators of oxygenation in a Cretaceous epicontinental sea. *Lethaia*, **26**, 49–63.

Briggs, D.E.G., Fortey, R.A. and Wills, M.A. (1993) How big was the Cambrian evolutionary explosion? A taxonomic and morphological comparison of Cambrian and Recent arthropods. *Linnean Society Symposium Series*, 33–44.

Clarkson, E.N.K. (1979) The visual system of trilobites. *Palaeontology*, **22**, 1–22.

Clarkson, E.N.K. (1988) The origin of marine invertebrate species: a critical review of microevolutionary transformation. *Proceedings of the Geologists' Association*, **99**, 153–171.

Clarkson, E.N.K., Harper, D.A.T. and Peel, J.S. (1995) The taxonomy and palaeoecology of the mollusc *Pterotheca* from the Ordovician and Silurian of Scotland. *Lethaia*, **28**, 101–114.

Doyle, P. and MacDonald, D.I.M. (1993) Belemnite battlefields. *Lethaia*, **26**, 65–80.

Fortey, R.A. (1975) Early Ordovician trilobite communities. *Fossils and Strata*, **4**, 331–352.

Harper, E.M. and Skelton, P.W. (1993) The Mesozoic Marine Revolution and epifauna: bivalves. *Scripta Geologica Special Issue*, **2**, 127–153.

Hollingworth, N.T.J. and Barker, M.J. (1991) Colour pattern preservation in the fossil record, taphonomy and diagenetic significance. In S.K. Donovan, (ed.) *The processes of fossilization*. Belhaven Press, London, pp. 105–119.

Jacobs, D.K. and Landman, N.H. (1993) *Nautilus* – a poor model for the function and behavior of ammonoids. *Lethaia*, **26**, 101–111.

Labandeira, C.C. and Sepkoski, J.J. (1993) Insect diversity in the fossil record. *Science*, **261**, 310–315.

Owen, A.W. (1985) Trilobite abnormalities. *Transactions of the Royal Society of Edinburgh: Earth Sciences*, **76**, 255–272.

Peel, J. S. (1991) Functional morphology, evolution and systematics of early Palaeozoic univalved molluscs. *Grønlands Geologiske Undersøgelse*, **161**, 1–116.

Raup, D.M. (1966) Geometric analysis of shell coiling: general problems. *Journal of Paleontology*, **40**, 1178–1190.

Ross, D.J. and Skelton, P.W. (1993) Rudist formations of the Cretaceous: a palaeoecological, sedimentological and stratigraphical review. *Sedimentology Review*, **1**, 73–91.

Runnegar, B. (1983) Molluscan phylogeny revisited. *Memoir of the Association of Autralasian Palaeontologists*, **1**, 121–144.

Runnegar, B. and Pojeta, J. (1974) Molluscan phylogeny: the palaeontological viewpoint. *Science*, **186**, 311–317.

Selden, P.A., Shear, W.A. and Bonamo, P.M. (1991) A spider and other arachnids from the Devonian of New York, and reinterpretation of Devonian Araneae. *Palaeontology*, **34**, 241–281.

Stanley, S.M. (1970) Relation of shell form to life habits in the Bivalvia. *Memoir of the Geological Society of America*, **125**, 1–296.

Vermeij, G.J. (1987) *Evolution and escalation. An ecological history of life*. Princeton University Press, New Jersey.

9 Vertebrates

Key Points

- Vertebrates are the main group of the Phylum Chordata, and they are characterized by a skeleton made from bone (apatite).
- The oldest vertebrates are Cambrian conodont animals, jawless fishes.
- Armoured fishes dominated Devonian seas and lakes.
- After the Devonian, the cartilaginous and bony fishes radiated in several phases.
- Tetrapods arose during the Devonian from rhipidistian fish ancestors, and fish-eating amphibians diversified in the Carboniferous.
- The first reptiles were small insect-eaters.
- Mammal-like reptiles dominated on land during the Permian and Triassic.
- These groups were heavily hit by the end-Permian mass extinction event, and diapsid reptiles took over in the Triassic.
- Dinosaurs were a hugely successful group for 160 Myr of the Mesozoic.
- Birds evolved from dinosaurs, and radiated particularly during the Tertiary.
- The first mammals were small insect-eaters of the latest Triassic, and the mammals achieved great diversity and abundance only after the extinction of the dinosaurs.
- Humans arose about 5 Ma, and fossil evidence points to repeated human migrations out of Africa.

Introduction

Vertebrates have a backbone consisting of vertebrae, and other bony elements, like a skull enclosing the brain and sense organs, and bones in the fins or limbs. Vertebrates are important in the present world because of the dominance by humans, and also because of the huge diversity and abundance of species of bony fishes, birds and mammals. Other groups, such as insects and microbes, are even more abundant and diverse, but vertebrates include the largest animals on land and in the sea. The secret of their success is the internal skeleton.

Origin of the vertebrates

Chordates

Vertebrates are the largest subgroup within the Phylum Chordata. Chordates possess a notochord, a flexible tough rod that runs along the length of the body in its upper part. In addition, chordates have a dorsal nerve cord (unlike the ventral nerve cord of most invertebrates)

and a tail (defined as a projection of the body behind the anus). The closest relatives of chordates among invertebrates are the echinoderms (see pp. 132–144), based on evidence of embryonic development, molecular phylogenetic studies, and, perhaps, on the fossil calcichordates (see pp. 143–144).

Other living chordates include such unlikely creatures as the sea squirts and the lancelet (see Box 9.1). Adult sea squirts (Figure 9.1a) are flexible bag-like creatures that are fixed to the seabed and feed by filtering particles of food from sea-water pumped through their mouth region. The clue to the true affinities of sea squirts is seen in the larvae, or young, which are like tiny tadpoles (Figure 9.1b), with a head region with an eye and gill slits, and a long tail for swimming which contains a notochord and a dorsal nerve cord.

The lancelet, or amphioxus, is more fish-like as an adult (Figure 9.1c). It is a slender 50-mm-long cigar-shaped animal, with a notochord, a dorsal nerve cord, and a throat region provided with gill slits. Water is pumped in through the mouth, and food particles and oxygen are extracted before the water is forced out through the gill slits. Amphioxus swims by beating its body from side to side, just as most fishes do, and the power comes from V-shaped muscle blocks, or myotomes, a typical chordate feature.

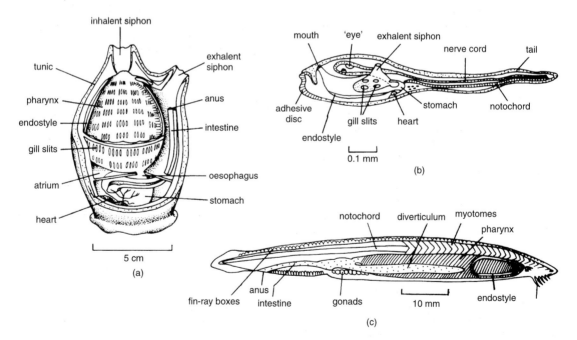

Fig 9.1 The sea squirt *Ciona* (a) as an adult and (b) in larval form. The larva shows obvious chordate features, such as a tail, a dorsal nerve cord, a head region, and gill slits in the pharynx, most of which are lost when the larva metamorphoses. An adult amphioxus *Branchiostoma* (c) is more clearly a chordate.

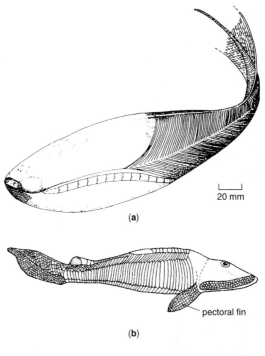

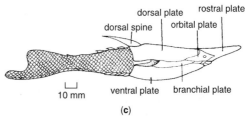

Fig 9.2 Early agnathan fishes. (a) *Sacabambaspis* from the mid Ordovician of Brazil, the oldest well-preserved fish; (b) the osteostracan *Hemicyclaspis* from the Devonian; (c) the heterostracan *Pteraspis*, also from the Devonian.

The skeleton

The skeleton of vertebrates is made from bone and cartilage. Bone consists of a network of collagen fibres on which needle-like crystals of apatite (calcium phosphate, $CaPO_4$) accumulate. Hence bone has a flexible component and a hard component, which explains why bones may undergo a great deal of strain before they break, and that breaks are not formed along simple brittle faces. Cartilage is usually unmineralized, and contains collagen and elastic tissues, and it is a flexible gristly tissue. In humans, most of the bones are laid down in the early embryo in the form of cartilage, and this progressively mineralizes, or ossifies, by deposition of apatite. In adult humans, cartilage remains in the ears, the nose, the ends of the ribs, and at the ends of some limb bones.

The first vertebrates probably had a cartilaginous skeleton. The oldest fish fossils, such as *Sacabambaspis* from the Ordovician of Brazil (Figure 9.2a), had the beginnings of a bony skeleton, but only on the outside of the body, and there is no trace preserved of an internal mineralized skeleton. The outer armour is made from sheets of tiny denticles, or scales, that lock together to form solid shields over the head and trunk region.

The internal skeleton of vertebrates, quite different from the external skeleton of invertebrates, has the unique property that it allows its possessors to grow very large. There is no need for vertebrates to shed their skeletons as they grow larger, as in arthropods, or to extend fleshy parts outside the skeleton to lay down new mineralized tissue, as in molluscs and graptolites, for example. The skeleton is maintained and rebuilt constantly within the body, and can act as a support for massive organisms.

Fishes

Jawless fishes

The first fishes date from the Late Cambrian, and the commonest group then, and in the Ordovician, were the conodont animals. Conodont jaw parts are common as fossils (see pp. 259–265), but their soft eel-like bodies are found only rarely. Fishes became common and diverse during the late Silurian and Devonian.

The first fishes lacked jaws, and are placed in the paraphyletic group Agnatha (see Box 9.2). In the Devonian there were several successful agnathan groups, some strongly armoured and some unarmoured. Osteostracans, such as *Hemicyclaspis* (Figure 9.2b), have a semi-circular head shield bearing openings on top for the eyes and nostrils, as well as porous regions round the sides which may have served for the passage of electrical sense organs, perhaps used in detecting other animals by their movements in the water. Heterostracans, such as *Pteraspis* (Figure 9.2c), were more streamlined in shape, and were perhaps more active swimmers. Both forms had their mouths underneath the head shield, and they probably fed by sieving organic matter from the sediment.

After the Devonian, the agnathan fishes almost disappeared, and today they are represented by 40 species of lampreys and hagfishes, eel-shaped animals that feed in a parasitic fashion, by attaching themselves to the sides of other fishes. Although lampreys and hagfishes have no jaws, their mouths are filled with tooth-bearing bones, and these are used to grip prey animals, and to rasp off lumps of flesh.

Box 9.2 Fish relationships

'Fishes' form a paraphyletic group, consisting of several distinct clades of swimming vertebrates.

Subphylum Vertebrata

Class Agnatha
A paraphyletic group of jawless fishes, including armoured and unarmoured Palaeozoic forms, and modern lampreys and hagfishes. Late Cambrian–Recent.

Class Placodermi
Heavily armoured fishes with jaws and a hinged head shield. Early–late Devonian.

Class Chondrichthyes
Cartilaginous fishes, including modern sharks and rays. Mid Devonian–Recent.

Class Acanthodii
Small fishes with many spines and large eyes. Mid Silurian–early Permian.

Class Osteichthyes
Bony fishes, with ray fins (Subclass Actinopterygii) or lobe fins (Subclass Sarcopterygii), the latter group including the ancestors of tetrapods. Mid Devonian–Recent.

The first jaws

All other vertebrates have jaws, and these probably evolved during the Silurian. There is no fossil evidence for how jaws arose, but studies of the anatomy of modern vertebrates suggest derivation from the strengthening bars of cartilage or bone between the gill slits, each of which consists of several elements, all linked by tiny muscles.

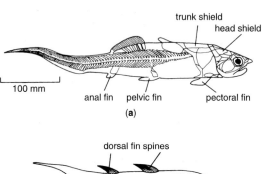

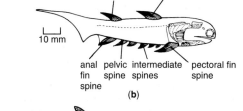

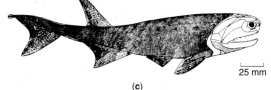

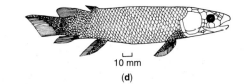

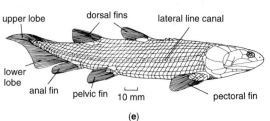

Fig 9.3 Jawed fishes of the Devonian. (a) The placoderm *Coccosteus*; (b) the acanthodian *Climatius*; (c) the actinopterygian bony fish *Cheirolepis*; (d) the lungfish *Dipterus*; (e) the rhipidistian *Osteolepis*.

Some of the oldest jaw-bearing fishes were the placoderms; for example, *Coccosteus* (Figure 9.3a) which had an armour of large bony plates over the head and shoulder region, as in the ostracoderms, and a more lightly armoured posterior region. Swimming was by sideways movements of this tail region. The edges of the jaws did not carry teeth, but instead sharp bony plates, which would have been just as effective in snapping at prey. Placoderms were fearsome predators, some of them (e.g. *Dunkleosteus* from the late Devonian of North America) reaching the incredible length of 10 m. This was the biggest animal that had lived until then, and perhaps this kind of giant predator explains the armour covering of so many Devonian fishes.

Other Devonian fishes were more modern in appearance. The first shark-like chondrichthyans, or cartilaginous fishes, came on the scene during the late Devonian. Acanthodians were small fishes, mostly in the range 50–200 mm in length, and they bore numerous spines, at the front of each fin, and in spaced rows on their undersides (Figure 9.3b). It is likely that acanthodians swam in huge shoals in open water, perhaps feeding on small arthropods and plankton. They escaped predators by rapid darting from side to side in their shoals, and perhaps their spininess made them difficult to swallow.

Bony fishes

The final fish group to appear during the Devonian were the osteichthyans, or bony fishes. There are two groups: those with ray-like fins, ancestors of most fishes today

---

Box 9.3 The fish beds of the Scottish Old Red Sandstone

The Old Red Sandstone Continent of northern Europe and Canada lay close to the equator, in hot tropical conditions, during the Devonian. Fishes lived in the shallow seas and in landlocked lakes around this continent. One of the best collecting areas is in the north of Scotland, where the first specimens came to light nearly 200 years ago.

The fish beds were laid down in large deep lakes (Trewin, 1986). Bulky armoured agnathans and placoderms fed on the bottom in shallow waters, while shoals of silvery acanthodians darted and swirled near the surface. Bony fishes, such as the actinopterygian *Cheirolepis*, the rhipidistian *Osteolepis*, and the lungfish *Dipterus*, moved rapidly through the plants near the water's edge seeking prey, and sometimes swam out through the deeper waters to new feeding grounds.

The fishes are usually found, beautifully preserved and nearly complete (Figure 9.4a), in dark-coloured siltstones and fine sandstones. These rocks were deposited in the deepest parts of the lake, probably in anoxic (low-oxygen) conditions (Figure 9.4b). There were repeated cycles of deposition, perhaps controlled by the climate. During times of high rainfall, great quantities of sand were washed into the lakes from the surrounding Scottish Highlands. Lake levels then fell during times of aridity, and in places the lakes dried out, leaving mud cracks and soils. Then flooding occurred, together with mass kills of fishes, perhaps as a result of eutrophication (oxygen starvation caused by decaying algae after an algal bloom) or following storms. In all, some 2–4 km of lake deposits accumulated over the 40–50 million years of the Devonian, and there are dozens of fish beds throughout this thickness.

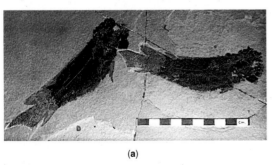

(a)

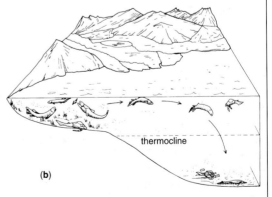

thermocline

(b)

Fig 9.4 The Old Red Sandstone lake in northern Scotland. (a) Typical preservation of two specimens of *Dipterus*. (b) Model of the lake. Sediment is fed in from surrounding uplands during times of heavy rainfall. Fishes inhabit shallow and surface waters, but carcasses may sink below the thermocline into cold relatively anoxic waters, where they sink to the bottom and are preserved in undisturbed condition in dark grey laminated muds.

from carp to salmon, and sea-horse to tuna; and the lobefins, which had thick muscular limb-like fins. Today, the lobefins are rare, being represented by only three species of lungfishes and the rare coelacanth. The coelacanth *Latimeria* is a famous 'living fossil', representative of a group that had been thought to have died out in the Mesozoic. However, in 1938, a coelacanth was fished out of deep waters off East Africa, and more have been caught since.

The ray-fins of the Devonian include *Cheirolepis* (Figure 9.3c), which had a flexible body covered with small scales, and a plated head. This was an active predator that may have fed on acanthodians. The Devonian lobefins are represented by some diverse groups, such as the lungfishes and the rhipidistians. The lungfish *Dipterus* (Figure 9.3d) was a long slender fish, which hunted invertebrates and fishes, and crushed them with broad grinding tooth plates. The rhipidistian *Osteolepis* (Figure 9.3e) was also long and slender, and was an active predator. These lobefins had muscular front fins, and could have used these to haul themselves over mud from pond to pond. Specimens of these fishes are known from the Devonian of many parts of the world (see Box 9.3).

Chondrichthyan evolution after the Devonian

During the Carboniferous, numerous extraordinary shark-like fishes arose, and these were clearly important marine predators. A second shark radiation took place in the Triassic and Jurassic. *Hybodus* (Figure 9.5a) was a fast-swimming fish, capable of accurate steering using its large pectoral (front) fins. The hybodontiforms had a range of tooth types, from triangular pointed flesh-tearing teeth to broad button-shaped crushers, adapted for dealing with molluscs.

Modern shark groups, the neoselachians, radiated dramatically during the Jurassic and Cretaceous to reach their modern diversity of 35 families. The neoselachians have wider-opening mouths than earlier sharks, and they have adaptations for gouging masses of flesh from their prey. Their body shape (Figure 9.5b) is more bullet-like than in their ancestors, and the pectoral fins are wider and more flexible. Neoselachians range in size from common dogfishes (0.3–1m long) to basking and whale sharks (17 m long), but the monster ones are not predators: they feed on krill which they filter from the water. The skates and rays, unusual neoselachians, are specialized for life on the sea-floor, having flattened bodies and broad pectoral fins for swimming (Figure 9.5c).

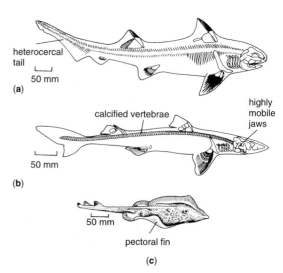

Fig 9.5 Sharks and rays, ancient and modern: (a) the Jurassic shark *Hybodus*; (b) the modern shark *Squalus*; (c) the modern ray *Raja*.

The radiations of bony fishes

The pulses of chondrichthyan radiation may relate to pulses of radiation of the much more numerous actinopterygian bony fishes. The first radiation (Devonian–Permian) consisted of the palaeonisciforms (Figure 9.6a), a paraphyletic group of bony fishes with large bony scales and heavy skull bones.

The second radiation of bony fishes, an assemblage termed the 'holosteans', occurred in the late Triassic and Jurassic. *Semionotus* (Figure 9.6b), a small form that has been found in vast shoals, had lighter scales than the palaeonisciforms, and a jaw apparatus that could be partly protruded, hence providing a wider gape.

The third, and largest, radiation of actinopterygian fishes, occurred in the late Jurassic and Cretaceous (Figure 9.6c), with the diversification of the teleosts. Teleosts are the most diverse and abundant fishes today, including 20 000 living species, such as eel, herring, salmon, carp, cod, anglerfish, flying fish, flatfish, seahorse and tuna. The huge success of this radiation may be the result of the remarkable teleost jaw apparatus. Palaeonisciforms opened their jaws like a simple trapdoor, holosteans could enlarge their gape a little, but teleosts can project the whole jaw apparatus like an extendable tube (Figure 9.6d). This came about because of a great loosening of the elements of the skull: as the lower jaw drops, the tooth-bearing bones of the upper jaw (the maxilla and premaxilla) move up and forwards. Rapid projection of a tube-like mouth allows many teleosts to suck in their prey,

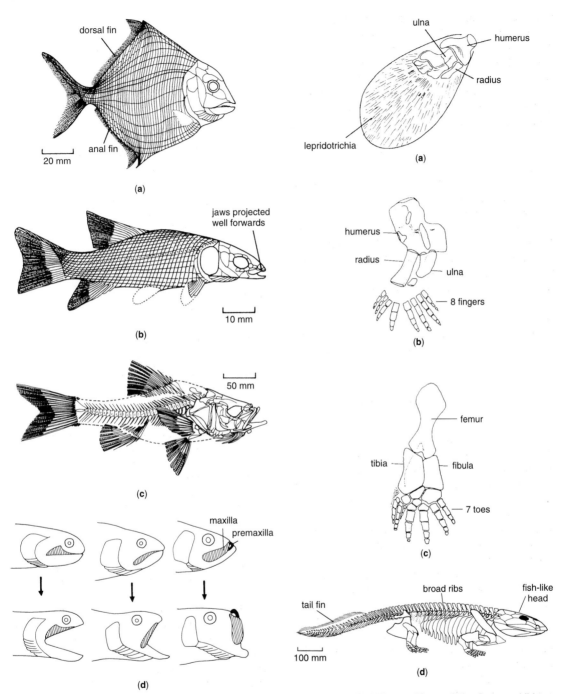

Fig 9.6 Evolution of the ray-finned bony fishes: (a) the Carboniferous 'palaeonisciform' *Cheirodus*, a deep-bodied form; (b) the Triassic 'holostean' *Semionotus*; (c) the Cretaceous teleost *Mcconichthys*; (d) evolution of actinopterygian jaws from the simple hinge of a 'palaeonisciform' (left) to the more complex jaws of a 'holostean' (middle), and the fully pouting jaws of a teleost (right).

Fig 9.7 Matching fins and legs of the first amphibians. The pectoral fin of the Devonian rhipidistian fish *Eusthenopteron* (a) shows bones that are probable homologues of tetrapod arm bones, such as in the Devonian amphibian *Acanthostega* (b). *Acanthostega* had eight fingers, and *Ichthyostega* had seven toes on its hindlimb (c). (d) The early amphibian *Ichthyostega*.

Box 9.4 The first tetrapods had seven or eight toes

The first good evidence of tetrapods comes from the late Devonian. *Ichthyostega* (Figure 9.7d), a 1-m-long animal with four limbs, shows land-living features in the strengthened rib cage and the clear separation between the head and shoulder girdle (these are joined in the rhipidistians), but retains a fish-like skull and a fin on its tail. The fish-like skull is seen in the amphibian *Acanthostega* (Figure 9.8), also from the late Devonian of Greenland.

New studies of *Ichthyostega* and *Acanthostega* (Coates and Clack, 1990), and of *Tulerpeton* from Russia, show that the first tetrapods had more than five fingers and toes; indeed as many as seven or eight (Figure 9.7b, c). This caused a major rethink of the classic story of the evolution of vertebrate limbs: five digits must have become standard only after the origin of tetrapods. The implications are wider, since the new evidence suggests that particular features of an organism may not all be preprogrammed in the genetic code of the developing embryo. Perhaps aspects of the developmental environment, rather than genetic programming, determine some details of adult structure, such as how many digits develop.

Fig 9.8 Skull of the late Devonian amphibian *Acanthostega*, showing the deeply sculpted bones and small teeth.

while others use the system to vacuum up food particles from the sea-floor, or to snip precisely at flesh or coral.

The three-phase radiation of bony fishes, palaeonisciforms, holosteans and teleosts, is paralleled by the three-phase radiation of sharks. It is impossible to say which set of evolutionary radiations came first: the bony fishes had to swim faster to escape their sharky predators, and the sharks had to swim faster to catch their bony fish prey.

Tetrapods

The origin of tetrapods

Tetrapods, the four-legged land vertebrates, arose from fishes during the Devonian. There has been some debate about the closest fish relatives of the tetrapods, although most attention has focused on the lobefins, because of their complex bony and muscular pectoral (front) and pelvic (back) paired fins (Figure 9.3d, e). It is possible to draw comparisons between the bones of the pectoral fin of a rhipidistian (Figure 9.7a) and those of the forelimb of an early amphibian (Figure 9.7b). The classic view has been that the fundamental tetrapod limb bore five digits (fingers or toes), but recent work shows that this is not the case (see Box 9.4).

In moving from a life in water to life on land, the first tetrapods faced major problems. The key problem was not necessarily the shift from breathing underwater, since early sarcopterygians almost certainly had *both* lungs and gills, and could already breathe air when necessary. The main problem was support: in water, an animal 'weighs' virtually nothing, but on land the body has to be held up from the ground, and the internal organs have to be supported in some way within a strong rib cage to prevent them from collapsing. In addition, reproductive, osmotic (water balance) and sensory systems had to adapt, but here the changes did not happen all at once. Amphibians are half-way to land life, but they retain many water-living adaptations.

The amphibians

Amphibians today are a minor group, consisting of 4000 species of mainly small animals that live in or close to water. Anurans (frogs and toads), known since the Triassic, have specialized in jumping: the hindlimbs are long and the hip bones reinforced to withstand the impact of landing. The head is broad, the jaws are lined with small teeth, and most feed by flicking a long sticky tongue out and trapping insects. The urodeles (salamanders and newts) date from the Jurassic, and consist of modest-sized long-bodied swimming predators. The third living amphibian group, the caecilians, are small limbless animals that look rather like earthworms, and live largely in soil and leaf litter in tropical lands. The oldest fossil form, with reduced limbs, is Jurassic in age.

All living amphibians appear to be closely related, forming a clade, the Lissamphibia (see Box 9.5), that is characterized by the structure of their tiny teeth. These kinds of teeth, and the broad semicircular skull shape, are seen in the 'temnospondyls', a paraphyletic group that was important in Carboniferous communities, and continued with reasonable success through the Permian and Triassic, finally dying out in the early Cretaceous. Temnospondyls had a low round-snouted skull (Figure 9.9a, b), and most of them appear to have operated like

---

Box 9.5 Classification of the amphibians

The amphibians are a paraphyletic group, since they exclude their descendants, the reptiles. Modern amphibians are clearly distinguishable from the diverse fossil groups.

Class Amphibia

Subclass Batrachomorpha

Order Nectridea
  Small slender aquatic forms. Early Carboniferous–late Permian.

Order Microsauria
  Terrestrial and aquatic long-bodied forms with deep skulls. Early Carboniferous–early Permian.

Order Temnospondyli
  Broad-snouted low-skulled amphibians, showing a range of sizes. Early Carboniferous–early Cretaceous.

Infraclass Lissamphibia
  Frogs, salamanders (newts), and caecilians (gymnophionans). Early Triassic–Recent.

Subclass Reptiliomorpha

Order Anthracosauria
  Narrow-skulled fish-eating amphibians. Early Carboniferous–late Permian.

Order Seymouriamorpha
  High-skulled terrestrial amphibians. Late Carboniferous–late Permian.

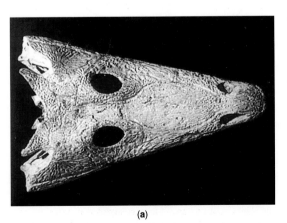

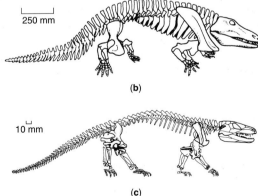

Fig 9.9 Fossil amphibians: (a) skull of the early Triassic temnospondyl *Benthosuchus*; (b) skeleton of the early Permian temnospondyl *Eryops*; (c) skeleton of the early Permian reptiliomorph *Seymouria*.

sluggish crocodiles, living in or near fresh waters, and feeding on fishes. Some temnospondyls became fully terrestrial, and others evolved elongate gavial-like snouts for catching rapidly swimming fishes. Some Carboniferous temnospondyls had tadpole young, just as modern amphibians do. This proves that they had a similar developmental pattern to modern amphibians, with an aquatic larval stage, the tadpole, which metamorphoses into the adult land-living form. Relatives of the temnospondyls included small forms, the aquatic nectrideans and the aquatic and terrestrial microsaurs.

The second amphibian lineage, the reptiliomorphs (see Box 9.5), included important groups in the Carboniferous and Permian, as well as the ancestors of reptiles, birds and mammals. Anthracosaurs had a longer narrower skull than the temnospondyls, but may have had similar lifestyles, hunting prey on land and in fresh waters. Some Permian reptiliomorphs, such as *Seymouria* (Figure 9.9c), were seemingly adapted to a

fully terrestrial life. *Seymouria* had long limbs and a relatively small skull, and probably hunted microsaurs and other small tetrapods.

Reign of the reptiles

Origin of the reptiles

The oldest-known reptile, *Hylonomus*, from the mid Carboniferous of Canada (Figure 9.10a, b) has been superbly well preserved inside ancient tree stumps, into which it crawled in pursuit of insects and worms, and then became trapped. *Hylonomus* looks little different from some amphibians of the time, such as the microsaurs, but it shows several clearly reptilian characters: a high skull, evidence for additional jaw muscles, and an astragalus bone in the ankle. One key reptilian

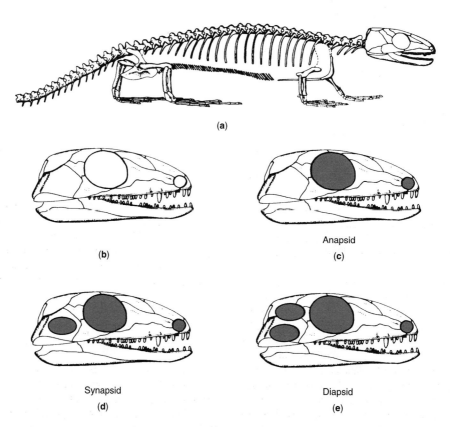

(a)

(b)

Anapsid

(c)

Synapsid

(d)

Diapsid

(e)

Fig 9.10   The earliest reptile, and early reptile evolution. (a), (b) The mid Carboniferous reptile *Hylonomus*, skeleton and skull. (c)–(e) The three major skull patterns seen in amniotes: anapsid, diapsid, and synapsid.

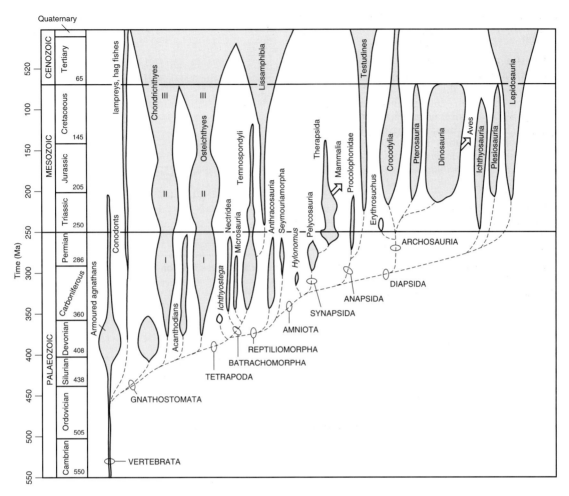

Fig 9.11 Phylogeny of the major groups of fishes and tetrapods.

character, and indeed a character of all Amniota, the clade consisting of reptiles, birds and mammals, is not known in *Hylonomus*. This is the cleidoic egg, the key to the success of these essentially fully terrestrial groups (see Box 9.6).

Reptiles radiated during the late Carboniferous, and the three main amniote lines became established. They are distinguished by the pattern of openings in the side of the skull (Figure 9.10c–e). The primitive state is termed the anapsid ('no hole') skull pattern, since there are no temporal openings. The two other skull patterns seen in amniotes are the synapsid ('same hole'), where there is a lower temporal opening, and the diapsid ('two hole') pattern, where there are two temporal openings. These temporal openings correspond to low-stress areas of the skull, and the edges serve as attachment sites for jaw muscles.

These three skull patterns diagnose the key clades

among amniotes (Figure 9.11). The Anapsida include various early forms such as *Hylonomus*, as well as some Permian and Triassic reptiles, and the turtles. The Synapsida include the mammal-like reptiles and the mammals, and the Diapsida include a number of early groups, as well as the lizards and snakes, and the crocodiles, pterosaurs, dinosaurs and birds.

The Anapsida

The oldest anapsids (see Box 9.7), such as *Hylonomus*, were small insect-eaters. During the Permian and Triassic, some unusual anapsids came on the scene. The most diverse of these were the procolophonids (Figure 9.13a), small animals with triangular skulls, and broad teeth adapted to a diet of tough plants and insects.

Box 9.6 The cleidoic egg: final break from the water

Amniotes (reptiles, birds and mammals), unlike amphibians, have broken with aquatic reproduction by enclosing their eggs within a tough semi-permeable shell, hence the term cleidoic ('closed'). The shell is usually hard and made from calcite, but some lizards and snakes have leathery egg shells. The shell retains water, preventing evaporation, but allows the passage of gases – oxygen in, and carbon dioxide out (Figure 9.12). The developing embryo is protected from the outside world, and there is no need to lay the eggs in water, nor is there a larval stage in development. Inside the egg shell is a set of membranes that enclose the embryo (the amnion), that collect waste (the allantois), and that line the egg shell (the chorion). Food is in the form of yolk.

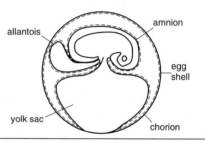

Fig 9.12 The cleidoic egg of amniotes in cross-section, showing the egg-shell and the extra-embryonic membranes.

The turtles appeared first in the late Triassic, being represented by *Proganochelys* (Figure 9.13b). Modern turtles have no teeth, but *Proganochelys* still had some on its palate. The skull is solid, and the body is covered above and below by a bony shell. Turtles live on land, in ponds (Figure 9.13c), and in the sea. Some marine turtles of the Cretaceous reached 3 m in length.

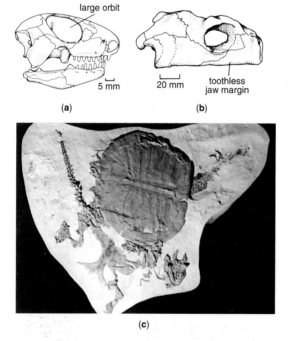

(a)

(b)

(c)

Fig 9.13 Fossil and recent anapsid reptiles. (a) Skull of the Triassic procolophonid *Procolophon*. (b) Skull of the Triassic turtle *Proganochelys*. (c) Example of the snapping turtle, from Miocene pond sediments filling an impact crater at Steinheim, Germany.

Box 9.7 Classification of the reptiles

Reptiles are a paraphyletic group of Amniota that excludes the reptile descendants, birds and mammals. The basic classification is founded on skull pattern (Figure 9.10).

Class Reptilia

Subclass Anapsida
No temporal openings. Various basal groups, such as procolophonids (Permian–Triassic).

Order Testudines (Chelonia)
   The turtles; bony carapace, retractable neck and limbs. Late Triassic–Recent.

Subclass Synapsida
One (lower) temporal opening.

Order Pelycosauria
   The sail-backed mammal-like reptiles and relatives. Late Carboniferous–early Permian.

Order Therapsida
   Mammal-like reptiles with differentiated teeth, and the mammals. Late Permian–Recent.

Subclass Diapsida
Two temporal openings.

Infraclass Archosauria
   Thecodontians, crocodilians, pterosaurs, dinosaurs and birds, characterized by an antorbital fenestra. Late Triassic–Recent.

Infraclass Lepidosauria
   Lizards, snakes and their ancestors. Late Triassic–Recent.

The rule of the synapsids

The first mammal-like reptiles, known from the late Carboniferous and early Permian, are grouped loosely as 'pelycosaurs'. Most of these were small to medium-sized insectivores and carnivores with powerful skulls and sharp flesh-piercing teeth. Some later pelycosaurs, such as *Dimetrodon* (Figure 9.14a), had vast sails supported on vertical spines growing up from the vertebrae, used in controlling body temperature. The pelycosaurs also include a number of groups that adapted to plant-eating, the first herbivorous land vertebrates.

Mammal-like reptiles radiated dramatically in the late Permian, as a new clade, the Therapsida. The most astonishing carnivores were the gorgonopsians (Figure 9.14b), with their large wolf-like bodies and massive sabre teeth which they probably used to attack the larger thick-skinned herbivores. The dicynodonts had bodies shaped like overstuffed sausages, and no teeth at all, or only two tusks (Figure 9.14c). They were successful herbivores, and some of the first animals to have a complex chewing cycle which allowed them to tackle a wide variety of plant foods. Late Permian therapsids are common in the continental sediments of the Karoo Basin in South Africa, and the Urals in Russia. At the end of the Permian, at the time of major mass extinction in the sea (see p. 30), most of these animals died out. The gorgonopsians disappeared, and the dicynodonts were nearly wiped out – mass slaughter on an unbelievable scale.

The cynodonts were an important Triassic synapsid group. The early Triassic form *Thrinaxodon* (Figure 9.15a) looked dog-like. In the snout area of the skull, there are numerous small canals which indicate small nerves serving the roots of sensory whiskers. If *Thrinaxodon* had whiskers, it clearly also had hair on other parts of its body, and this means insulation and temperature control. Cynodonts evolved along several

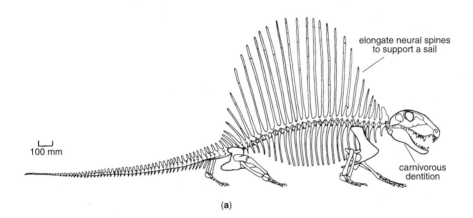

100 mm

elongate neural spines
to support a sail

carnivorous
dentition

(a)

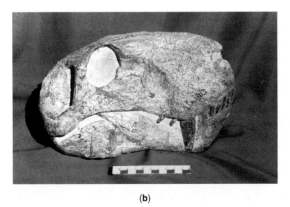

(b)

(c)

Fig 9.14 Synapsids of the Permian. (a) the carnivorous pelycosaur *Dimetrodon*; (b) the carnivorous gorgonopsian *Lycaenops*; (c) the herbivorous dicynodont *Dicynodon*.

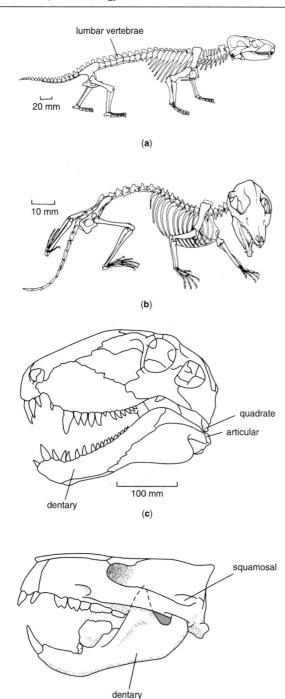

lumbar vertebrae

20 mm

(a)

10 mm

(b)

quadrate

articular

100 mm

dentary

(c)

squamosal

dentary

(d)

Fig 9.15 Transition to the mammals. (a) The early Triassic cynodont *Thrinaxodon*. (b) The early Jurassic mammal *Megazostrodon*. (c, d) Skulls of an early mammal-like reptile (c) and a mammal (d) to show the reduction in elements in the lower jaw, and the switch of the jaw joint.

lines during the Triassic, and gave rise to mammals, such as *Megazostrodon* (Figure 9.15b), in the late Triassic and early Jurassic.

The transition from mammal-like reptile to mammal is marked by an extraordinary shift of the jaw joint into the middle ear. Reptiles typically have six bones in the lower jaw, and the articular bone articulates with the quadrate in the skull (Figure 9.15c). In mammals, on the other hand, there is a single bone in the lower jaw, the dentary, which articulates with the squamosal. The reptilian articular–quadrate jaw joint became reduced in Triassic cynodonts, and moved into the middle ear passage. That is why we have three tiny ear bones, the hammer, anvil and stirrup, which transmit sound from the ear drum to the brain, while reptiles have only one, the stirrup or stapes.

The diapsids take over

The third major amniote group, the diapsids (see p. 206), were initially small to medium-sized carnivores that never matched the abundance of the synapsids. Things began to change during the Triassic, perhaps as a result of the end-Permian extinction event, which had such a devastating effect on therapsid communities. Small and large meat-eaters such as *Erythrosuchus* (Figure 9.16a) appeared – one of the first of the archosaurs, a group that was later to include the dinosaurs, pterosaurs, crocodiles and birds. Archosaurs have an additional skull opening between the orbit and the naris, termed the antorbital fenestra, whose function is unclear.

During the Triassic, some archosaurs became large carnivores, others became specialized fish-eaters, others adopted a specialized grubbing herbivorous lifestyle, yet others were small two-limbed fast-moving insectivores (the crocodilians and dinosaurs), and some became proficient flyers (the pterosaurs). It took another mass extinction event, near the beginning of the late Triassic (about 225 Ma) to set the new age of diapsids fully in motion. Most of the synapsids died out then, as did various basal archosaur groups. Many new kinds of land tetrapods then radiated: the dinosaurs, pterosaurs, crocodilian, and lizard ancestors, as well as the turtles, modern amphibians and true mammals.

The pterosaurs were proficient flapping flyers (Figure 9.16b), with a lightweight body, narrow hatchet-shaped skull, and a long narrow wing supported on a spectacularly elongated fourth finger of the hand. The bones of the arm and finger support a tough flexible membrane that could fold away when the animal was at rest, and stretch out for flight. Pterosaurs were covered with hair, and were almost certainly endothermic. Some later

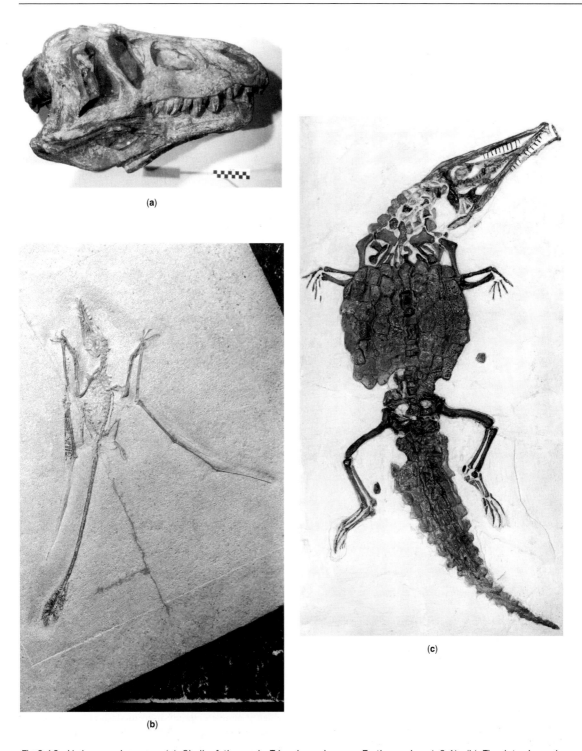

Fig 9.16 Various archosaurs. (a) Skull of the early Triassic archosaur *Erythrosuchus* (×0.1). (b) The late Jurassic pterosaur *Rhamphorhynchus*, showing the elongated wing finger on each side, and the long tail with its terminal 'sail' made from skin (×0.3). (c) The late Jurassic crocodilian *Crocodilemus*, showing the skeleton and armour covering (×0.2).

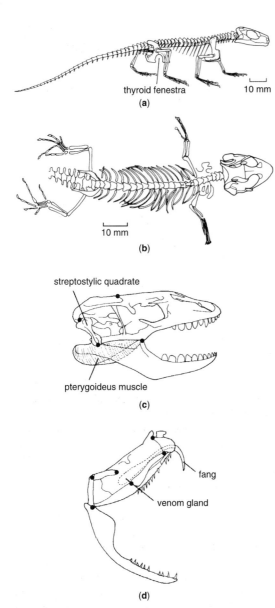

thyroid fenestra 10 mm
(a)

10 mm
(b)

streptostylic quadrate

pterygoideus muscle
(c)

fang

venom gland
(d)

Fig 9.17 Lepidosaurs. (a) The late Triassic sphenodontid *Planocephalosaurus*. (b) The late Jurassic lizard *Ardeosaurus*. (c, d) Skulls of a modern lizard and snake, showing the points of mobility that permit wide jaw open-

bony plates. Crocodilians were more diverse and abundant during the Jurassic and Cretaceous than they are now (Figure 9.16c). Some even became highly aquatic in adaptations, to the extent of having paddles instead of hands and feet, and a deep tail fin to speed their swimming. The modern crocodilian types – crocodiles, alligators and gavials – all arose in the late Cretaceous.

The second major diapsid clade, the lepidosaurs, represented today by lizards and snakes, radiated in the late Triassic. The key forms then were sphenodontids, which were snub-nosed lizard-sized animals (Figure 9.17a) that fed on plants and insects. The group dwindled after the Jurassic, except for a single living representative, *Sphenodon*, the tuatara of New Zealand, a famous 'living fossil'. The first true lizards are known from the mid and late Jurassic (Figure 9.17b), and they show characteristic mobility of the skull: the bar beneath the lower temporal opening is broken, the quadrate is mobile, and the snout portion of the skull can tilt up and down (Figure 9.17c). This process of loosening of the skull was taken even further in the snakes, a group known first in the Early Cretaceous. Snakes have such mobile skulls that they can open their jaws to swallow prey animals that are several times the diameter of the head (Figure 9.17d).

The age of dinosaurs

The dinosaurs were the most important of the new groups, both in terms of their abundance and diversity, and in terms of the vast size reached by some of them. The first dinosaurs were modest-sized bipedal carnivores. After the late Triassic extinction, a new group of herbivorous dinosaurs, the sauropodomorphs, radiated dramatically, some like *Plateosaurus* (Figure 9.18a) reaching lengths of 5–7 m during the late Triassic. Later sauropodomorphs were mainly large and very large animals, some of them, such as *Brachiosaurus* (Figure 9.18b), reaching lengths of 23 m or more, and heights of 12m. These giant dinosaurs pose fascinating biological problems (see Box 9.8).

The theropods include all the carnivorous dinosaurs, and in the Jurassic and Cretaceous the group diversified to include many specialized small and large forms. *Deinonychus* (Figure 9.19a) was human-sized, but immensely agile and intelligent (it had a bird-sized brain). Its key feature was a huge claw on its hind foot which it almost certainly used to slash at prey animals. *Tyrannosaurus* (Figure 9.19b) is famous as the biggest land predator of all time, reaching a body length of 14 m, and having a gape of nearly 1 m. The theropods and sauropodomorphs share the primitive reptilian hip pat-

pterosaurs were much larger than any known bird, such as *Pteranodon* with a wing span of 5–8 m, and *Quetzalcoatlus* with a wing span of 11–15 m. Most pterosaurs fed on fishes caught in coastal seas, but others were insectivorous.

Early crocodilians were largely terrestrial in habits, but walked on all fours, and had an extensive armour of

Fig 9.18 Sauropodomorph dinosaurs: (a) the late Triassic *Plateosaurus*; (b) the late Jurassic sauropodomorph *Brachiosaurus*.

tern, in which the two lower elements point in opposite directions, the pubis forwards and the ischium backwards (Figure 9.19b). They also share derived characters of the skull and limbs which shows they form a clade, the Saurischia.

All other dinosaurs share a unique hip pattern in which

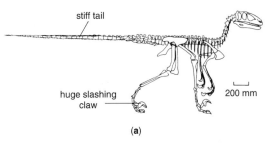

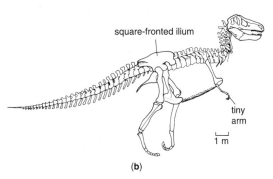

Fig 9.19 Cretaceous theropod dinosaurs: (a) *Deinonychus* and (b) *Tyrannosaurus*.

the pubis has swung back and runs parallel to the ischium (Figure 9.20a), and these are termed the Ornithischia, all of which were herbivores. Two groups of armoured ornithischians are the stegosaurs and the ankylosaurs. *Stegosaurus* (Figure 9.20a) has a row of bony plates along the middle of its back, which may have had a temperature-control function. *Euoplocephalus* (Figure 9.20b) is a massive tank-like animal with a solid armour of small plates of bone set in the skin over its back, tail, neck and skull; it even had a bony eyelid. The tail club was a useful defensive weapon.

Most ornithischians were ornithopods, bipedal forms, initially small, but later often large. In the late Cretaceous, the hadrosaurs were successful fast-moving plant-eaters. Many of them have bizarre crests on top of their heads which may have been used for species-specific signalling, and their duck-billed jaws are lined by multiple rows of grinding teeth (Figure 9.21). Close relatives of the ornithopods were the ceratopsians ('horn-faces'), like *Centrosaurus* (Figure 9.20c), which had a single long nose-horn and a great bony frill over the neck.

There has been a continuing debate about whether the dinosaurs were warm-blooded (endothermic) or not. Evidence for warm-bloodedness is strongest for the small active predators such as *Deinonychus* which might have required the added stamina and speed. However,

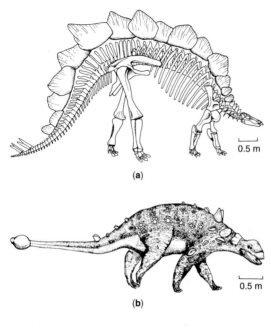

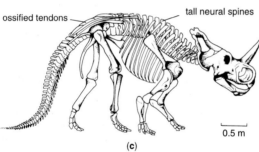

Fig 9.20 Armoured ornithischian dinosaurs from the Jurassic (a) and Cretaceous (b, c): (a) *Stegosaurus*, (b) *Euoplocephalus* and (c) *Centrosaurus*.

endothermy is costly in terms of the extra food required as fuel, and it is not clear whether the larger dinosaurs could have eaten fast enough. Indeed, larger dinosaurs would have maintained a fairly constant core body temperature simply because of their size, whether they were endothermic or not.

Dinosaur reproductive habits have also come under scrutiny recently. Discoveries of eggs and nests in North America and Mongolia have shown that many dinosaurs practised parental care. They laid their eggs in earth nests scooped in the soil, and returned to feed the young when they hatched out. Some of the most spectacular finds are unhatched eggs with the tiny bones of the dinosaur embryos still inside.

Box 9.8 Palaeobiology of the largest animals ever

When the monster sauropods of the late Jurassic were first discovered in the 19th century, many palaeontologists thought that they were too big to have lived fully on land. It was assumed that the sauropods lived in lakes, supporting their bulk in the water, and feeding on waterside plants. However, new evidence shows that life on land was quite possible, and indeed the long neck of *Brachiosaurus* (Figure 9.18b) made it a super-giraffe, a dinosaur that could feed on leaves from very tall trees, well out of the range of any other animal.

More recently, some palaeontologists suggested that sauropods might have been able to gallop at speed, and hoist themselves up on their hind legs to reach even higher. These ideas are not likely since both activities would be likely to break their legs. Experiments with modern bones show how much strain they can withstand before they break. It is a simple matter to work out the breaking strain of dinosaur leg bones, and a galloping *Brachiosaurus* would collapse in a pile of broken leg bones.

Fig 9.21 Skull of the late Cretaceous hadrosaur *Edmontosaurus* (×0.1).

Dragons of the deeps

During the Mesozoic, several reptile groups ruled the waves. The ichthyosaurs (Figure 9.22a) were fish-shaped animals, entirely adapted to life in the sea, but almost certainly evolved from land-living diapsids. Ichthyosaurs had a long thin snout lined with sharp teeth, and they fed on ammonites, belemnites and fishes. Exquisite preservation of many specimens shows the dorsal fin and the paddle outlines. Ichthyosaurs swam by

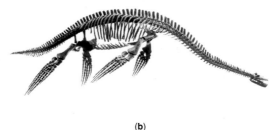

(a)

(b)

Fig 9.22 Jurassic marine reptiles: (a) *Stenopterygius* (×0.05) and (b) *Cryptoclidus* (×0.05).

beating the body and tail from side to side, and they used the front paddles for steering.

The second major marine reptile group were the plesiosaurs. Most plesiosaurs had long necks and small heads (Figure 9.22b), but the pliosaurs were larger, and had short necks and large heads. Plesiosaurs fed mainly on fishes, using the long neck like a snake to dart after fast-moving prey, and they swam by beating their paddles in a kind of 'flying' motion. The extraordinary diversity of tetrapod predators in the sea came to an end 65 million years ago during the great KT mass extinction (see p. 301) which saw the end of the dinosaurs and pterosaurs too.

Bird evolution

One of the most famous fossils is *Archaeopteryx*, the oldest known bird (Figure 9.23). The first specimen was found in Upper Jurassic sediments in southern Germany in 1861, and was hailed as the ideal 'missing link' or proof of evolution in action. Here was an animal with a beak, wings and feathers, so it was clearly a bird, but it still had a reptilian bony tail, claws on the hand, and teeth. Since 1861, six more skeletons have come to light, the last two in 1987 and 1992.

Archaeopteryx was about the size of a magpie, and it fed on insects. The claws on its feet and hands suggest that *Archaeopteryx* could climb trees, and the wings are clearly those of an active flying animal. This bird could fly as well as most modern birds, and flying allowed it to catch prey that were not available to land-living relatives. The skeleton of *Archaeopteryx* is very like that of *Deinonychus* (Figure 9.19a), especially in the details of the arm and hind limb, showing that birds are small flying theropod dinosaurs.

Birds remained rare until the late Cretaceous when new marine forms radiated. These birds still had teeth, but the bony tail was reduced to a short knob as in modern birds, and they had other modern features. Modern groups of birds appeared in the latest Cretaceous and early Tertiary, including flightless ratites and ancestors of water birds, penguins, and birds of prey. The perching birds, consisting today of 5000 species of song birds, radiated only in the Miocene.

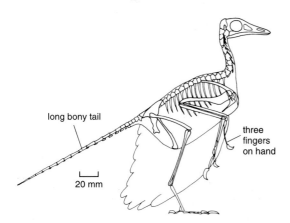

long bony tail

three fingers on hand

20 mm

Fig 9.23 The oldest bird, *Archaeopteryx* from the late Jurassic.

Rise of the mammals

Primitive forms

The first mammals, small insect-eaters in the late Triassic and early Jurassic (Figure 9.15b), probably hunted at night. Mammals remained small through most of the Mesozoic, and they failed to make an impact as long as the dinosaurs existed. Several lines of insectivorous, car-

---

Box 9.9 Classification of the mammals

Modern mammals fall into three groups: the monotremes, marsupials and placentals, each characterized by their breeding modes. Various primitive groups are omitted.

Class Mammalia

  Subclass Monotremata
    Females lay eggs, and newborn young grow in pouch. Early Cretaceous–Recent.

  Subclass Marsupialia
    Young are born live, but continue development in pouch. Late Cretaceous–Recent.

  Subclass Eutheria (placental mammals)
    Young are born live at an advanced stage, having been nourished by a placenta, while in the womb. Main orders only are listed. Late Cretaceous–Recent.

  Order Edentata
    (armadillos, tree sloths, anteaters). Palaeocene–Recent.

  Order Carnivora
    (dogs, bears, cats, hyaenas, seals). Palaeocene–Recent.

  Order Insectivora
    (hedgehogs, moles, shrews). Late Cretaceous–Recent.

  Order Primates
    (monkeys, apes, humans). Late Cretaceous–Recent.

  Order Chiroptera
    (bats). Palaeocene–Recent.

  Order Rodentia
    (mice, rats, squirrels, porcupines, beavers). Palaeocene–Recent.

  Order Lagomorpha
    (rabbits and hares) Palaeocene–Recent.

  Order Artiodactyla
    (pigs, hippos, camels, cattle, deer, giraffes, antelopes). Eocene–Recent.

  Order Perissodactyla
    (horses, rhinos, tapirs). Eocene–Recent.

  Order Cetacea
    (whales and dolphins). Eocene–Recent.

  Order Proboscidea
    (elephants). Eocene–Recent.

---

nivorous and herbivorous forms appeared. Some of them adapted to climbing trees, but only one or two of these clades survived the KT extinction. Three of the clades that did survive into the Tertiary were the monotremes, marsupials and placentals – the modern groups (see Box 9.9).

Monotremes today are restricted to Australasia, being represented by the platypus and the echidnas. These mammals are unique in still laying eggs, as the cynodont ancestors of mammals presumably did. The young hatch out as tiny helpless creatures, and feed on mother's milk until they are large enough to live independently.

Marsupials

Marsupial young are born tiny and helpless, and have to feed on maternal milk in a pouch for many months, but egg-laying has been abandoned. The oldest marsupial fossils come from the late Cretaceous of the Americas. The group radiated successfully in South America during the Tertiary, and included several lines of insectivores, carnivores and herbivores, many of which were remarkably like unrelated placental mammals elsewhere. Some

forms were dog-like, and *Thylacosmilus* (Figure 9.24a) independently evolved all the characters of the placental sabre-toothed cats of Europe and North America.

In Australia, the marsupials diversified even more, after reaching that continent by the Oligocene, and they radiated to parallel placental mammals in functions and body forms, except of course for the unique kangaroos. In the Pleistocene, there were abundant and diverse faunas of large marsupials, including giant kangaroos and the hippopotamus-sized herbivorous *Diprotodon* (Figure 9.24b).

Basal placental mammals

Placental mammals produce young that are retained in the mother's womb much longer than is the case in marsupials, and they are nourished by blood passed through the placenta. The oldest fossils of placental mammals are late Cretaceous in age, but the clade only began to radiate substantially after the extinction of the dinosaurs. During the Palaeocene there were 20 lineages of placental mammals, including the ancestors of all modern groups, as well as some extinct orders.

Most cladograms (Figure 9.25) separate the South

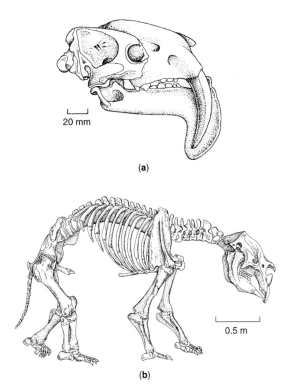

(a)

(b)

Fig 9.24 Extinct marsupials: (a) the sabre-tooth *Thylacosmilus* from South America; (b) the giant herbivore *Diprotodon* from Australia.

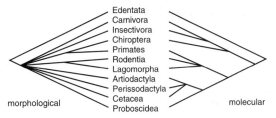

Fig 9.25 Cladogram of the placental mammals, based on morphology.

American edentates from the other placentals. The edentates today include armadillos, tree sloths and ant-eaters, all of which had a spectacular fossil record, particularly in the Pliocene and Pleistocene. There were giant armadillos such as *Glyptodon* (Figure 9.26a), and giant ground sloths such as *Mylodon* (Figure 9.26b) which reached a length of 6 m and fed on coarse leaves from the tree tops. The ground sloths survived in South America until 11 000 years ago, and their subfossil remains include clumps of reddish hair and caves full of fresh dung that occasionally ignites spontaneously.

Varied Old World placentals

The Order Carnivora (cats, dogs, hyaenas, weasels and seals) are characterized by sharp cheek teeth (carnassials) which are used for tearing flesh. The cats have a long history during which dagger-toothed and sabre-toothed forms evolved many times. The sabre-tooths such as *Smilodon* (Figure 9.27a) preyed on large thick-skinned herbivores by cutting chunks of flesh from their

bodies. The sabre-toothed adaptations were evolved independently by some South American marsupials (cf. Figure 9.24a). Early dogs such as *Hesperocyon* (Figure 9.27b) were light fast-moving animals, close to the ancestry of modern dogs and bears. Some carnivores related to raccoons and weasels entered the sea during the Oligocene, and gave rise to the seals, sealions and walruses. Early forms such as *Allodesmus* (Figure 9.27c) had broad paddle-like limbs and fed on fish.

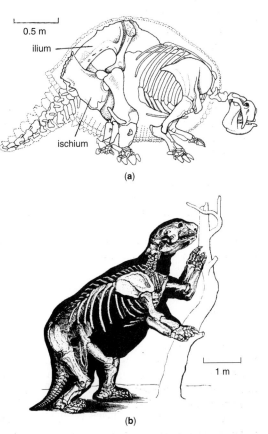

(a)

(b)

Fig 9.26 Pleistocene edentates from Argentina: (a) *Glyptodon* and (b) *Mylodon*.

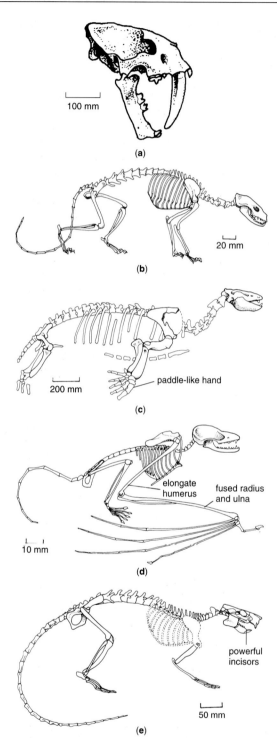

(a)

(b)

paddle-like hand

(c)

elongate humerus   fused radius and ulna

10 mm

(d)

powerful incisors

50 mm

(e)

Fig 9.27 Diverse mammals. (a) the Pleistocene sabre-toothed cat *Smilodon*; (b) the Eocene dog *Hesperocyon*; (c) the Miocene 'seal' *Allodesmus*; (d) the Eocene bat *Icaronycteris*; (e) the Eocene rodent *Paramys*.

The Order Insectivora is composed of dozens of species of hedgehogs, moles and shrews, all small animals with long snouts which feed on insects. The oldest Insectivora are late Cretaceous in age, and for the most part the fossil forms probably looked like the modern ones. One exception is the giant spiny hedgehog, *Deinogalerix*, which was half a metre long.

The next group in the cladogram (Figure 9.25) is a clade consisting of the Chiroptera and Primates. The Chiroptera, or bats, are diverse today, consisting of about 1000 species. The earliest bats, such as *Icaronycteris* from the Eocene (Figure 9.27d), show the typical wing structure in which the flight membrane is supported on four fingers of the hand which spread out. The feet turn backwards, and *Icaronycteris* could have hung upside down. It also has large eyes, and the ear region is modified for echolocation. *Icaronycteris*, like most modern bats, hunted insects at night, using its large eyes to pick up movements, and sending out high-pitched squeaks to detect its prey by the echoes they made. The primates include monkeys, apes and humans, and will be described in more detail below (see p. 219–221).

Another clade (Figure 9.25) is composed of the Rodentia and Lagomorpha. Rodents are the largest group, consisting of over 1700 species of mice, rats, squirrels, porcupines and beavers. They owe their success to their powerful gnawing teeth: the front incisors are deep-rooted and grow continuously, so that thcy can be used to grind wood, nuts and husks of fruit. The oldest rodents, such as the ischyromid *Paramys* (Figure 9.27e) already had the front grinders, and this ability to chew materials ignored by other animals triggered several phases of rapid radiation. Beavers, porcupines and cavies radiated in the Miocene. The cavies include a giant Miocene guinea pig as large as a pygmy rhinoceros. Rabbits and their relatives (Order Lagomorpha) have never been as diverse as the rodents. Fossil forms in the Oligocene have elongate hindlimbs used in jumping.

The ungulates

The remaining placental mammal orders are mainly medium-sized to large plant-eaters which have classically been grouped together as the ungulates, but it is not clear whether they form a single clade or not (Figure 9.25). There are several living and extinct ungulate groups, some of which were unique to South America.

The artiodactyls arose in the Eocene (Figure 9.28a), and the group includes pigs, hippos, camels, cattle, deer, giraffes and antelopes, all with an even number of toes (two or four). Pigs and hippos share ancestors in the

(a)

Oligocene, at the same time as vast herds of oreodonts fed on the spreading grasslands of North America. Oreodonts are related to the camels and the ruminants. The first camels were long-limbed and lightly built North American animals: it was only later that camels moved to Africa and the Middle East and evolved adaptations for living in conditions of drought.

Most artiodactyls today are ruminants, i.e. they pass their food into a fore-stomach, regurgitate it (chew the cud), and swallow it again. The multiple digestive process allows ruminants to extract all the nourishment from their plant food, usually grass, and to pass limited waste material (compare the homogenous excrement of cattle with the fibrous undigested droppings of horses, which do not ruminate). Ruminants became successful after the mid Miocene, when a great variety of deer, cattle and antelopes appeared. These animals usually have horns or antlers, seen in spectacular style in the Irish deer, *Megaloceros* (Figure 9.28b). The head gear is used

(b)

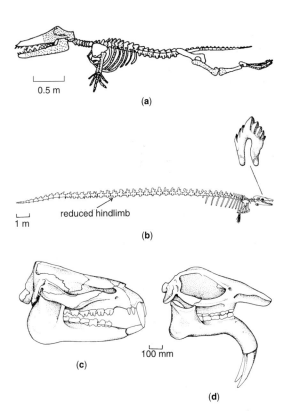

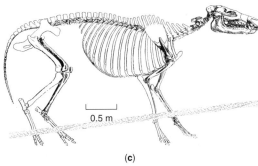

(c)

Fig 9.28 Diverse ungulates. (a) The small four-toed artio-dactyl *Messelobunodon*, showing the complete skeleton and a mass of chopped plant material in the stomach area, from the oil shale deposit of Messel, Germany (×0.15). (b) The Pleistocene giant Irish deer *Megaloceros*. (c) The Miocene horse *Neohipparion*.

Fig 9.29 Whales and elephants. (a) The early Eocene whale *Ambulocetus*. (b) The late Eocene whale *Basilosaurus*. (c, d) Skulls of the Eocene proboscidean *Moeritherium* (c) and the Miocene proboscidean *Deinotherium* (d).

in all cases for displays and fights between males seeking to establish territories and win mates.

The second major ungulate group, the Order Perissodactyla, consisting of horses, rhinos and tapirs, all have an odd number of toes (one, three or five). The evolution of the horses is a classic (see p. 295–296). The first horse, *Hyracotherium*, was a small woodland-living animal which had four fingers and three toes, and low teeth used for browsing on leaves. During the Oligocene and Miocene, horses (Figure 9.28c) became adapted to the new grasslands which were replacing the forests, and they became larger, lost toes, and evolved deep-rooted cheek teeth for grinding tough grass.

Tapirs and rhinoceroses are probably related. Eocene and Oligocene rhinos were modest-sized running animals, not much different from some of the early horses. Tapirs later became a rare group, restricted to Central and South America and south-east Asia. The rhinoceroses flourished for a while, producing monsters such as the Miocene *Indricotherium*, the largest land mammal of all time, 5.5 m tall at the shoulder, and weighing 15 tonnes. The whales, Order Cetacea, probably belong in this group. The oldest whale, Ambulocetus (Figure 9.29a), retains limbs, and proves the ancestry of the group from Palaeocene terrestrial meat-eaters. By the late Eocene, whales such as Basilosaurus (Figure 9.29b) had become very large, at lengths of 20 m or more. Basilosaurus had a long thin body, like a mythical sea-serpent, and a relatively small skull armed with sharp teeth. It was probably a fish-eater, like the toothed whales today. The baleen whales, the biggest of all modern whales, arose later, and they owe their success to their ability to filter vast quantities of small crustaceans, krill, from polar sea waters.

The African and Indian elephants of today (Order Proboscidea) are a sorry remnant of a once-diverse group. Early elephants such as Moeritherium (Figure 9.29c) were small hippo-like animals that probably fed on lush plants in the ponds and rivers of Africa. Later, many lines of proboscideans diverged, distinguished by an astonishing array of tusks, which are modified teeth. Some had tusks in the upper jaw (as in modern elephants), others had tusks in the lower jaw, and others had tusks in both as in Gomphotherium (Figure 9.29d). The Pleistocene mammoths were abundant in cold northern Ice Age climates, but died out as the ice retreated 10 000 years ago.

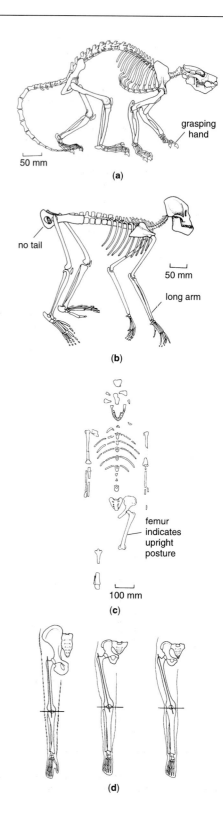

Fig 9.30 Primates. (a) The Palaeocene *Plesiadapis*. (b) The Miocene ape *Proconsul*. (c) The Pliocene hominid *Australopithecus afarensis*, known as 'Lucy'. (d) Comparison of the hindlimb of an ape (left), Lucy (middle), and a modern human (right).

The line to humans

Early primates

The Order Primates is one of the oldest of the modern placental mammal groups: it arose in the latest Cretaceous. For most of their history, the primates have been a rare and rather obscure group. All primates share a number of features that give them agility in the trees (mobile shoulder joint, grasping hands and feet, sensitive finger pads), a larger than average brain, good binocular vision, and enhanced parental care (one baby at a time, long time in the womb, long period of parental care, delayed sexual maturity, long lifespan).

Early relatives of the primates, such as *Plesiadapis*

(Figure 9.30a), were squirrel-like animals which may have climbed trees and fed on tough leaves. Various basal primates radiated in the Palaeocene, Eocene and Oligocene, and gave rise to the modern lemurs, lorises and tarsiers.

True monkeys arose in the Oligocene, and they diverged into two groups: the New World monkeys of South America, and the Old World monkeys of Africa, Asia and Europe. The New World monkeys have flat noses and prehensile tails which may be used as extra limbs in swinging through the trees. The Old World monkeys have narrower projecting noses and non-prehensile tails, or no tails at all.

The apes arose from the Old World monkeys before the end of the Oligocene, and the group radiated in Africa in the Miocene. Even early forms, such as *Proconsul* (Figure 9.30b), have no tail, and a relatively

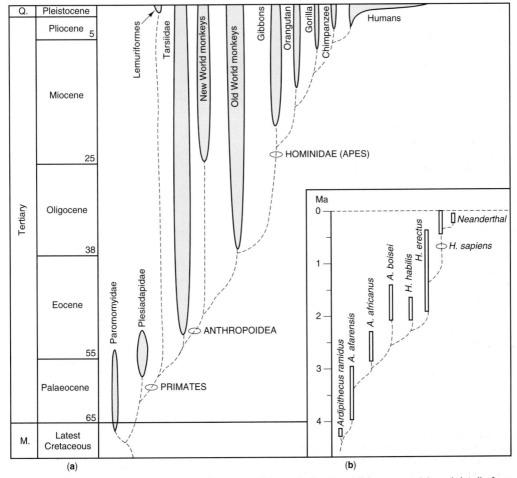

Fig 9.31 Phylogeny of the primates, showing some of the main fossil and living groups (a), and detail of one view of human evolution (b). Abbreviations: *A., Australopithecus; H., Homo;* M., Mesozoic; Q., Quaternary.

large braincase, indicating high intelligence. These apes ran about on the ground, and along low branches, on all fours, and fed on fruit. The apes spread out from Africa into the Middle East, Asia and southern Europe by the mid Miocene, and gave rise to some of the modern ape groups at that time. Fossil and molecular evidence on phylogeny (see p. 49) suggests that the gibbons of southeast Asia are the most primitive living apes, having branched off about 20 Ma, followed by the orangutan 15.5 Ma (Figure 9.31). The focus of ape (and human) evolution remained in Africa.

Gorillas, chimpanzees and humans appear to be very closely related, sharing many anatomical characters, and with more than 99% of their protein structure identical. Gorillas seem to have diverged first, about 9 Ma, and the ancestors of humans and chimps separated about 5–7 Ma.

Human evolution

Humans are proud of their large brains, but it seems that the other key human character of bipedalism evolved first. This may have arisen because of a major environmental change in Africa in the late Miocene. Much of Africa had been covered with lush forests in which the ancestral apes flourished, but climates then became arid and the East African Rift Valley began to open up, separating the lush forests in the west from the arid grasslands in the east. Tree-living apes (chimps and gorillas) retreated west, and the remaining apes (our ancestors) remained in the eastern grasslands. They had to stand upright to look for enemies, to permit them to run long distances in search of food, and to free their arms for carrying food. Humans evolved from bipedal ape-like ancestors which had no special high brain power.

The oldest human remains are teeth, jaws and an arm, called *Ardipithecus ramidus* from Ethiopia, found in 1993 in rocks dated at 4.4 Ma. These remains are insufficient to detect how this hominid walked. The oldest clear evidence of bipedalism consists of human tracks in volcanic ash from Tanzania, dated at 4.1–3.9 Ma. The oldest substantial skeletons, of *Australopithecus afarensis*, come from rocks dated at about 3.2 Ma; these also show clear evidence for advanced bipedalism, but still an ape-sized brain. The famous skeleton of a female *A. afarensis*

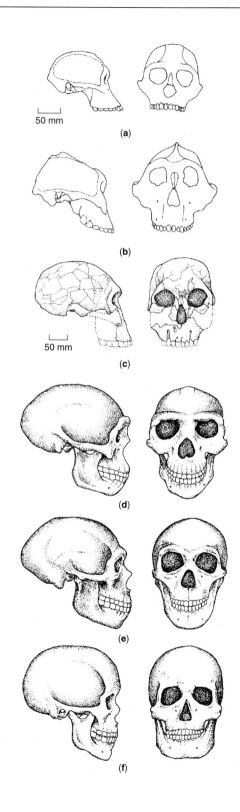

Fig 9.32 Skulls of fossil humans in front (top) and side (bottom) views: (a) *Australopithecus africanus*; (b) *Australopithecus boisei*; (c) *Homo habilis*; (d) *Homo erectus*; (e) *Homo sapiens*, Neanderthal race; (f) modern *Homo sapiens*.

from Ethiopia, called Lucy by its discoverer Don Johanson in the 1970s (Figure 9.30c), has a rather modern humanoid pelvis and hindlimb (Figure 9.30d). The pelvis is short and horizontal, rather than long and vertical as in apes, the thigh bone slopes in towards the knee, and the toes can no longer be used for grasping. Lucy's brain, however, is small: only 415 cm^3 for a height of 1–1.2 m, i.e. not much different from a chimpanzee.

The human genus *Australopithecus* continued to evolve in Africa from about 3–1.5 Ma, giving rise to further small species, and some large robust ones (Figure 9.32a, b). The larger australopithecines reached heights of 1.75 m, but their brain capacities did not exceed 520 cm^3, a rather ape-like measure. The leap forward to modern human brain sizes only came with the origin of a new human genus, *Homo*. The first species, *H. habilis* (Figure 9.32c), lived in Africa from 2.4 to 1.5 Ma, and had a brain capacity of 630–700 cm^3 in a body only 1.3 m tall. *H. habilis* may have used tools. It is a remarkable fact that, for over 1 Myr, three or four different human species lived side by side in Africa.

So far, the focus of human evolution had been entirely in Africa, but a new species, *Homo erectus*, which arose around 1.9 Ma in Africa, spread to China, Java and central Europe. *H. erectus* had a brain size of 850–1100 cm^3 (Figure 9.32d) in a body up to 1.6 m tall, and there is clear evidence that this early human species had semi-permanent settlements, a basic tribal structure, knew the use of fire for cooking, and made tools and weapons from stone and bone.

Modern peoples

Truly modern humans, *Homo sapiens*, may have arisen as much as 400 000 years ago, and certainly by 100 000 years ago, in Africa, having evolved from *H. erectus*. It seems that all modern humans arose from a single African ancestor, and that the *H. erectus* stocks in Asia and Europe died out. *H. sapiens* spread to the Middle East and Europe by 90 000 years ago. The European story is particularly well known, and it includes a phase, from 120 000 to 35000 years ago, when Neanderthal man occupied much of Europe from Russia to Spain, and from Turkey to southern England. Neanderthals had large brains (average, 1400 cm^3), heavy brow ridges (Figure 9.32e), and stocky powerful bodies. They were a race of *H. sapiens* adapted to living in the continuous icy cold of the last Ice Ages, and had an advanced culture, which included communal hunting, the preparation and wearing of sewn animal-skin clothes, and religious beliefs.

The neanderthals disappeared as the ice withdrew to the north, and more modern humans advanced across Europe from the Middle East. This new wave of colonization coincided with the spread of *Homo sapiens* over the rest of the world, crossing Asia to Australasia before 30 000 years ago, and reaching the Americas 11 500 years ago, if not earlier, by crossing from Siberia to Alaska. These fully modern humans, with brain sizes averaging 1360 cm^3 (Figure 9.32f), brought more refined tools than those of the neanderthals, art in the form of cave paintings and carvings, and religion. The nomadic way of life began to give way to settlements and agriculture about 10 000 years ago.

Further reading

Benton, M. J. (1989) *On the trail of the dinosaurs*. Kingfisher, London.

Benton, M. J. (1997) *Vertebrate palaeontology* 2nd edition. Chapman & Hall, London.

Benton, M. J. (1991) *The reign of the reptiles*. Kingfisher, London.

Carroll, R. L. (1987) *Vertebrate paleontology and evolution*. Freeman, San Francisco.

Feduccia, A. (1996) *The age of birds*, 2nd edition. Harvard University Press, Cambridge, Mass.

Gould, S. J., (ed.) (1993) *The book of life*. Hutchinson, London.

Kemp, T. S. (1982) *Mammal-like reptiles and the origin of mammals*. Academic Press, London.

Lewin, R. (1993) *Human evolution*, 3rd edition. Blackwell Scientific Publications, Oxford.

Norman D. B. (1986) *Illustrated encyclopedia of dinosaurs*. Salamander, London.

Savage, R. J. G. and Long, M. R. (1986) *Mammal evolution*. British Museum (Natural History), London.

References

Coates, M. I. and Clack, J. A. (1990) Polydactyly in the earliest known tetrapod limbs. *Nature*, **347**, 66–69.

Trewin, N. H. (1986) Palaeoecology and sedimentology of the Achanarras fish bed of the Middle Old Red Sandstone, Scotland. *Transactions of the Royal Society of Edinburgh: Earth Sciences*, **77**, 21–46.

10 Fossil plants

Key Points

- Plants may be preserved as permineralized tissues, coalified compressions, cemented casts, or as hard parts.
- Fungi have a long fossil record, but they are not true plants.
- Green algae, and their relatives, are close to the origin of green plants.
- Plants moved on to land in the Silurian, a move enabled by the evolution of vascular and woody tissues, waterproof cuticles and stomata, and durable spores.
- Various non-seed-bearing plants arose during the Devonian, but tree-like lycopsids, equisetaleans, and progymnosperms became established only during the Carboniferous. These formed the great 'coal forests'.
- The gymnosperms (seed-bearing plants) radiated in several phases, during the Carboniferous–Permian (e.g. medullosans, cordaites and cycads), and the Mesozoic (e.g. conifers, ginkgos, bennettitaleans and gnetales).
- The angiosperms (flowering plants) radiated dramatically during the Cretaceous, and they owed their success to fully enclosed and protected seeds, flowers and double fertilization.

Introduction

The study of fossil plants falls into two disciplines: palaeobotany, which concentrates on macroscopic (visible with the naked eye) plant remains, and palynology, which is mainly the study of pollen and spores. Palynology is usually treated as a branch of micropalaeontology (see Chapter 11), since palynologists use microscopes, and since much of the work is aimed specifically at biostratigraphical correlation, often for commercial purposes. This chapter concentrates on the history of whole plants, based on the study of leaves, roots, wood, flowers, fruits and seeds, but evidence from spores and pollen is included when it is relevant to an understanding of the whole plant.

The fossil record of plants is rich, and a great deal of information is available about the main stages in plant evolution. Many fossil localities show conditions of exquisite preservation of plant fossils, and this has allowed very detailed microscopic study of the cellular structure of ancient leaves, seeds and wood. True plants, or metaphytes, are considered in this chapter, together with their closest algal relatives, and the Fungi.

Plant preservation

Plant parts are usually preserved as compression fossils in fine-grained clastic sediments, such as mudstone, siltstone or fine sandstone, although 3D preservation may occur in exceptional situations. Many of these modes of preservation match those identified for animal fossils (see pp. 8–17). There are four main modes of plant preservation (Schopf, 1975): cellular permineralization, coalified compression, authigenic preservation, and hard-part preservation (Figure 10.1).

Plant fossils preserved by cellular permineralization, or petrifaction, may show superb microscopic detail of the tissues (Figure 10.1), but the organic material has gone. The plant material was invaded throughout by minerals in solution, such as silicates, carbonates or iron compounds, which precipitated to fill all spaces and replace some tissues. Examples of cellular permineralization are seen in the Precambrian Gunflint Chert, the Devonian Rhynie Chert, and the Triassic wood of the Petrified Forest, Arizona. The most studied examples of permineralized plant tissues have come from coal balls. Coal balls are calcareous specimens preserved in Carboniferous rocks in association with seams of bituminous coal, and huge collections have been made this century in North America

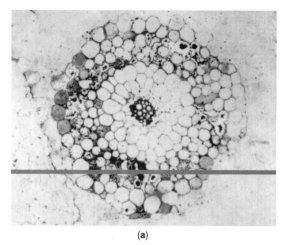

(a)

(b)

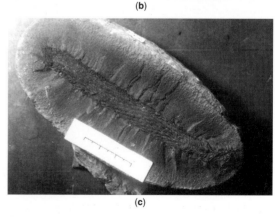

(c)

Fig 10.1 Different modes of plant preservation. (a) Permineralization: cross-section of a silicified plant stem from the Rhynie Chert (early Devonian, Scotland) (×50). (b) Coalified compression: leaves of *Annularia*, late Carboniferous, Wales (×0.7). (c) Authigenic preservation: a mould of *Lepidostrobus*, late Carboniferous, Wales (×0.5).

and Europe. They are studied by means of acetate peels which can reveal astounding detail.

The second common kind of plant preservation is the coalified compression, produced when masses of plant material lose their soluble components and are compressed by accumulated sediments. The non-volatile residues form a black coaly material, made from broken leaves, stems and roots, and with rarer flowers, fruits, seeds, cones, spores and pollen grains. The coalified compressions may be found within commercially workable coal beds, or as isolated coalified remains in clastic sediments such as siltstones and fine sandstones (Figure 10.1b).

The third mode of plant preservation, authigenic preservation or cementation, involves casting and moulding. Iron or carbonate minerals become cemented around the plant part, and the internal structure commonly degrades. The cemented minerals produce a faithful mould of the external and internal faces of the plant specimen, and the intervening space may be filled with further minerals, producing a perfect replica, or cast, of the original stem or fruit. Some of the best examples of authigenic preservation of plants are ironstone concretions, such as those from Mazon Creek in Illinois (Figure 10.1c).

The fourth typical mode of plant preservation is the direct preservation of hard parts. Some microscopic plants in particular have mineralized tissues in life, which survive unchanged as fossils. Examples are the coralline algae, with calcareous skeletons, and the diatoms, with their silicified cell walls.

Fungi

The Fungi, represented by familiar moulds and mushrooms, are not true plants. They form a separate kingdom that is apparently more closely related to multicelled animals (Metazoa) than to multicelled plants (Metaphyta) (see p. 66). Fungi are classified into a number of phyla on the basis of reproductive patterns. In some cases, there are specialized reproductive structures, which may be identified in well-preserved fossils.

There are a number of records of possible Precambrian fungi, but most of these are dubious. The first good fossils of fungi are found in permineralized deposits of Devonian and Carboniferous age. In these, the fungi appear to have acted as decomposers, feeding on decaying plant material, or as parasites, infesting the tissues of living plants. In the early Devonian Rhynie Chert, for example, fungal remains include mats of hyphae, branching tissue strands, some of them bearing

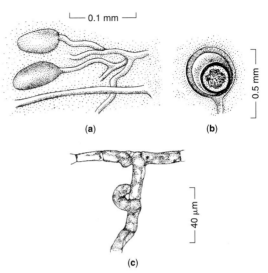

Fig 10.2 Examples of fossil fungi. (a, b) *Palaeomyces*, a possible oomycete fungus from the early Devonian Rhynie Chert of Scotland, showing branching non-septate hyphae terminated by enlarged vesicles (a) and a resting spore (b). (c) *Palaeancistrus*, with basidiomycete-like clamp connections, from the late Carboniferous of North America.

reproductive structures (Figure 10.2a, b), similar to those of modern oomycete Fungi.

Coal balls have yielded information on other fungal groups. A Carboniferous fungus, *Palaeancistrus* (Figure 10.2c), shows extensive developments of hyphae in a mat-like structure, or mycelium, with specialized hooked terminations on marginal hyphae, called clamp connections. These are characteristic of another living fungal group, the Basidiomycotina. After the Carboniferous, there are sporadic records of fungi of different groups. Particularly abundant finds come after the radiation of flowering plants, when fungi of various groups adapted to parasitize the roots, stems and leaves of the new plant group, especially in humid tropical conditions.

Terrestrialization of plants

Relationships of green plants

Palaeobotanists have long sought the origins of true plants, metaphytes, among the Chlorophyta, the green algae, and molecular evidence confirms their close relationship (see p. 68). Broader cladistic studies (Kenrick, 1994) have now shown that a number of groups, traditionally classified as 'algae', are close outgroups of land plants: Charophyceae (including charophytes), Micromonadophyceae (including dinoflagellates, see p. 247), and Chlorophyta.

These algal groups, together with land plants, form a larger clade, termed the Chlorobionta, or green plants, which are all characterized by the possession of chlorophyll *b*, and similarities of their flagellate cells and chloroplasts (Kenrick, 1994). The origin of the group dates back at least to 900 Ma, with finds of possible green algae in the Bitter Springs Chert of Australia (see p. 68). Chlorobiont evolution is hard to track in detail in these early stages because of the rarity of diagnostic fossils, but finds improve with the diversification of land plants in the Silurian and Devonian.

Adaptations to life on land

It is likely that various algae and fungi adopted partly terrestrial lifestyles perhaps during later parts of the Precambrian, but these organisms were probably small in size, and they lacked many of the adaptations required for a fully terrestrial existence. Key adaptations for life on land are as follows:

1. spores with durable walls to resist desiccation;
2. surface cuticle over leaves and stems to prevent desiccation;
3. stomata, or controllable openings, to allow gas exchange through the impermeable cuticle;
4. vascular conducting system to pass fluids through the plant;
5. lignification of tracheids to resist collapse. The cellulose cell walls of the conducting tubes, or tracheids, of vascular plants are invested with lignin, the tough polymer that makes up all woody tissues, providing strength and waterproofing.

These key adaptations relate to the problems a water plant must overcome when moving on to land. In water, a plant may absorb nutrients and water all over its surface, but on land, all such materials must be drawn from the ground, and passed round the tissues internally. Land plants typically have specialized roots that draw moisture and nutrient ions from the soil, which are passed through water-conducting systems that connect all cells. The system is powered by transpiration, a process powered by the evaporation of water from leaves and stems. As water passes out of aerial parts of the plant, fluids are drawn up into the water-conducting system hydrostatically.

Water loss is a second key problem for plants on land. Whereas, in water, fluids may pass freely in and out of a plant, a land plant must be covered with an impermeable

covering, the waxy cuticle. Gaseous exchange and water transport are then facilitated in many land plants by specialized openings, the stomata (singular: stoma), often located on the underside of leaves. Typically, stomata open and close depending on carbon dioxide concentration, light intensity and water stress.

The third problem of life on land is support. Water plants simply float, and the water renders them neutrally buoyant. Most land plants, even small ones, stand erect in order to maximize their uptake of sunlight for photosynthesis, and this requires some form of skeletal supporting structure. All land plants rely on a hydrostatic skeleton, and some groups have evolved additional structural support through lignification of certain tissues in the wood and cortex.

Mosses, liverworts and hornworts: first land plants?

Bryophytes consist of three distinctive groups. Liverworts and hornworts are flattened branching structures, some of which show differentiation into upright

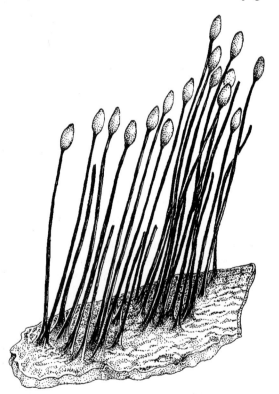

Fig 10.3 *Sporogonites*, an early Devonian bryophyte, seemingly showing numerous slender sporophytes (20 mm tall) growing from a basal gametophyte portion.

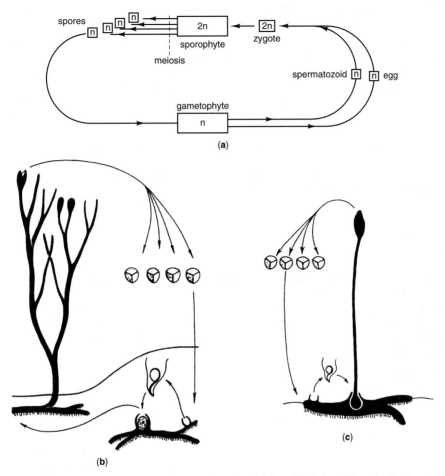

Fig 10.4 The origin of vascular land plant (tracheophyte) life cycles. (a) Simplified plant life cycle showing alternation of phases. (b) Life cycle of a hypothetical tracheophyte, with dominant sporophyte phase and reduced gametophyte, in comparison with the life cycle of a hypothetical bryophyte (c), where the dominant phase is the gametophyte, and the sporophyte is a reduced dependent structure.

stems and leaves. Mosses are upright plants with slender stems and spirally arranged leaves. These three are commonly grouped together as the bryophytes, but cladistically (Kenrick, 1994), they form three successive outgroups to a clade including all other land plants.

Because of this relationship, bryophytes share some of the land-plant characters noted above with the vascular plants, but their solutions to the problems of life on land are not always the same. Bryophytes do not have lignin, but they do possess a cuticle over leaves and stems. Many hornworts and mosses have stomata, but they are absent in liverworts. Some of the larger mosses and liver-worts have a vascular conducting system. Some bryophytes have the unusual ability to dry up completely, and then to rehydrate when rain falls, and continue as normal.

The fossil record of the bryophytes is patchy. This is often attributed to low preservation potential, but fossil specimens are also difficult to distinguish from other simple land plants. The oldest recorded fossil bryophytes are Silurian and Devonian in age, although interpretations are uncertain (Kenrick, 1994). For example, *Sporogonites* (Figure 10.3) from the Lower Devonian of Belgium, has been interpreted as a part of the flattened portion of a liverwort with, growing from it, the slender-stemmed spore-bearing phases of the plant. This specimen indicates a key difference in reproductive cycles between bryophytes and vascular plants (see Box 10.1).

In typical vascular plants, the green plant that we see is known as the sporophyte, the phase in the life cycle of the plant that produces spores, while the gametophyte, the phase that produces sperms and eggs, is very small

(Figure 10.4b). The opposite is the case in bryophytes, where the visible mosses and liverworts are haploid gametophytes, and the sporophyte is a small plant that depends for nourishment on the larger gametophyte (Figure 10.4c). Hence, in *Sporogonites* (Figure 10.3), numerous sporophytes appear to be growing from a portion of the larger flattened gametophyte phase.

First vascular plants in the Silurian and Devonian

Devonian land plants have been known for a long time, but recent work has pushed the records of vascular land plants well back into the Silurian. Vascular plants are characterized by the possession of tracheids, true vascular conducting systems. Lignin and stomata are typical of vascular plants, but may not have been present in the earliest forms.

The oldest vascular plant is *Cooksonia* from the mid Silurian of southern Ireland, a genus that survived until the end of the early Devonian. *Cooksonia* (Figure 10.5) is composed of cylindrical stems that branch in two at various points and which are terminated by cap-shaped sporangia, or spore-bearing structures, at the tip of each branch. The specimens of *Cooksonia* range from tiny Silurian examples, as little as a few millimetres long, to larger Devonian forms up to 65 mm long. Extraordinary anatomical detail has been revealed by studies of specimens of these tiny plants that have been freed from the

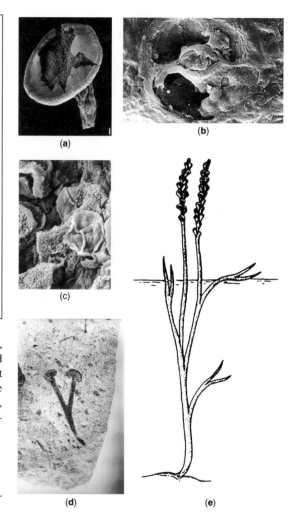

Fig 10.5 Early vascular plants. (a–d) The oldest land plant, *Cooksonia* from the Silurian to early Devonian. Early Devonian examples from Wales, showing a complete sporangium at the end of a short stalk (a), a stoma (b), and spores (c). The sporangium is 1.6 mm wide, the stoma is 40 μm wide, and the spores are 35 μm in diameter. (d) *Cooksonia caledonica*, a late Silurian form, about 60 mm tall. (e) *Zosterophyllum*, a zosterophyllopsid from the early Devonian of Germany, 150 mm tall.

rock by acid digestion, and then mounted in resin. The sporangia have been dissected to reveal that they were packed with spores. The vascular conducting tissues of early Devonian examples have thickened walls, and there are stomata on the outer surfaces of the stems (Edwards *et al.*, 1992).

Cooksonia is a member of the Rhyniopsida, the basal group of vascular plants, the Tracheophyta. Rhyniopsids are known most fully from the early Devonian Rhynie

Chert of north-east Scotland which has preserved numerous plants and arthropods exquisitely in silica. Some of the Rhynie rhyniopsids reached heights of 180 mm. They consisted of groups of vertical stems supported on horizontal branching structures which probably grew in the mud around small lakes.

Several other groups of vascular land plants arose in the early Devonian. *Zosterophyllum* (Figure 10.5e), a zosterophyllopsid, shares many features with the rhyniopsids, but has numerous lateral sporangia, instead of a single terminal one, on each vertical stem. Later in the Devonian, some basal tracheophytes became taller, as much as 3 m, the size of a shrub, and these indicate the future evolution of some vascular plants towards large size.

Evolution of vascular land plants

Recent studies (Crane, 1989; Kenrick, 1994) have allowed palaeobotanists to clarify the pattern of relationships of the major land plant groups (Figure 10.6). All green land plants belong to a larger group, the Chlorobionta, which includes chlorophyte and charophycean algae (see p. 226). The Embryophytina is a subclade that includes the three bryophyte groups (liverworts hornworts, and mosses; see pp. 226–228) and the Tracheophyta. The tracheophytes (see Box 10.2) are the vascular plants, characterized by vascular canals with secondary thickening, and including the rhyniopsids and lycopodiopsids (lycopsids and zosterophyllopsids) as basal groups. Next up the main axis of the cladogram are the horsetails (equisetopsids) and ferns (filicopsids), but their exact relationships cannot yet be determined. The progymnosperms are the sister group of the seed-bearing plants.

The seed-bearers are traditionally divided into gymnosperms, a paraphyletic group, and angiosperms, the clade of flowering plants. The long-term mystery of angiosperm origins (called the 'abominable mystery' by Charles Darwin) has been resolved to some extent:

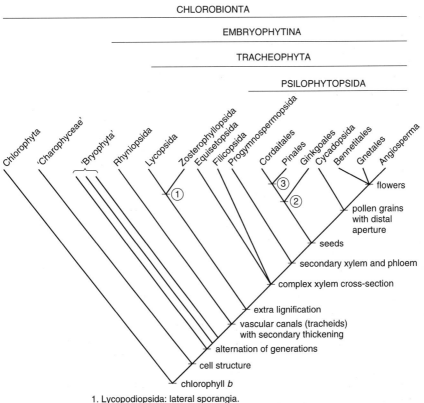

1. Lycopodiopsida: lateral sporangia.
2. Seed-bearing structures borne on short shoot.
3. Coniferopsida: seed-bearing structures grouped in a 'flower'.

Fig 10.6 Cladogram showing the postulated relationships among the major groups of vascular land plants. Some synapomorphies that define particular nodes are indicated.

Box 10.2 Classification of the tracheophytes (vascular land plants)

Tracheophytes are the vascular land plants, and include modern ferns, horsetails, conifers, and flowering plants, as well as numerous extinct groups. The basal groups are distinguished in terms of branching patterns and sporangial morphology.

Division Tracheophyta
Class Rhyniopsida
Simple vascular plants with dichotomously branching stems and terminal sporangia. Mid Silurian–early Devonian.

Class Horneophytopsida.
Small group of simple vascular plants. Early Devonian.

Class Trimerophytopsida.
Psilophyton and relatives. Early-mid Devonian.

Superclass Lycopodiopsida
Class Lycopsida
Small to large plants with lateral sporangia and (usually) small leaves. Late Silurian–Recent.

Class Zosterophyllopsida
Simple vascular plants with dichotomously branching stems and lateral kidney-shaped sporangia. Early to late Devonian.

Class Equisetopsida
Horsetails; vertical stems with jointed structure and a whorl of fused leaves at the nodes; sporangia grouped in umbrella-like structure. Late Devonian–Recent.

Class Filicopsida
Ferns; dichotomously-branching flat leaves which uncurl as they develop; sporangia are grouped in clusters on the underside of leaves. Mid Devonian–Recent.

Superclass Psilophytopsida
Class Progymnospermopsida
Fern-like plants which did not produce seeds, but had gymnosperm-like wood and produced woody tissue. Mid Devonian–late Permian.

'Seed ferns'
Order Medullosales
Primitive seed plants with large pollen grains and unusual stem anatomy. Early Carboniferous–Permian.

Order Glossopteridales
Tree-like and bush-like seed plants. Late Carboniferous–late Triassic.

Class Coniferopsida.
Early Carboniferous–Recent.

Order Cordaitales.
Trees with strap-shaped parallel-veined foliage. Early Carboniferous–late Permian.

Class Pinales
Conifers; trees with resin canals, and needle- or scale-like leaves. Early Carboniferous–Recent.

Class Ginkgoales
Trees with seed-bearing shoots and with fan-shaped or more divided leaves. Late Triassic–Recent.

Class Cycadopsida
Bushy to tree-like plants with leaf traces that girdle the stem; frond-like leaves; seeds attach to megasporophyll stalk below a leaf-like structure. Early Carboniferous–Recent.

Plants with flowers
Class Bennettitales
Bushy to tree-like plants with sterile scales between the seeds; frond-like leaves; flower-like cones with enclosing structures that surround ovules and pollen sacs. Late Triassic–late Cretaceous.

Class Gnetales
Leaves opposite each other, and vessels in the wood; male and female cones are flower-like. Late Triassic–Recent.

Division Angiosperma
Ovules are enclosed in carpels, within a flower, and fertilization is double (involving two sperm nuclei). Early Cretaceous–Recent.

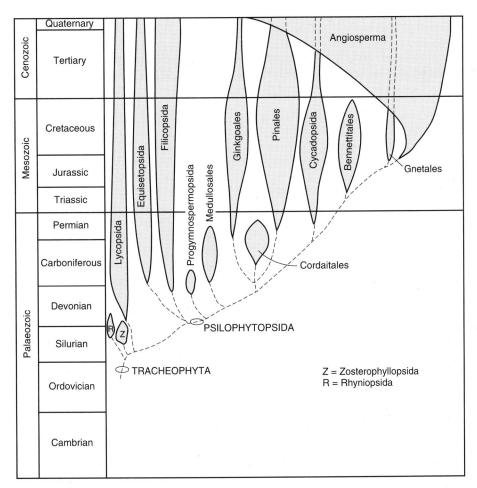

Fig 10.7   Phylogenetic tree of the main groups of vascular land plants. The pattern of postulated relationships is based on the cladogram (Figure 10.6), and details of known stratigraphic range and species diversity are added.

either bennettitaleans or gnetaleans are the sister group of angiosperms. Successively more distant outgroups are the cycads, the conifers + ginkgos, and the medullosans.

A phylogeny of tracheophytes (Figure 10.7) shows the broad stratigraphic range and relative abundance of each group at different points in plant history. The phylogeny highlights the three major bursts of land-plant evolution: in the Devonian (e.g. rhyniopsids, zosterophylls and other basal vascular plants), in the Carboniferous and Permian (e.g. lycopsids, ferns, horsetails, seed ferns), and in the Cretaceous (e.g. angiosperms).

The great coal forests

Lycopsids small and large

The clubmosses, Class Lycopsida, arose at the same time as the rhyniopsids and other dichotomously branching plants, but they are distinguished by having their sporangia arranged along the sides of vertical branches, and by having numerous small leaves attached closely around the stems.

Low herbaceous lycopsids existed throughout the

Devonian and Carboniferous, and they showed considerable variation in leaf and sporangium shape, and in the nature of the spores. From the late Devonian onwards, most lycopsids produced two kinds of spores, small and large (microspores and megaspores), which developed within terminal cones. The lycopsids are represented today by a few small herbaceous forms.

During the Carboniferous, several lycopsid groups achieved giant size, and these are the dominant trees seen in reconstruction scenes of the great coal swamps of that period. The best-known is *Lepidodendron,* a clubmoss that reached 50 m or more in height. Fossils of *Lepidodendron* have been known for 200 years, since they are commonly found in association with commercial coal occurrences in North America and Europe. At first, the separate parts – roots, trunk, bark, branches, leaves, cones and spores – were given different names, but over the years they have been assembled to produce a clear picture of the whole plant (Figure 10.8).

The giant lycopsids were adapted to the wet conditions of the coal swamps, but these habitats receded at the time of a major arid phase in the latest Carboniferous and early Permian. *Lepidodendron* and its like died out. Medium-sized lycopsids, about 1 m high, existed during the Mesozoic, but truly arborescent forms never evolved again.

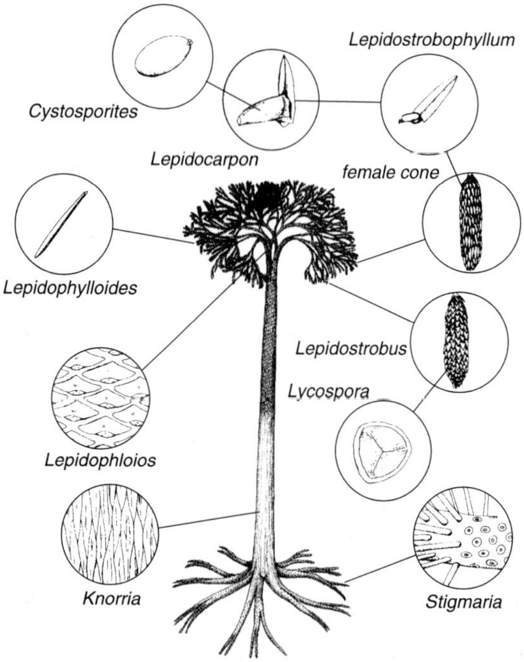

The horsetails

The horsetails, or equisetopsids (sometimes called sphenopsids or arthrophytes) are familiar to gardeners as small pernicious weeds. Their upright green shoots, with a characteristic jointed structure, are linked by underground rhizome systems. The sporangia are grouped into bunches of five or ten, below an umbrella-like structure, a unique feature of the group. The horsetails are a small group today, consisting of two genera, and 15 species, most of them small, but some reaching a height of 10 m. The early history of the group shows much greater diversity.

The horsetails arose during the Devonian, but flourished in the Carboniferous swamps, just as did the lycopsids. One form, *Calamites* (Figure 10.9a) reached nearly 20 m in height, but shows the jointed stems, and whorls of leaves at the nodes, of modern smaller horsetails. The

Fig 10.8 Reconstructing the arborescent lycopsid *Lepidodendron*, a 50-m-tall tree from the Carboniferous coal forests of Europe and North America. No complete specimen has ever been found, but complete root systems, *Stigmaria*, and logs from the tree trunk are relatively common. The details of the texture of the bark, the branches, leaves, cones, spores and seeds are restored from isolated finds.

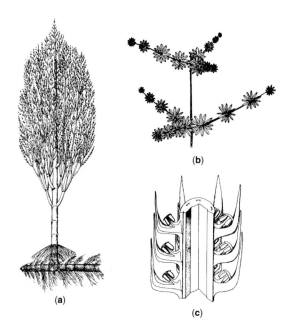

Fig 10.9 Giant Carboniferous horsetails. (a) *Calamites*, a 10-m-tall tree. (b) *Annularia*, portion of a terminal shoot bearing 10-mm-long leaves. (c) *Palaeostachya*, diagrammatic cross-section of a cone-like structure, 15 mm in diameter, bearing small numbers of megaspores.

trunk of *Calamites* generally arose from a massive underground rhizome. The leaves formed radiating bunches at nodes along the side branch (Figure 10.9b), and there were usually two types of cones, some bearing megaspores (Figure 10.9c), and others bearing microspores. The giant horsetails disappeared at the end of the Carboniferous, as did the arborescent lycopsids. Some modest tree-like forms up to 2 m tall existed in the Permian and Triassic, but later horsetails were mainly small plants living in damp boggy areas.

The ferns

Ferns, Class Filicopsida, are familiar plants today, typically with long leaves or fronds, each composed of feathery side branches, which uncurl as they develop. Fern-like plants are known in the Devonian and Carboniferous, and undisputed ferns are known in abundance from the Carboniferous onwards. As with the lycopsids and horsetails, some of the Carboniferous ferns, such as *Psaronius* (Figure 10.10), were tree-like. The fronds were borne on a vertical trunk, and they show all the features of their smaller modern tropical relatives. Other Carboniferous and Permian ferns were smaller herbaceous plants.

The ferns showed a second burst of evolutionary radiation during the Jurassic and Cretaceous, and they are the dominant plants in some Jurassic floras. Again, there were tree-like forms, as well as the more familiar low-growing ferns seen today.

Plant ecology of the Coal Measures

Early reconstructions of Carboniferous vegetation tend to show crowds of ferns, horsetails, tree ferns and clubmosses growing in dense profusion around vegetation-filled lakes. However, detailed studies have shown that the floodplain vegetation consisted almost exclusively of clubmosses such as *Lepidodendron* and *Sigillaria*, with rarer examples of horsetails such as *Calamites* (Thomas and Cleal, 1993). Pteridosperms (see p. 234), conifers and ferns were adapted to drier conditions, and they occupied elevated locations such as levees, the banks of sand thrown up along the sides of rivers. There are hints of extensive dry upland vegetations during the Carboniferous, but the fossil record of these is barely preserved.

Towards the end of the Carboniferous, the floodplain vegetation of Europe and North America changed, probably as a result of a slight drying of the environment. The clubmosses were replaced to some extent by the dry-land

Fig 10.10   The tree ferns *Psaronius*, a 10-m-tall fern from the late Carboniferous of North America.

ferns and seed ferns. These boggy habitats virtually disappeared in Europe and North America by the end of the Carboniferous, but persisted to the end of the Permian in China.

Progymnosperms: the missing link?

One of the greatest developments in the evolution of plants was the seed – the characteristic feature of the dominant modern plant groups, the gymnosperms and angiosperms. These two groups also show an advance in their woody tissues that permits the growth of very large trees; their lignified tracheids, vascular canals, can develop in a secondary system. Progymnosperms (e.g. *Archaeopteris* in the mid to late Devonian) look superficially like tree ferns, but their trunks show the development of secondary woody tissues and growth rings. Some progymnosperm fertile structures show retention of megaspores in a protective sheath, and these may be precursors of ovules and seeds.

Seed bearing plants

The origin of seeds

The first plants with seeds are known from late Devonian rocks, and seed-bearers rose to prominence during the Carboniferous. After the end of the Carboniferous, and the extinction of arborescent lycopsids, ferns and horsetails, seed-bearing plants (or gymnosperms) took an increasingly dominant role in floras around the world.

Seeds in gymnosperms are naked, i.e., they are not enclosed in ovaries as they are in flowering plants (angiosperms). Seeds follow from the fertilization of an ovule, the structure containing the egg. The gymnosperm ovule (Figure 10.11) consists of a large food store, the megasporangium, and an outer protective layer, the integument, with an open end through which the pollen grains enter. The pollen grains settle on the ovule, and may send pollen tubes into the tissues of the ovule, through which sperm head for the fertile female structures, the archegonia. Upon fertilization, the ovule becomes a seed, containing a viable embryo that develops within the seed coat, and feeding on the nutritive material that composed the bulk of the ovule.

Ovules and pollen are believed to have arisen from different kinds of spores found in non-seed-bearing plants. In primitive plants, spores are shed, and fertilization generally occurs externally on an independent plant, the gametophyte (see p. 227). Many non-seed-bearers show differentiation of spores, with some sporangia producing small numbers of megaspores and others producing large numbers of tiny microspores. These are interpreted as precursors of the female ovules and the male pollen. The seed habit came from the retention of one functional megaspore and the female gametophyte in the sporangium, which became surrounded by the integu-

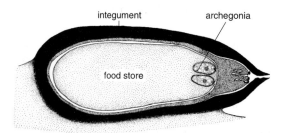

Fig 10.11 A typical gymnosperm seed, the ovule of *Pinus*, the pine, showing the archegonia (fertile female structures) surrounded by a substantial food store. Sperm enter through a narrow gap in the protective integument, and pass through pollen tubes to the archegonia.

ment, and adaptations to trap microspores (pollen) from other sporangia on the same plant, or on another plant.

Gymnosperms are said to have owed their success in the Carboniferous to the fact that they retained their ovules, and that the developing embryo had extra protection from the parent plant. In addition, the free-living gametophyte phase was eliminated, and water was not required for the sperm to swim through, as in more primitive groups, so that gymnosperms could inhabit dry upland habitats. Gymnosperms may have had adaptive advantages in certain situations as a result of seed-bearing, but it would be wrong to assume that they prevailed in all situations. Ferns, horsetails, lycopsids, and more primitive plants such as mosses, continued to diversify, especially in damp situations, and they continued their successful evolution without the 'benefit' of seeds.

Seed ferns

The seed ferns, or pteridosperms, have been regarded traditionally as a major gymnosperm class, but they share no unique characters, and it is clear that they are a polyphyletic assemblage of gymnosperms of varied affinities that were important during the late Palaeozoic and Mesozoic.

The Carboniferous and Permian seed ferns belong to a variety of groups, such as the Medullosales, which looked superficially like tree ferns, but bore ovules and pollen. Another group of late Palaeozoic seed ferns, the Glossopteridales, include *Glossopteris* (Figure 10.12), a 4-m-tall tree with radiating bunches of tongue-shaped leaves. This seed fern was the typical member of the famous *Glossopteris* flora which characterized Gondwana, the southern-hemisphere continents, from the late Carboniferous to late Permian. The Glossopteridales existed through the Triassic, and a number of other groups of seed ferns of uncertain affinities radiated during the Triassic and Jurassic.

Conifers

The conifers are the most successful of the gymnosperms, having existed since the late Carboniferous, and being represented today by over 500 species, including the largest living organisms of all time, the 110-m-tall redwoods (*Sequoia sempervirens*). Conifers have a variety of adaptations to dry conditions, including the possession of narrow needle-like leaves with thick cuticles and sunken stomata, all adaptations to minimize water loss. The seeds are contained in tough scales

Fig 10.13   The early conifer *Cordaites*, about 25 m tall.

with a single fertile scale at the tip, showing apparent intermediate stages, to the cones of modern conifers, where all or most scales are fertile.

Modern conifers radiated in the late Triassic and Jurassic, possibly from ancestors among the Voltziales. The main families, the Podocarpaceae (southern podocarps), Taxaceae (yews), Araucariaceae (monkey puzzle), Cupressaceae (cypresses and junipers), Taxodiaceae (sequoia, redwood and bald cypress), Cephalotaxaceae, and Pinaceae (pines, firs and larches) are distinguished by leaf shape and features of the cones. Podocarps and yews do not have cones.

Fig 10.12   The seed fern *Glossopteris*, a 4-m-tall tree, from the late Permian of Australia.

Diverse gymnosperm groups

grouped in spirals into cones, usually borne at the end of branches, while the pollen-producing cones are usually borne on the sides of branches.

The Cordaitales of the Carboniferous and Permian are a distinctive group of early conifers which had strap-shaped parallel-veined leaves (Figure 10.13). Some Cordaitales were tree-like, and bore their leaves (sometimes up to 1 m long) in tufts at the ends of lateral branches. The Voltziales, of late Carboniferous to Jurassic age, are represented by abundant finds of leaves and cones. The cones show a variety of structures, some

Compared to the conifers, the other gymnosperm groups did not radiate so widely. The ginkgos are represented today by one species, *Ginkgo biloba*, the maidenhair tree, a native of China, but seen today as a typical urban tree in parts of Europe and North America. Ginkgos were more diverse in the Mesozoic. Leaf shape varies from the fan-shaped structure in the modern form, to deeply dissected leaves in some Mesozoic taxa (Figure 10.14a, b). Catkin-like pollen organs and bulbous stalked ovules are borne in groups on separate male and female plants. The leaves in the modern *Ginkgo* are deciduous, i.e., they are

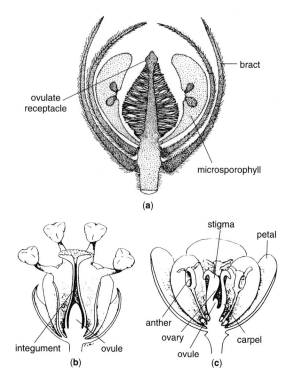

Fig 10.15   Evolution of the angiosperm flower. (a) Cone of the Jurassic bennettitalean *Williamsoniella*, showing the female fertile structure, the ovule, contained in a central receptacle, and surrounded by the male fertile structures, the microsporophylls. (b) Flower of the gnetale *Welwitschia*, showing the central ovule, and surrounding male elements. (c) Flower of the angiosperm *Berberis*, showing the same pattern, but with the seed enclosed in a carpel.

shed in winter, and this may have been a feature of ancient ginkgos.

The cycads, represented today by 10 tropical and sub-tropical genera, are trees with a stem that ranges in length from a small tuber to a palm-like trunk up to 18 m tall. The leaves are provided with deep-seated leaf traces that partially girdle the stem. *Leptocycas* (Figure 10.14c) from the late Triassic of North Carolina has a 1.5-m-tall trunk, showing a few traces of attachment sites of leaves that had been lost as the plant grew, and a set of nine or ten long fronds near the top of the trunk. Many other cycads show marked leaf bases along the entire length of the trunk. Cycad fronds are typically composed of numerous parallel-sided leaflets attached to a central axis in a simple frond-like arrangement, but others had undivided leaves.

The bennettitaleans, or Cycadeoidales, were a Mesozoic group of bushy plants with frond-like leaves very similar to those of cycads. Some bennettitaleans had

Fig 10.14   Diverse gymnosperms. Leaves of the modern ginkgo, *Ginkgo biloba* (a) and of the Jurassic ginkgo, *Sphenobaiera paucipartita* (b). (c) *Leptocycas gigas*, reconstruction of a 1.5-m-tall cycad from the late Triassic of North America. (d) *Cycadeoidea*, reconstruction of a 2-m-tall bennettitalean, from the Cretaceous of North America.

a trunk up to 2 m tall, with bunches of long fronds at the top of the trunk and on subsidiary branches. Other bennettialeans, such as *Cycadeoidea* (Figure 10.14d), had an irregular ball-like trunk covered in leaf bases, representing former attachment sites of fronds, and with a tight tuft of long feathery fronds on top. Classic dinosaur scenes of Jurassic and Cretaceous age often picture one or other of these bennettialeans in the background.

The gnetales have a patchy fossil record, with two late Triassic examples, a few in the Cretaceous and Tertiary, and three living genera. The group has gained prominence among botanists since gnetales appear to represent the closest approach to angiosperms among the gymnosperms. In particular, gnetales may have their ovules and pollen organs in cones that are rather flower-like (Figure 10.15b).

Flowering plants

Flowers and angiosperm success

The angiosperms are by far the most successful plants today, with 250 000 species and dominance of most habitats on land. Most of the food plants used by humans are angiosperms. The angiosperms arose during the Mesozoic, and radiated dramatically during the mid Cretaceous. Characteristics of angiosperms are as follows:

1. The ovules are fully enclosed within carpels (Figure 10.15c). It is believed that carpels are modified leaves that grew around the ovules, and provided a secure protective covering. In angiosperm development, the carpels grow around the ovules and fuse, although in some magnolias the carpels are not completely fused when fertilization takes place.
2. Most angiosperm ovules have two integuments, or protective casings.
3. Most angiosperms have pollen grains with a double outer wall separated by columns of tissue.
4. Angiosperms have a flower (Figure 10.15c), a structure that is composed of whorls of sepals and petals in dicots. The flower includes the carpels and stamens. The structure of flowers is not standard in all angiosperms.
5. Angiosperms all show double fertilization, i.e., two sperm nuclei are involved in fertilization. One unites with the egg nucleus, while the other fuses with another nucleus that divides to form the food supply for the developing embryo. Double fertilization has been described also in Gnetales.

6. Most angiosperms have water-transporting vessels rather than just xylem tracheids. However, this feature is absent in some magnolids and hamamelidids, and is present in gnetales.
7. Most angiosperms have a net-like pattern of veins in their leaves.

Most of these characters are regarded as typical of angiosperms, but many are not unique to angiosperms, nor are they present in all angiosperms. The only one that seems to be an acceptable apomorphy of the group is the possession of carpels around the ovule (character 1).

Flowers are certainly the most obvious feature of angiosperms, but several gymnosperm groups are also characterized by flowers. Bennettialeans (Figure 10.15a) and Gnetales (Figure 10.15b) have flower-like structures with the ovule in the centre, and around them structures resembling petals.

The secret of the success of the angiosperms may be the flower and the fully enclosed ovule. The carpels protect the ovule from fungal infection, desiccation and the unwelcome attentions of herbivorous insects. Double fertilization is said to offer the advantage that the parent plant does not invest energy in creating a large food store, as in gymnosperms (Figure 10.11) until fertilization of the ovule is assured. Pollen is produced within the anthers which are typically borne on long filaments arranged around the centrally placed ovary or ovaries. Pollen grains are transported by animals (often insects) or by the wind, to the stigma, and from it the pollen grains send pollen tubes to the ovules through which the sperm pass.

The petals, often brightly coloured, the special fragrances, and the supplies of nectar (sugar water) are all adaptations of angiosperms to ensure fertilization by insects. Some gymnosperms show hints of this pattern: the living gnetalean *Welwitschia* has a flower with 'petals' (Figure 10.15b), and it secretes a nectar-like pollination drop as tempting food for its insect pollinator. It is clear that the evolution of angiosperm characters was paralleled by the evolution of major new groups of insects that fed from flowers and pollinated the flowers (see Box 10.3).

The first angiosperms

There has been heated discussion among palaeobotanists over the past century about the oldest angiosperm fossil, and about the closest relatives of angiosperms. The oldest generally accepted angiosperms are early Cretaceous

Box 10.3 A new career for insects: pollination

Pollinating insects existed before the Cretaceous and the radiation of the angiosperms, but their role was minor, feeding at flowers of some of the advanced gymnosperms. During the Cretaceous, however, there is striking evidence for angiosperm–insect co-evolution (Figure 10.16). Groups of beetles and flies that pollinate various plants were already present in the Jurassic and early Cretaceous, but the hugely successful butterflies, moths, bees, and wasps are known as fossils only from the Cretaceous and Tertiary.

The composition of insect faunas changed during the Cretaceous and Tertiary. One group to evolve substantially at that time was the Hymenoptera (bees and wasps). The first hymenopterans to appear in the fossil record, the saw-flies (Xyelidae), had been present since the Triassic. Some fossil specimens have masses of pollen grains in their guts, a clear indication of their preferred diet. The sphecid wasps, which arose during the early Cretaceous, had specialized hairs and leg joints that show they collected pollen. Other wasps, the Vespoidea, and the true bees appear to have arisen in the late Cretaceous.

The first angiosperms may not have had specialized relationships with particular insects, and may have been pollinated by several species. More selective plant–insect relationships probably grew up during the late Cretaceous with the origin of vespoid wasps which today pollinate small radially symmetrical flowers. These kinds of specialized relationships are shown by increasing adaptation of flowers to their pollinator in terms of flower shape, and the food rewards offered, and of the pollinator to the flower. Evidence from the late Cretaceous plant record shows that angiosperms similar to roses then had specialized features that catered for pollinators that fed on nectar as well as pollen.

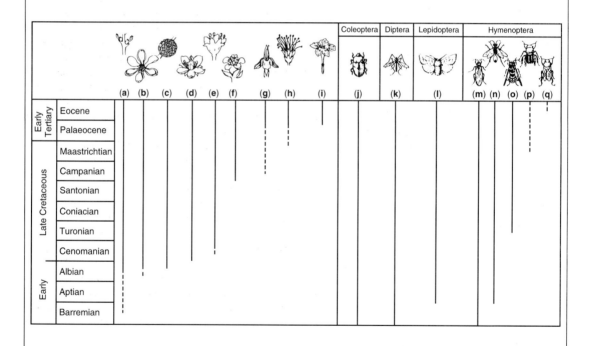

Fig 10.16   The coevolution of floral structures, and of pollinating insects during the entire span of the Cretaceous, and the early part of the Tertiary. Some of the major floral types are small simple flowers (a), flowers with numerous parts (b), small unisexual flowers (c), flowers with parts arranged in whorls of five (d), flowers with petals, sepals and stamens inserted above the ovary (e), flowers with fused petals (f), bilaterally symmetrical flowers (g), brush-type flowers (h), and deep funnel-shaped flowers (i). Pollinating insects include beetles (j), flies (k), moths and butterflies (l), and various groups of wasps and bees (m–q): Symphyta (m), Sphecidae (n), Vespoidea (o), Meliponinae (p), and Anthophoridae (q).

(a)

(b)

Fig 10.17 Fossil angiosperm remains from North America. (a) Flower of an early box-like plant, *Spanomera*, from the mid Cretaceous of Maryland (×10). (b) Leaf of the birch, *Betula*, from the Eocene of British Columbia (×1).

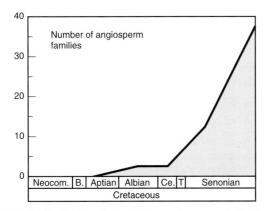

Fig 10.18 Rapid radiation of the angiosperms during the Cretaceous, shown by a rise in the number of angiosperm families, from 0 at the beginning of the Cretaceous to more than 35 by the end of the period. Abbreviations: B., Barremian; Ce., Cenomanian; Neocom., Neocomian; T., Turonian.

in age, but a late Triassic angiosperm reproductive structure, associated with angiosperm-like pollen and leaves was reported in 1993. This identification is controversial (Crane et al., 1995).

The oldest flowers are early Cretaceous in age, and fossils have been reported from Europe, North America and East Asia. Some of these rare and remarkable fossils show soft parts of flowers, such as sets of fleshy stamens with pollen grains inside, and evidence of fivefold symmetry of the flower parts, typical of modern angiosperms. Even more spectacular fossil specimens of flowers are known from the beginning of the Tertiary, where specimens are preserved in lithographic limestones, in chert, and in amber (Figure 10.17). Other more commonly preserved fossil evidence for the first angiosperms consists of pollen, leaves, fruits and wood.

The radiation of the angiosperms

Angiosperms radiated to a diversity of 35 families by the end of the Cretaceous (Figure 10.18). The success of the angiosperms in the mid Cretaceous may have been driven by environmental stresses. The early angiosperms lived in disturbed ephemeral habitats, such as river beds and coastal areas, and they were opportunists that could spread quickly when conditions were right. In addition, the specialized reproductive systems of angiosperms perhaps promoted rapid speciation, especially in terms of the increasing matching of flower and pollinator.

Angiosperms are typically divided into two major subgroups, the monocotyledons and the dicotyledons. The monocots – lilies, gingers, grasses, palms, and their relatives – have one cotyledon (food-storage area of the seed), flower parts arranged in threes, and parallel leaf venation patterns. The dicots (i.e. all other flowering plants) have two cotyledons, flower parts often in fours and fives, net-like venation patterns on the leaves, and specialized features of the vascular tissues in the wood. These two groups are readily distinguishable in the Cretaceous. During the Tertiary, more and more of the modern families appeared, so that at least 250 of the 400 or so extant families of angiosperms have a fossil record of some kind.

Angiosperms and climate

Angiosperms are highly sensitive indicators of palaeoclimates on land, and they provide the best tool at present for estimating temperatures, rainfall patterns, and measures of seasonality. The key to the use of angiosperms

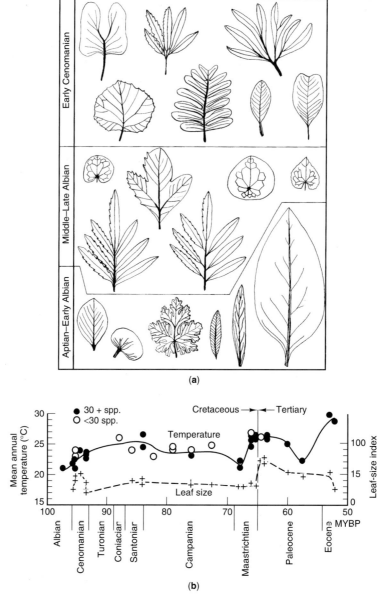

(a)

(b)

Fig 10.19 The evolution of angiosperm leaf shape and palaeoclimate. (a) Samples of typical leaf shapes from North American floras spanning the mid Cretaceous, showing variations in length, margins and shapes. The average leaf size declines, suggesting an increase in temperature. (b) The leaf-size index (percentage of entire-margined species and average leaf size) for low-latitude North American floras through the late Cretaceous shows fluctuations. These are interpreted as the result of changes in temperature.

in this way is the fact that so many modern taxa may be traced well back into the Tertiary and Cretaceous, and palaeobotanists assume that adaptations that are observed today had the same functions in the past.

Studies on North American late Cretaceous angiosperm leaves have shown how precise these climatic estimates may be. Upchurch and Wolfe (1987) established ways of assessing temperatures and rainfall measures from key leaf features, including the following:

1. *Leaf size*: the largest leaves are found in tropical rain forest, and size diminishes as temperature and moisture decline.
2. *Leaf margins*: in tropical areas, most angiosperm leaves have entire (unbroken) margins, whereas in temperate areas, there are many more toothed-margined leaves.
3. *Drip tips*: leaves from tropical rain-forest species have elongated tips to allow water to clear the leaf during major downpours.
4. *Deciduousness*: the proportion of deciduous trees (those that shed all their leaves simultaneously in winter, or during the dry season, i.e. at times of low growth rate) to evergreens is highest in temperate zones, while tropical trees are more likely to retain their leaves since they grow more continuously.
5. *Lianas*: certain angiosperms in tropical forests grow as long rope-like plants that hang down from tall trees (see any Tarzan film), but such plants are uncommon in temperate forests.
6. *Vessels in wood*: in areas subject to freezing or drying, the vascular canals possess adaptations to prevent air filling the canals when water is in short supply; the canals are narrow and densely packed.
7. *Growth rings*: in areas subject to highly seasonal climates, wood grows rapidly during the warm or wet season, and slows or stops growing when conditions are cold and/or dry. The variation in growth ring style indicates the degree of seasonality.

Abundant assemblages of leaves have been recovered from several hundred late Cretaceous localities in North America, and these together show changes in leaf shapes of the sort that reflect palaeoclimates (Figure 10.19a). Measurements of the leaf-shape characters noted above, based on floras of numerous species, can give a clear plot of palaeotemperature change in North America during the late Cretaceous (Figure 10.19b). Temperatures remained around 20–25 °C, with slight variations, until the last 5 Myr of the Cretaceous, when there was a dramatic rise in temperature to 27°C, then a drop, and a further rise in the early Tertiary.

Further reading

Friis, E. M., Chaloner, W. G. and Crane, P. R. (1987) *The origins of angiosperms and their biological consequences.* Cambridge University Press, Cambridge.

Mauseth, J. D. (1991) *Botany; an introduction to plant biology.* Saunders College Publishing, Philadelphia.

Stewart, W. N. and Rothwell, G. W. (1993) *Paleobotany and the evolution of plants*, 2nd edition. Cambridge University Press, Cambridge.

Thomas, B. A. and Spicer, R. A. (1995) *Evolution and palaeobiology of land plants*, 2nd edition. Chapman & Hall, London.

References

Crane, P. R. (1989) Colonisation of the land. In K. C. Allen and D. E. G. Briggs (eds) *Evolution and the fossil record.* Belhaven, London, pp. 153–187.

Crane, P. R., Friis, E. M. and Pedersen, K. R. (1995) The origin and early diversification of angiosperms. *Nature*, **374**, 27–33.

Edwards, D., Davies, K. L. and Axe, L. (1992) A vascular conducting strand in the early land plant *Cooksonia. Nature*, **357**, 683–685.

Kenrick, P. (1994) Alternation of generations in land plants: new phylogenetic and palaeobotanical evidence. *Biological Reviews*, **69**, 293–330.

Schopf, J. M. (1975) Modes of plant fossil preservation. *Review of Palaeobotany and Palynology*, **20**, 27–53.

Thomas, B. A. and Cleal, C. J. (1993) *The Coal Measures forests*. National Museum of Wales, Cardiff.

Upchurch, G. R., jr and Wolfe, J. A. (1987) Mid-Cretaceous to Early Tertiary vegetation and climate: evidence from fossil leaves and woods. In E. M.Friis, W. G. Chaloner and P. R. Crane (eds) *The origins of angiosperms and their biological consequences.* Cambridge University Press, Cambridge, pp. 75–105.

11 Microfossils

- Micropalaeontology is the study of micro-organisms or the microscopic parts of larger organisms.
- Protists (unicellular eukaryote organisms) include both animal-like and plant-like groups.
- Acritarchs and dinocysts are palynomorphs with important biostratigraphical applications.
- Foraminifera (single-celled animal-like protozoans) are mainly benthic, commonly with calcareous tests.
- Radiolarians (animal-like protozoans) and diatoms (plant-like protozoans), both siliceous, are important rock formers.
- Ostracodes (microscopic crustacean arthropods) are mainly benthic, and occupied a wide range of environments.
- Conodonts occurred together as apparatuses for the grasping, grinding and shearing of food; the conodont animal was a jawless vertebrate.
- Ichthyoliths (vertebrate microfossils including fish scales and teeth) are biostratigraphically useful.
- Spores and pollen are resistant reproductive parts of plants, and reflect the development of the pteridophyte, gymnosperm and angiosperm floras; precise zonal schemes are available for the Upper Palaeozoic, Mesozoic and Cenozoic.

Introduction

Micropalaeontology is a multidisciplinary science. Microfossils are the microscopic remains, commonly less than a millimetre in size, of either micro-organisms or the disarticulated or reproductive parts of larger organisms. Micropalaeontology has thus attracted the attention of botanists, zoologists, biochemists and microbiologists, together with, of course, palaeontologists and geologists. The taxonomic groups included as microfossils are very varied, but they are united by their method of study; all require the use of an optical microscope, although more recently both scanning and transmission electron microscopes have significantly enhanced their investigation.

Microfossils include material derived from most of the major animal and plant groups, as well as microbes themselves. The broad classification adopted here is conventional and operational: the group is divided into the prokaryotes (mainly bacteria), the protists (unicellular eukaryote organisms with a variety of tests and cysts), microinvertebrates (mainly the ostracodes and scolecodonts), microvertebrates (mainly the conodonts and various other microscopic parts of fishes), and spores and pollen (microscopic reproductive organs of plants). Prokaryotes are the focus of Chapter 4 and are not discussed here further.

The abundance and durability of many microfossil groups renders them ideal for biostratigraphical correlation. Most groups are widely used in geological exploration by oil and mining companies; many groups are common in small rock samples such as well chips or cores. Moreover, irreversible changes in colour in particular microfossils when heated has made them a useful tool for assessments of thermal maturation and the prediction of likely sites for oil and gas.

Micro-organisms have made a phenomenal contribution to the evolution of the planet. Many groups, such as the coccolithophores, diatoms, foraminifera and radiolaria, are rock-forming organisms. The prokaryotic cyanobacteria fundamentally changed the planet's atmosphere from anoxic to oxygenated during the Precambrian. Moreover, recent research suggests that carbonate mudmounds, for example the late Ordovician mudbanks in central Ireland, the north of England and Sweden, the early Carboniferous Waulsortian mounds in Ireland and elsewhere together with the early Cretaceous mudmounds in the Urgonian limestones of the Alpine belt, were precipitated by microbes such as bacteria.

The influence of micro-organisms may also be more subtle. Coccolith-producing organisms (e.g. *Emiliania*) can, during blooms, manufacture massive amounts of calcium carbonate; this material may be subducted and recycled through volcanoes as CO_2. The build-up of this greenhouse gas probably maintained warm climates during the last 200 million years.

The extraction and retrieval of microfossils from rocks and sediments usually requires a number of specialized preparation techniques only attempted in purpose-built laboratories (Aldridge, 1990). Most processes start with the disaggregation or solution of the rock with a variety of chemicals, specific for particular microfossil groups. The microfossils are isolated by settling in heavy liquids or by electromagnetic separation, and then picked and mounted on a slide for microscope study. Many groups, such as algae and foraminifera, may also be studied in thin section.

Protista

The protists are single-celled organisms with nuclei and organelles, including both autotrophs, which metabolize inorganics such as CO_2, and heterotrophs, which process other organics. Many of these organisms were very abundant in the geological past and, as today, most were probably primary producers at the base of the food chain. The protist fossil record probably illustrates only a small sample of the entire catalogue of these groups; despite their abundance in modern lakes and oceans, relatively few are potential fossils.

Organisms such as *Carosphaeroides* in the Bitter Springs Chert, Australia (850 Ma) and *Latisphaera* in the Beck Spring Dolomite, California (850 Ma) may represent the first coccoid green algae and possibly some of the first eukaryotes. A range of protists such as the acritarchs, chitinozoans, coccolithophores and diatoms dominated the phytoplankton at various stages from the late Precambrian to the present, whereas the foraminiferans and radiolarians were important parts of the zooplankton (Figure 11.1). Apart from a role as a primary food source, the marine phytoplankton functions as a major carbon sink, initially removing CO_2 from the atmosphere as carbonate ions.

Calcareous algae

Many carbonate-producing organisms are termed 'calcareous algae', specifically the rhodophytes (red algae) and chlorophytes (green algae). The organisms are multicellular and have a calcified thallus. After death, many

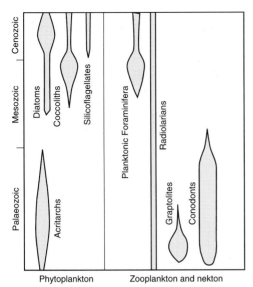

Phytoplankton   Zooplankton and nekton

Fig 11.1   Distribution of the main nektonic and planktonic microfossil groups throughout geological time.

thalli disintegrate into microscopic carbonate particles in a variety of shallow-water environments (Figure 11.2). The blue-green algae (cyanophytes) or Cyanobacteria are not true algae and are discussed elsewhere (see Chapter 4); the coccolithophores (chrysophytes) are dealt with later in this chapter.

The rhodophytes have a red pigment and secrete calcium carbonate, mainly magnesian calcite, within and between the cell walls. Two life modes occur: encrusting, as in the solenoporaceans which lack reproductive structures, and erect, as in the corallinaceans. Both are commonly associated with reef environments. The group has a long history and has been an important component of ancient reef ecosystems such as the Silurian reefs of the Swedish island of Gotland and the Devonian reefs of Australia and Europe (Figure 11.3). The solenoporaceans range in age from the Cambrian to Eocene, whereas the corallinaceans appeared during the Jurassic and are extant.

The chlorophytes include two main groups: the codiaceans and the dasyclads, most typical today of nearshore settings with abnormal salinities. The group appeared during the late Precambrian and is common in ancient and modern backreef, lagoonal and tidal flat environments, particularly in skeletal sands.

Many of the genera are long-ranging. *Girvanella* was common in shallow-water environments, and had a wide distribution from the Cambrian to Cretaceous; it has a series of fine filaments, twisted together into massive clumps. *Renalcis* is compound and hemispherical, forming large mounds in shallow water from the Cambrian to the Carboniferous.

Acritarchs

The acritarchs are a mixed bag of hollow organic-walled microfossils, not readily accommodated in any other groups. The acritarchs are thus polyphyletic and contain a wide range of diverse forms, probably representing the cyst stages in the life cycles of various groups of planktonic algae. At first it was thought acritarchs were the eggs of planktonic invertebrates, but they were later assigned to the phytoplankton. Although many more taxa have been described since, and their value in biostratigraphical correlation has been proved, particularly in the Lower Palaeozoic, uncertainty still surrounds the origin and affinities of the group: although some

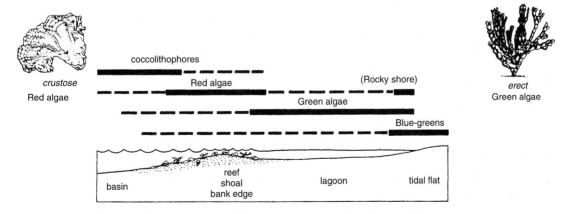

Fig 11.2   Environmental ranges of main algal groups.

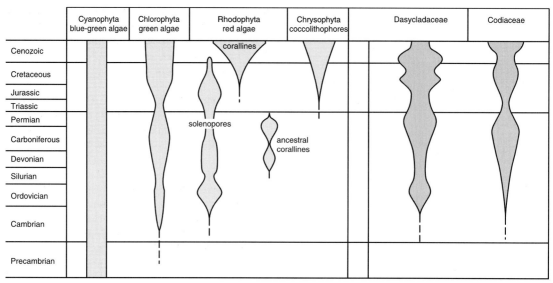

Fig 11.3 Stratigraphical ranges of main algal groups.

acritarch-like cysts have been assigned to the Chlorophyta, the majority are unclassified.

The composition and broad morphology of the acritarchs suggest relationships with the dinocysts; like the dinocysts, acritarchs are also often found in clusters. Clearly the group had a similar life cycle to that of the dinoflagellates; during the next phase of the organism's life, or reproductive cycle, material escaped through encystment structures. A number of escape structures have been described, including median splits, pylomes and cryptopylomes.

Acritarchs consist of vesicles of condensed fatty acids, ranging in shape from spherical to cubic and in size from 50 to 100 µm, although many are preserved as flattened films in black shales. There is a huge variety of basic shapes (Figure 11.4). Acritarchs can have single- or double-layered walls, and the wall structure is often useful taxonomically. The central cavity or chamber can be closed or open externally through a pore, slit or pylome. The opening, or epityche, presumably allowed the escape of the motile stage and may be modified with a hinged flap.

Externally the acritarch may be smooth or, for example, have a granulate or microgranulate ornament. Commonly the vesicle may be modified by various extensions or processes projecting outwards from the vesicle wall. If an acritarch has a set of similar processes, it is termed homomorphic, whereas if it has a variety of different projections it is termed heteromorphic.

Over 400 genera of acritarchs are known, defined mainly on vesicle shape and ornament. All acritarchs are aquatic, with the vast majority found in marine environments. The classification of the group is based on the wall structure, the shape of the body vesicle, pylome type, and the nature of the extensions and processes.

Distribution and evolution

Acritarchs had a wide geographic range, apparently mainly controlled by latitude, although the entire group ranged from the poles through the tropics. The wide distribution of the group is similar to that of the dinoflagellates, and strongly suggests that acritarchs were also members of the phytoplankton. Acritrachs have been of considerable value in regional stratigraphical correlations.

The acritarchs include some of the oldest documented fossils, with a history of over 1500 million years, although the group was not common until 1000 Ma, when sphaeromorphs dominated. During the early Cambrian radiation of the group, spinose morphs such as *Baltisphaeridium* and *Micrhystridium*, together with the crested *Cymatiosphaera*, appeared; significantly, these armoured vesicles evolved during the expansion of marine predators. By the late Cambrian–early Ordovician, acritarch palynofacies were dominated by three main groupings: the *Acanthodiacrodium*, *Cymatiogalea* and *Leiofusa* groups.

The acritarchs declined at the end of the Devonian, and are rare in Carboniferous–Triassic rocks. Nevertheless, the group staged a weak recovery during the Jurassic and continued through the Cretaceous and Tertiary.

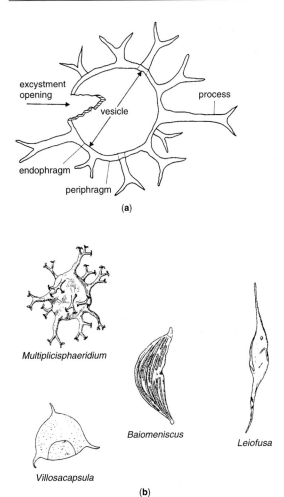

excystment
opening

process

vesicle

endophragm

periphragm

(a)

Multiplicisphaeridium

Baiomeniscus

Leiofusa

Villosacapsula

(b)

Fig 11.4 (a) Descriptive morphology of the acritarchs and (b) some acritarch morphotypes: *Multiplicisphaeridium* (×800), *Baiomeniscus* (×200), *Leiofusa* (×400) and *Villosacapsula* (×400).

Dinoflagellates

The dinoflagellates, or 'whirling whips', comprise a group of microscopic algae with organic-walled cysts. The life history of these organisms thus oscillates between a motile and a cyst stage; the cysts usually range in size from 20 to 150 μm. The motile phase is either flexible and unarmoured or rigid and armoured with a network of plates, the theca; the arrangement of thecal plates comprises the dinoflagellate tabulation. The plates of a dinoflagellate theca are arranged from the apex to antapex as follows: apical, precingular, cingular, postcingular and antapical (Figure 11.5). There are a number of other plates with further specialized terms and together the plates are commonly labelled in a precise way. The motile phase is rarely fossilized. In contrast, the cysts are chemically resistant and relatively common (Figure 11.6). Their morphology crudely shadows that of the theca, and comparable structures are prefixed by the term 'para'.

Cysts have a paratabulation which is useful taxonomically: the peridiniacean cysts have seven precingular and five postcingular paraplates, whereas the gonyaulacaceans have six precingular and six postcingular paraplates.

The dinoflagellates are one of the more abundant members of the living phytoplankton, forming an important base to the food chains of the oceans. However, dinoflagellate blooms or red tides can cause asphyxiation of other marine groups by producing neurotoxins. Mass mortalities of Cretaceous bivalves in Denmark and Oligocene fishes in Romania have been associated with fossil red tides.

There are three main cyst types. The proximate cyst is developed directly against the theca itself and has a similar configuration. A chorate cyst is smaller than the theca; the cysts are contained within the theca, interconnected by various appendages and spines, related to the external tabulation of the theca. In the cavate morphs there is a gap between the cyst and the theca at the two poles.

Distribution and evolution

To date, the oldest dinoflagellate cyst is probably *Arpylorus* from the late Silurian of Tunisia; the cyst has feeble paratabulation and a precingular archeopyle. Oddly, there are no further definite records until the early Triassic, when *Sahulidinium* appears off north-west Australia; some authors have suggested that a number of Palaeozoic acritarch taxa may in fact be dinoflagellates. Multiplated forms such as *Rhaetogonyaulax* and *Suessia* appear in dinocyst floras ranging from Australia to Europe. *Nannoceratopsis* cysts with characteristic archeopyles and tabulation are common in early Jurassic floras, while *Ceratium*-like forms appeared first during the late Jurassic and diversified in the Cretaceous. Many precise zonation schemes for Mesozoic and Cainozoic strata, particularly the Jurassic–Tertiary marine successions, are based on dinocyst distributions. The global biodiversity of the group began a steady decline during the Eocene.

Chitinozoans

Chitinozoans are small bag-shaped hollow vesicles with smooth or ornamented surfaces, consisting of organic material similar in composition to the graptolite rhabdo-

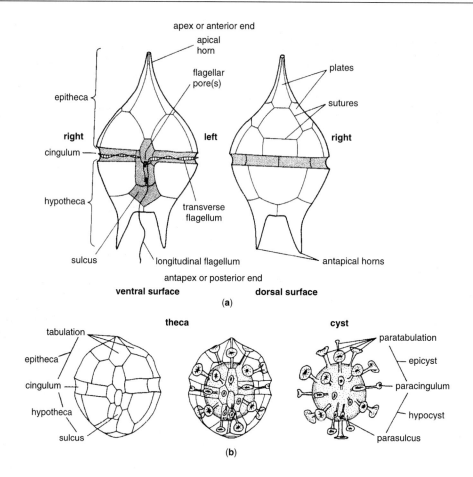

Fig 11.5   Descriptive morphology of (a) dinoflagellate and (b) dinoflagellate (left), unpeeled (middle) to reveal corresponding cyst (right).

some (Figure 11.7). The vesicle encloses a chamber which ranges in shape from spherical through ovoid to cylindrical and conical forms. The chamber opens through a pseudostome at the oral end, either directly or at the end of a neck with collar. The pseudostome is closed by an operculum which may be supported by the prosome. The base of the vesicle may be flat or extended as a copula. There are over 50 genera of chitinozoans.

Distribution and affinities

The group is most common in fine-grained sediments, usually those deposited in anoxic environments and associated with pelagic macrofauna such as graptolites and nautiloids, together with acritarchs. In some lithologies, such as black slates, chitinozoans are the only fossils preserved. These associations, together with their

widespread geographical range, suggest that chitinozoans were also pelagic. The group has proved useful for both local and regional correlations, mainly in Lower Palaeozoic rocks.

The precise affinity of the group remains uncertain. Chitinozoan vesicles were probably tightly sealed, and they occur as chains and clusters, both of which suggest they may have contained eggs or acted as dormant cysts. The group has been associated with annelids, echinoderms, gastropods and graptolites, but it is more likely that chitinozoans were the products of some soft-bodied vermiform animal developed during a pelagic life stage.

Two main groups have been established, based on the morphology of the operculum. The Simplexoperculati with relatively simple opercula are the older, and the more complex opercula of the Complexoperculati appeared later.

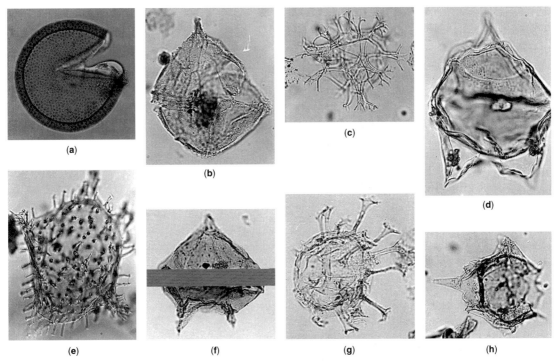

Fig 11.6   A prasinophyte (a) and some dinoflagellate taxa: (a) *Tasmanites* (Jurassic), (b) *Cribroperidinium* (Cretaceous), (c) *Spiniferites* (Cretaceous), (d) *Deflandrea* (Eocene), (e) *Wetzeliella* (Eocene), (f) *Lejeunecysta* (Eocene), (g) *Homotryblium* (Eocene), (h) *Muderongia* (Cretaceous), (b, c, h, i, f ×425; a, d, e ×250).

Evolution

Chitinozoans appeared and diversified during the early Ordovician to achieve an abundance continuing until the end of the Silurian. Chitinozoans were rare during the Devonian and finally became extinct in the earliest Carboniferous. Through time, the group developed smaller self-contained chambers with an increased complexity of ornament and a greater degree of apparent coloniality (Figure 11.8).

Calcareous nannoplankton

Nannofossils are usually smaller than 50 μm; in practice, the upper size limit for the group is taken at 63 μm, the smallest standard mesh size for sieves. Although the nannoplankton contains organic-walled and siliceous forms, the calcareous groups are most common in living floras and dominate the fossil record. Coccolithophores are the most dominant members of the fossil calcareous

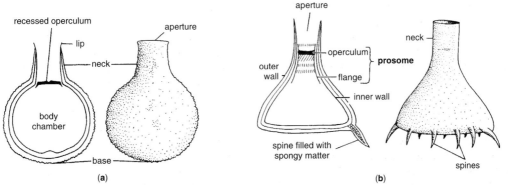

Fig 11.7   Descriptive morphology of the chitinozoans: (a) Simplexoperculate, *Lagenochitina*; (b) Complexoperculate, *Ancyrochitina*.

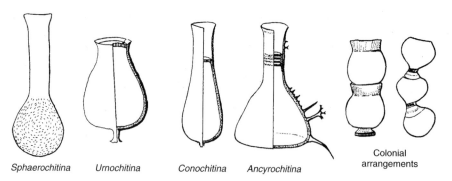

Sphaerochitina   *Urnochitina*   *Conochitina*   *Ancyrochitina*   Colonial arrangements

Fig 11.8 Some chitinozoan morphotypes: *Sphaerochitina* (×160), *Urnochitina* (×160), *Conochitina* (×80), *Ancyrochitina* (×240) and colonial arrangements (×40).

nannoplankton, represented by abundant isolated calcareous plates, coccoliths. Noncircular calcareous plates are rejected from the coccolithophores and usually termed nannoliths. These may be related to coccolith-bearing organisms, but in view of their diversity in form, the group may contain a number of quite different microbes secreting broadly similar calcareous plates. As a whole, calcareous nannoplankton first appeared during the late Triassic and is abundant in the surface waters of modern oceans.

Coccolithophores

Coccolithophores are unicellular algae, autotrophic in dietary mode, usually ranging in size from 10 to 50 μm, and globular, fusiform or pyriform in shape (Figure 11.9). The group has been assigned to the Phylum Haptophyta together with the closely related non-calcifying algae. They have golden-brown photosynthetic pigments and, in motile phases, two flagellae together with a third flagellum-like structure, the haptonema. The shell is composed of distinctive calcitic platelets or coc-

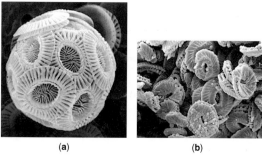

(a)   (b)

Fig 11.9 Some coccolith morphotypes: (a) coccospheres of the living *Emiliana huxleyi*, currently the most common coccolithophore (×6500); (b) late Jurassic coccolith limestone (×2000).

coliths, produced intra-cellularly; they migrate to the cell surface and are embedded in a layer of gelatinous, organic scales. Commonly the shell or coccosphere consists of 10–30 discrete platelets, although some forms have many more. Many taxa consist of similar coccoliths, but some have a variety of plate morphologies. There are two fundamentally different types of coccoliths: heterococcoliths have a radial array of relatively few (usually 20–50) complex-shaped crystal units, whereas holcococcoliths have several hundred, smaller, less complex-shaped crystals.

Haptophyte life cycles are poorly known; the few well-documented examples suggest most have alternating haploid and diploid stages. Coccoliths have a complex life cycle consisting of at least two main phases which were described initially as two different species: the *Crystallolithus hyalinus* stage may be haploid, flagellate and coated by minute coccoliths; the *Coccolithus pelagicus* stage may be diploid, nonflagellate and coated by placolith coccoliths. Coccoliths may be secreted during both or neither of the stages. In general terms, heterococcoliths are apparently formed during the diploid, usually non-motile, stage whereas holcococcoliths are formed during the motile, haploid stage.

Classification is based largely on coccolith form, which can present problems. Each coccolithophore can have a variety of coccoliths of different shapes, and moreover during ontogeny a single species can have a series of different holococcolith and heterococcolith plate morphologies. A biological classification based on the entire organism is not feasible since specimens are nearly always disarticulated; rather a parataxonomy has been devised for the individual elements of the coccolithophore. The intricate and distinctive form of coccoliths makes them ideal for morphological classification while they have a good fossil record. Although a number of types of classifications are available, all broadly based on the shapes and structure of the coccoliths, data from cytology and molecular genetics strongly supports classifications based on morphological criteria.

Together with diatoms, dinoflagellates and picoplankton, coccolithophores are the most abundant phytoplankton in modern oceans. Greatest diversity is developed within the tropics. Dependence on sunlight for photosynthesis restricts the group to the photic zone, with a depth range of 0 to about 150 m; however most are found clustered at depths of about 50 m, avoiding the most intense rays. Within wave-mixed surface waters there is normally only a slight vertical stratification of assemblages; quite different assemblages occur beneath the thermocline.

Rare coccoliths are known in the late Triassic, and are more common during the Jurassic and Cretaceous; the group peaked in the late Cretaceous, and the Chalk was almost entirely composed of nannofossils. Only a few species survived the end-Cretaceous extinction event to radiate again during the Cenozoic. There are at least 20 families of coccolith-bearing haptophytes.

Diatoms

The diatoms are unicellular autotrophs included among the chrysophyte algae; they are characterized by large green-brown chloroplasts. Both individuals, and loosely integrated colonies of diatoms, occur in a range of aquatic environments from saline to freshwater and across a range of temperatures, being particularly common in the Antarctic plankton. Both benthic and planktonic life modes occur, although within the plankton one group, the Centrales, preferred marine environments, whereas the other, the Pennales, is more common in freshwater lakes.

The diatom cell is contained within a siliceous skeleton or frustule comprising two unequally sized valves or thecae (Figure 11.10). The smaller hypotheca fits into the larger epitheca; the valve plates and congula of both valves interface with the congulum of the epitheca covering that of the hypotheca to form a connective seal.

Two main divisions are recognized: the Centrales, as the name suggests, have round valves with pores radiating in concentric rows from the valve centre; and the Pennales have more elliptical valves with the pores arranged in pairs. The latter are usually characterized by a median gash or raphe.

During reproductive fission, both the parent valves are used as the epitheca by the offspring, which then constructs its own hypotheca; this process occurs a number

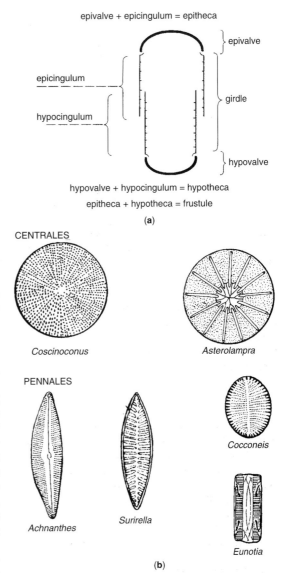

Fig 11.10 (a) Descriptive morphology of the diatoms, (b) some diatom morphotypes. *Coscinoconus* (×250), *Asterolampra* (×400), *Cocconeis* (×360), *Achnanthes* (×150), *Surirella* (×200) and *Eunotia* (×400).

of times each day, progressively reducing the size of the frustule. A stage of sexual reproduction kicks in to restore the growth momentum of the individual.

Diatom evolution

Both diatom frustules and, more commonly, endospores are preserved in the fossil record. A late Jurassic assemblage from western Siberia including *Stephanopyxis* may be the oldest known diatom flora. The first diverse floras appeared during the mid Cretaceous with almost ten families recorded; the group further diversified after this. Nearly 100 genera of centric diatoms are recorded from the late Cretaceous. Some of the first pennate diatoms appeared during the early Tertiary, colonizing freshwater environments for the first time; the group reached an acme during the Miocene.

Diatomites

Pure accumulations of diatom frustules are very porous, often 80%, with a density of about 0.5. These diatomites, also termed kieselguhrs and tripolis, are widely used as purifiers, for filtering many types of drink, medicines and water. Over 2 million tonnes are extracted each year for commercial use. Diatomites can form deposits up to 500 m thick. Modern sedimentation rates suggest that 4–5 mm of diatomaceous ooze is deposited over 1000 years, currently occupying over 10% of the ocean floor today. Major commercial deposits occur in the Miocene of the Ardèche, France, and in the Pliocene and Pleistocene of Cantal, France; other deposits occur in Spain, Germany and Russia.

Foraminifera

Foraminifera are testate protozoans, common in a wide variety of Phanerozoic sedimentary rocks and of considerable biostratigraphical and palaeoenvironmental value. The foraminiferans are grouped in the Sarcodina, together with familiar living forms such as *Amoeba*. Many pioneer studies in micropalaeontology were based on the relatively abundant, characteristic and widespread tests of these unicellular organisms.

Morphology and classification

Although many different classifications of foraminiferans have been published, the morphology of the animal and its shell is the main basis for identification of species and higher categories. Most foraminifera have a mineralized shell or test comprising chambers, interconnected through holes or foramina. The test may be composed of a number of materials; four main categories have been documented (Figure 11.11). Agglutinated tests comprise fragments of extraneous material bound together by a variety of cements; the debris may be siliciclastic, such as quartz or mica grains, or calcareous, recycling fragments of coccoliths or other forams. Calcareous tests may be porcellaneous (imperforate) consisting of small,

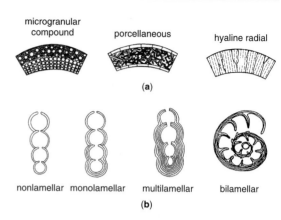

Fig 11.11   Main types of foraminiferan test walls: (a) The composition and structure of test walls and (b) lamellar construction.

randomly orientated crystals of high-magnesium calcite. Hyaline (perforate) tests have two main structural modes: the radial tests are made up of minute calcite crystals with their c-axes normal to the test surface, whereas granular forms consist of microcrystals of calcite with variable orientations. Both modes usually have a multilamellar structure and perforations. Finally, microgranular tests consist of tightly packed, equidimensional grains of crystalline calcite. Most members of this group are late Palaeozoic in age.

The gross morphology of a foraminiferan test is governed by the shape and arrangement of the animal's chambers. The group has evolved a wide range of test symmetries (Figure 11.12), from simple uniserial and biserial forms to more complex planispiral and trochospiral shapes. Chambers also come in a wide spectrum of shapes, from simple spherical compartments through tubular to clavate forms (Figure 11.13). Moreover, the shape and position of the aperture may vary. Surface ornament may include ribs and spines, or be merely punctate or rugose.

Life modes

The foraminifera adopted two main life modes, benthic and planktonic. The majority are benthic epifaunal types; they are either attached, or cling to the substrate or crawl slowly over the seabed by extending their protoplasmic pseudopodia. Infaunal types live within the top few centimetres of sediment. Most benthic forms have a restricted geographic range. Planktonic foraminiferans are most diverse in tropical equatorial regions, and may be extremely abundant in fertile areas of the oceans, particularly where upwelling occurs.

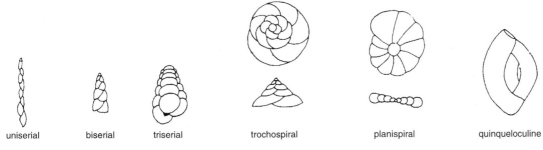

Fig 11.12   Main types of foraminiferan chamber construction.

Fig 11.13   Some genera of foraminifera: (a) *Textularia*, (b) *Cribrostomoides*, (c) *Milionella*, (d) *Spriorlina*, (e) *Brizalina*, (f) *Pyrgo*, (g) *Elphidium*, (h) *Nonion*, (i) *Cibicides*, (j) *Globigerina*, (k) *Globorotalia*, (l) *Elphidium* (another species) (all ×50–100).

Box 11.1 Foraminifera in modern environments

The ratio of agglutinated : hyaline : porcellaneous foram tests has been used extensively to differentiate among a range of modern environments. Ternary plots of the relative frequencies of test type have been assigned to fields for hypersaline and marine lagoons, estuaries and open shelf seas (Figure 11.14). Fossil faunas may be plotted on these templates, and estimates for palaeosalinity are possible for ancient environments.

Measures of the ratio of benthic to planktonic foraminifera are also useful in environmental studies. In general terms, the percentage of benthic taxa declines rapidly below depths of about 500 m in modern seas and oceans. Data from living assemblages have been used to interpret palaeoenvironments with diverse fossil foraminifera faunas. For example, microfossil analysis of the upper part of the Chalk of the Anglo-Paris Basin has suggested water depths of between 600 and 800 m during the mid Cretaceous on the basis of the high proportions of planktonic foraminifera; however, by the late Cretaceous, water depths of about 100 m are suggested by the rich benthic fauna.

Evolution and geological history

The earliest foraminifera are known from the Lower Cambrian, represented by simple agglutinated tubes assigned to *Bathysiphon*, a living benthic genus (Figure 11.15). More diverse agglutinated forms appeared during the Ordovician, and microgranular tests evolved during the Silurian; however, it was not until the Devonian that multichambered tests probably developed. Nevertheless, Carboniferous assemblages have a variety of uniserial, biserial, triserial and trochospiral agglutinated tests. Around the Devonian–Carboniferous boundary, the first partitioned tests displaying multilocular growth modes appeared. Two microgranular families, the Endothyridae and the Fusulinidae, dominated Carboniferous assemblages, and the porcellaneous Miliolinidae achieved importance in the Permian. The Fusulinidae, despite a high diversity during the late Permian, became extinct at the end of the Palaeozoic, while the Endothyridae and the Miliolinidae were very much reduced in number.

Although Triassic assemblages were generally impoverished, the stage was set for a considerable radiation during the Jurassic. Two hyaline groups, the benthic Nodosariidae and planktonic Globigerinidae, diversified, while the agglutinates, Lituolitidae and Orbitolinidae continued. The planktonic foraminifera underwent a period of diversification in the Cretaceous, but this culminated in the near extinction of the group in the late Maastrichtian. Two further periods of diversification took place, during the Palaeocene–Eocene and the Miocene.

Radiolaria

The radiolarians are marine unicellular plankton with delicate exoskeletons, usually composed of a framework of opaline silica (Figure 11.16). Most radiolarians feed on bacteria and phytoplankton, and occupy levels in the water column from the surface to the abyssal depths, although most live in the photic zone, commonly associated with symbiotic algae.

The radiolarian ectoplasm secretes the test and holds symbiotic zooxanthellae, whereas the endoplasm contains the nucleus and other inclusions. The group has two types of pseudopodia: the axopodia are rigid and not ramified, whereas the filipodia are thin ramified (branched) extensions of the ectoplasm.

The radiolarian skeleton or test consists of isolated or networked spicules, composed of opaline silica and forming sponge-like structures or trabeculae. Two main groups are recognized on the basis of skeletal structure and arrangement of perforations: the nassellarians develop a lattice from bar-like spicules, each end having a bundle of spicules, which together are enclosed in the cephalis, where chambers may be defined by perforate septa; the spumellarians, however, have a radial symmetry based on a spherical body plan.

Although some records suggest an origin in the late Cambrian or earlier, the radiolarians are first common during the early Ordovician when they occur in chert sequences often associated with other deep-water oceanic assemblages; the albaillellarians and the enactinids were the dominant forms, although after the Devonian spumellarians with sponge-like tests were more prominent. Spumellarians remained important during the Triassic, with genera such as *Capnuchosphaera*, although the nassellarians had appeared. Spumellarians continued domination through the Jurassic, Cretaceous and early Tertiary. Late Tertiary forms evolved thinner skeletons, perhaps because of increased competition with the diatoms for mineral resources.

Radiolarian oozes cover about 2.5% of modern ocean floors, accumulating at rates of 4–5 mm per 1000 years.

Box 11.1 (Cont.)

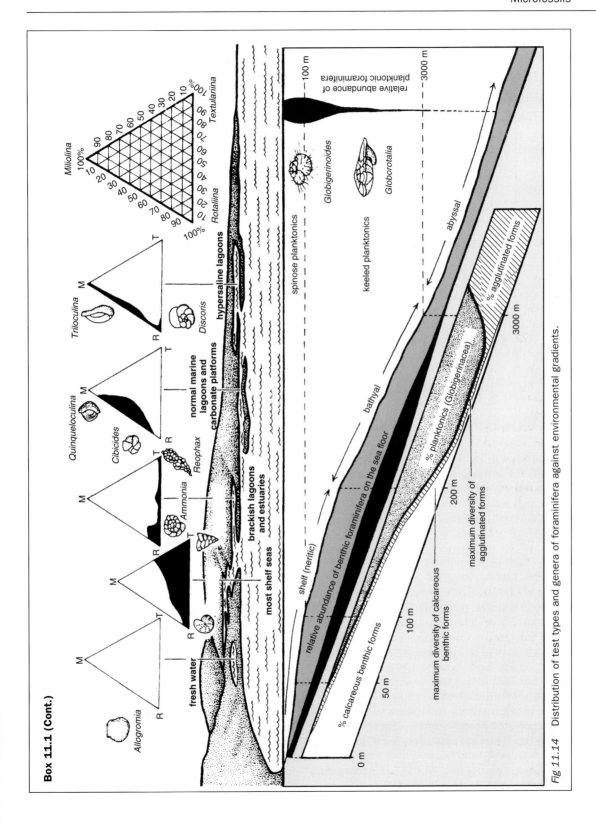

Fig 11.14   Distribution of test types and genera of foraminifera against environmental gradients.

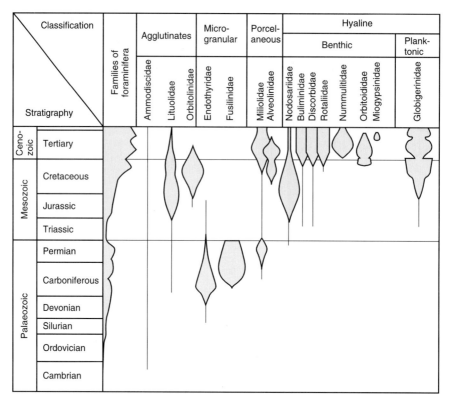

Fig 11.15   Stratigraphical ranges of the main foraminiferan groups.

Radiolarian cherts and radiolarites commonly occur in oceanic facies preserved in mountain belts and commonly associated with ophiolite sequences at major plate tectonic boundaries.

Tintinnids

The tintinnids are one of the few groups of ciliated protozoans with a fossil record. Fossil tintinnids, or calpionellids, are common in late Jurassic and early to mid Cretaceous pelagic limestones of the Alpine Belt in Europe and in rocks of similar age and facies in North Africa, the Caribbean and Mexico. These animals are minute, ranging from about 60–280 µm in diameter. Living forms are part of the zooplankton, providing a food source for larger members of the plankton. The soft tissue is enclosed within a lorica, often ten times larger than the cell itself. Modern forms have an organic lorica, whereas fossil forms had a calcareous test (Figure 11.17).

Two families of fossil tintinnid have been recorded, together ranging in age from the late Jurassic to the mid Cretaceous.

Microinvertebrates

Etched sediment residues commonly contain fragments or parts of invertebrates. For example, Palaeozoic residues often have graptolite debris such as the prosicula, the jaws of annelid worms or scolecodonts, and phosphatic microbrachiopods such as the acrotretides and mollusc-like tentaculitids. Ostracodes are, however, the most common microinvertebrates and they are found in a wide range of facies.

Ostracodes

Ostracodes are crustacean arthropods (see also Chapter 8), abundant and widespread in aquatic environments. They are the commonest fossil arthropods, ranging in age from the Cambrian to the present day. They have small bivalved carapaces, hinged along the dorsal margin (Figure 11.18); most are about 1 mm in size although some grow to over 10 mm. The bivalved carapace completely covers the entire animal when closed and in some

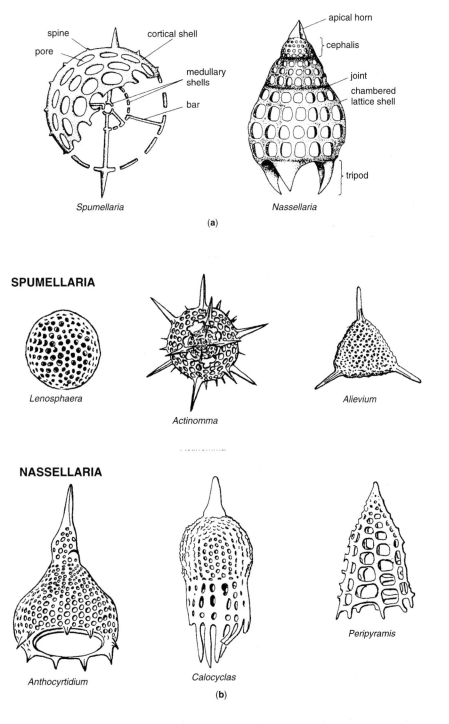

Fig 11.16 Descriptive morphology of (a) the radiolarians and (b) some radiolarian morphotypes. *Lenosphaera* (×100), *Actinomma* (×240), *Alievium* (×180), *Anthocyrtidium* (×250), *Calocyclas* (×150) and *Peripyramis* (×150).

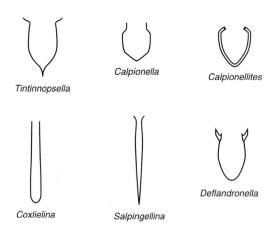

Tintinnopsella

Calpionella

Calpionellites

Coxlielina

Salpingellina

Deflandronella

Fig 11.17 Morphology of some tintinnids in cross-section from limestones (×100–200).

species is perforated by tiny pore canals, some containing sensory setae. Most ostracodes are benthic, swimming, crawling or burrowing at the sediment–water interface in muds or silts with abundant organic material; a few, such as the myodocopids, are planktonic and some are commensal or parasitic.

Ostracodes have weak segmentation, with a poorly defined head, thorax and sometimes an abdomen; the animal is contained within a twin-valved carapace united dorsally by an elastic ligament and a variably developed hinge; growth is by periodic ecdysis (moulting). Following each moult phase, the carapace initially develops as a pair of chitinous valves enclosing the animal; typically, most of the carapace is then calcified, with the exception of the dorsal margin, which remains as a chiti-

nous ligament forcing the valves apart when the internal adductor muscles relax.

The central muscle scars vary across the class (Figure 11.19) from complex patterns in the Leperditicopida and Podocopida to a single scar in some members of the Palaeocopida.

Articulatory structures are variably developed along the hinge line. Three main types of hinge are known. Adont hinges lack teeth, but have a long median element on the left valve that fits a socket on the right valve. The merodont hinge has anterior and posterior terminal elements on the right valve, fitting respective sockets on the left valve. Amphidont hinges have short terminal elements with the median element subdivided into longer (posterior) and shorter, tooth-like (anterior) parts.

The carapace is perforated by canals holding setae that communicate with the exterior. The body is suspended within the carapace, attached by muscles; it is equipped with seven pairs of appendages, three in front of the mouth and four behind. The appendages are specialized, acting as sensory organs, limbs for the capture and processing of food, locomotion and general cleaning and housekeeping within the carapace. The animal has a digestive system, sophisticated genitalia and a nervous system; a median eye or a pair of lateral eyes is present.

Sexual dimorphism is common, and often reflected in the shape and special features of the ostracode carapace. Males commonly have a greater length to height ratio than the females. In some benthic Palaeozoic ostracodes the female had a brood pouch as part of the carapace wall; these ostracodes, the supposed females, are often called heteromorphs, while the males and juveniles, lacking the brood pouch, are called tecnomorphs.

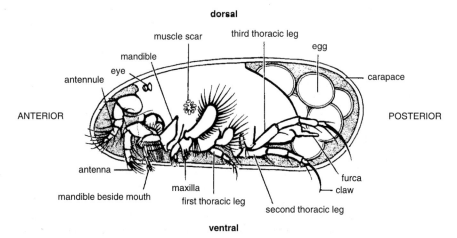

Fig 11.18 Descriptive morphology of the ostracode animal.

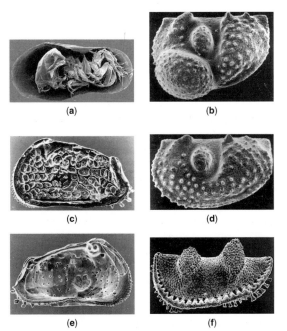

(a)　　　　　　　　(b)

(c)　　　　　　　　(d)

(e)　　　　　　　　(f)

Fig 11.19 Some ostracode genera: (a) left valve of male living *Limnocythere* showing details of appendages (×30); (b), (d). left valves of female and male heteromorphs of *Beyrichia* (Silurian) (×18), (c), (e) external and internal views of left valve of living *Patagonacythene* (×30), (f) palaeocopid *Kelletina* (Carboniferous) (×30).

Evolution of ostracodes

Ostracodes appeared first during the early Cambrian; the bradoriids (archaeocopides) were an unusual group with a large, weakly calcified carapace, highest at the posterior part of the valves. The group was short-lived, disappearing during the latest Cambrian–earliest Ordovician. The later history of the group shows a number of trends: the evolution of small size, perhaps simpler muscle systems and shorter hinge lines. The functional significance of these changes is not immediately clear.

Large Leperditicopida and Palaeocopida appeared during the Ordovician and dominated ostracode faunas until the Devonian, when deep-water environments were characterized by the small, sometimes spiny, myodocopids. The marked faunal turnover at the end of the Permian is emphasized by the virtual extinction of the main Palaeozoic group, the palaeocopids; records of the group in the Lower Triassic are rare. The Podocopida dominated Mesozoic and Cainozoic ostracode faunas. Although early Jurassic assemblages are of low diversity, podocopids such as the platycopines, cypridaceans and cytheraceans radiated steadily during the Jurassic.

By the Cenozoic, the cypridaceans dominated lake environments, whereas the cytheraceans were established in marine settings.

Microvertebrates

Conodonts are the most common microvertebrates in Palaeozoic strata. They are routinely extracted by dilute acids from a range of carbonate lithologies. However, various microremains, such as scales and teeth, of more typical fishes are potentially useful biostratigraphically, and are proving important in both morphological and palaeoenvironmental reconstructions.

Conodonts

Conodont elements are phosphatic tooth-like microfossils, ranging in age from Cambrian to Triassic. The function of the conodont elements as teeth or possibly lophophore supports, within an eel-like fish, has been established only recently (see pp. 197–198). The mineralized conodont elements were all arranged in some form of apparatus; many naturally occurring assemblages reflect the arrangement of original apparatuses. The group is useful biostratigraphically, exclusively marine, and most abundant in nearshore calciclastic facies. In addition, since colour changes of the elements can be related to changing temperature, conodonts are important indicators of thermal maturation (see Chapter 2).

Conodont morphology

On the basis of growth modes there are three main groups of conodonts: protoconodonts, such as *Hertzina*, are simple cones with deep basal cavities; paraconodonts, such as *Furnishina* and *Westergaardina*, are mainly simple cones; and euconodonts or true conodonts are more complex, with cones, bars and blades. The protoconodonts are almost certainly unrelated to true conodonts; they are probably chaetognaths or arrow worms, invertebrate cousins of the echinoderms and hemichordates.

Euconodonts occur as three broad types of element, consisting of laminae of apatite. These grew outwards from an initial growth locus. White matter often occurs between, or cross-cuts, lamellae; this material is like vertebrate bone. The three main morphotypes of conodont element have been used in the past as the basis of a crude single element or form taxonomy (Figures 11.21, 11.22).

Box 11.2 Ontogeny and phylogeny in Ordovician ostracodes

The moult series of ostracodes provides an excellent opportunity for the study of the relationship between ontogeny and phylogeny. Olempska's detailed study (1989) of *Mojczella* through the condensed Lower–middle Ordovician limestones of the Holy Cross Mountains, Poland, demonstrates phyletic gradualism in these microarthropod lineages (Figure 11.20). In broad terms, during ontogeny, the two lateral crests, C1 and C2, tend towards coalescence. In earlier populations, the crests are separated throughout ontogeny, but in later populations this separation is only preserved in the earliest growth stage; intermediate morphologies are seen in stratigraphically intermediate populations.

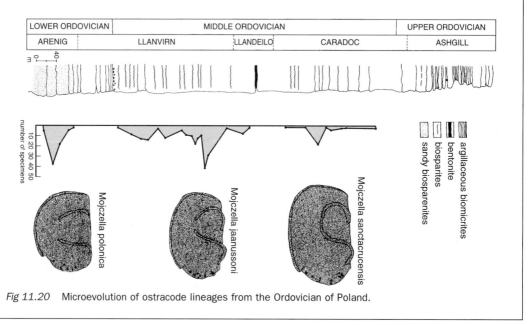

Fig 11.20   Microevolution of ostracode lineages from the Ordovician of Poland.

The cones or coniform elements are the simplest with the base surmounted by a cone-like cusp, tapering upwards, sometimes ornamented with ridges or costae. Bars or ramiform elements consist of an elongate blade-like ridge with up to four processes developed posterior, anterior or lateral to the cusp. Platforms or pectiniform elements have a wide range of shapes with denticulate processes extending both anteriorly and posteriorly from the area of the basal cavity; some also have primary lateral processes. The cusp is attenuated, whereas the base may be expanded to form a platform with denticles on its upper surface. The basal cavity of some elements was probably covered by plates. In many elements the basal cavity was open whereas in others the cavity was filled by the basal body of the element in the form of cartilage- or dentine-like material.

At first individual elements were given a Linnean binomen based on their external shape, since entire animals were once thought to contain only one type of element. Multielement taxonomy now forms the basis for the description, identification and classification of the group. The first-established species name for elements of a complete apparatus is used for the entire assemblage.

Rare examples of associated elements, or apparatuses (Figure 11.23a), showed that the conodonts occurred naturally as symmetrical associations of paired elements, with commonly seven or eight pairs. Natural conodont assemblages are arranged in three main zones; the pectiniform (P-elements), makelliform (M-elements) and symmetry transition series (S-elements). P-elements occur posteriorly, whereas the M-elements form the anterior part of the assemblage. These assemblages may contain two or more different types of elements; generally bars and platforms occupy P positions, whereas bars and cones are found in M and S positions. The P, M and S positions are more finely tuned with alphabetical subscripts, e.g. P_a and P_b elements. Unimembrate apparatuses have a single type of element whereas multimembrate assemblages have two or more types of elements.

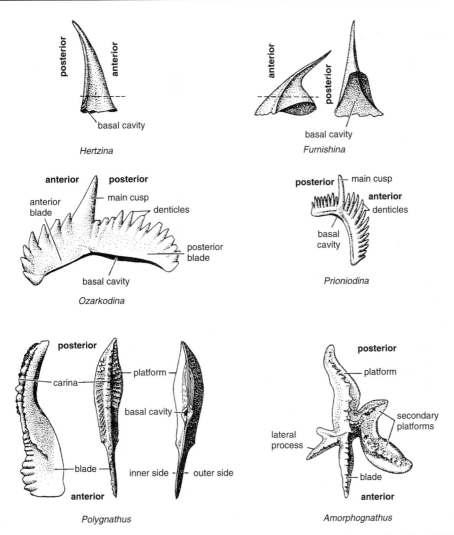

Fig 11.21 Descriptive morphology of the main types of conodont element: *Hertzina* (×40), *Furnishina* (×40), *Ozarkodina* (×40), *Prionidina* (×20), *Polygnathus* (×40) and *Amorphognathus* (×40).

The conodont animal

Conodont elements were recognized by Pander in 1856 and are variably common in rocks ranging in age from late Cambrian to late Triassic. Surprisingly, only recently has the conodont animal been identified and described. Nevertheless, there have been a number of past attempts to identify the conodont organism. Most authorities agreed that the conodont animal was soft-bodied, bilaterally symmetrical and nektonic. Past attempts to identify the conodont animal have since been shown to have been mistaken: they include worm-like creatures that had eaten some conodonts, and even a loosely constructed creature that looked like a toilet roll with a tooth-like element arranged outside.

In the early 1980s, during an investigation of the Carboniferous shrimps of the Granton Shrimp Bed, near Edinburgh, an eel-like animal, containing an assemblage of conodonts near its anterior end (Figure 11.23b) was discovered. Detailed examination of the animal revealed the elements were *in situ* and, this time, had not been merely eaten by the animal (Aldridge *et al.*, 1993a). One the greatest mysteries in palaeontology had been solved. Ten specimens of the conodont animal from the Granton Shrimp Bed are now known. The animals are elongate, up to 55 mm in length, with a short, lobate head and a caudal fin supported by rays. The head had two large eyes.

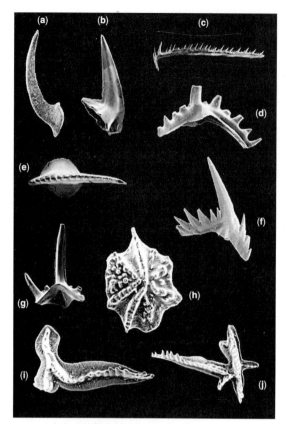

Fig 11.22 Conodont elements: (a) coniform, lateral view, (b) coniform, lateral view, (c) ramiform, lateral view, (d) ramiform, lateral view, (e). straight blade, upper view, (f) arched blade, lateral view, (g) ramiform, posterior view, (h) platform, upper view, (i) platform, upper view, (j) platform, upper view (×20–35).

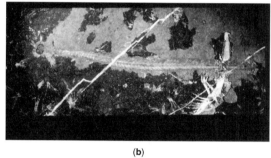

Fig 11.23 (a) Natural assemblage of conodonts from the Carboniferous of Illinois (×24). (b) The conodont animal from the Carboniferous Granton Shrimp Bed, Edinburgh with anterior to the left (×1.5).

Immediately behind the eyes, ventrally, is the feeding apparatus consisting of an assemblage of conodont elements, and lacking any preserved surrounding tissue. The Granton specimens have an ozarkodinid-type apparatus; the comb-like ramiform elements comprise the anterior basket and probably captured prey whereas, posteriorly, the pectiniform elements acted as a food processor.

Biostratigraphy

Detailed biostratigraphical schemes based on conodonts have been established for many parts of the Palaeozoic and Triassic. For example, over 20 conodont zones have been established for the Ordovician System whereas the Upper Devonian is the most congested interval, with over 30 biozones. In north-west Europe the Carboniferous is routinely correlated on the basis of conodont zones.

Remarkable precision is now available in some zonal schemes. This has permitted the development of models for global environmental change during the early Silurian (Figure 11.24) tied to a tight conodont zonation (Aldridge *et al.*, 1993b). Two oceanic states are recognized: primo episodes with oxygenated cool oceans with a good vertical circulation and adequate supplies of nutrients, and secundo episodes with warm stratified oceans with deep saline levels and poor nutrient supplies. Sudden changes between episodes would involve changes in vertical circulation and nutrient supply, probably promoting extinction events.

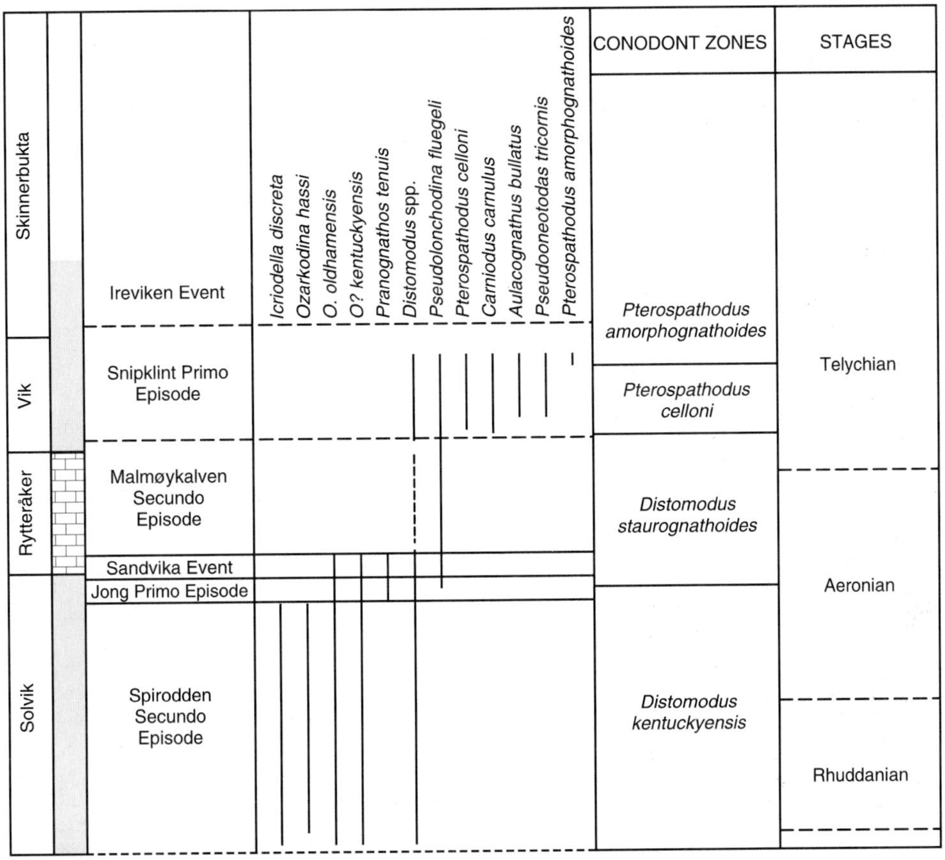

		Events/Episodes	Icriodella discreta	Ozarkodina hassi	O. oldhamensis	O? kentuckyensis	Pranognathos tenuis	Distomodus spp.	Pseudolonchodina fluegeli	Pterospathodus celloni	Carniodus carnulus	Aulacognathus bullatus	Pseudooneotodas tricornis	Pterospathodus amorphognathoides	CONODONT ZONES	STAGES
Skinnerbukta		Ireviken Event													*Pterospathodus amorphognathoides*	Telychian
Vik		Snipklint Primo Episode													*Pterospathodus celloni*	
Rytteråker		Malmøykalven Secundo Episode													*Distomodus staurognathoides*	Aeronian
		Sandvika Event														
		Jong Primo Episode														
Solvik		Spirodden Secundo Episode													*Distomodus kentuckyensis*	Rhuddanian

Fig 11.24 Alternation of p (primo) and s (secundo) oceanic states correlated with part of the Lower Silurian succession of the Oslo Region, Norway.

Biogeography

There have been many studies on the provincialism of the diverse Palaeozoic conodont faunas (e.g. Bergström, 1990). Cambrian faunas show a distinction between low-latitude warm-water associations of the North American Midcontinent Faunal Region and the high-latitude cool-water Atlantic Faunal Region. During the early Ordovician, six discrete provinces are recognized statistically; but at the end of the Ordovician most typical Atlantic taxa disappeared, and the majority of Silurian forms originated in the Midcontinent region.

Dzik (1990) has documented in graphic detail the evolution of conodont lineages at high latitudes during the Ordovician. Conodonts evolved independently and there are only a few incursions from lower-latitude faunas. Towards the end of the Ordovician, high-latitude cold-water faunas migrated into lower latitudes. Thus, late Ordovician equatorial Midcontinent assemblages originated in polar and subpolar regions and themselves

formed the foundation for the Silurian fauna. During the mid and late Palaeozoic, conodonts were mainly restricted to tropical latitudes. Devonian and Carboniferous faunas show some biogeographic differentiation among shelf associations.

Palaeoecology

Conodonts occur in a wide range of marine and marine-marginal environments, although the group is most common in nearshore carbonate facies, commonly in the tropics. Several environment-related conodont palaeocommunities have been identified in many parts of the Palaeozoic. In view of the large sample sizes available, multivariate techniques such as principal component and correspondence analyses have been widely used. A sequence of depth-related associations have been decribed in the Missourian (Upper Carboniferous) rocks of Kansas (Figure 11.25). In general terms, the first

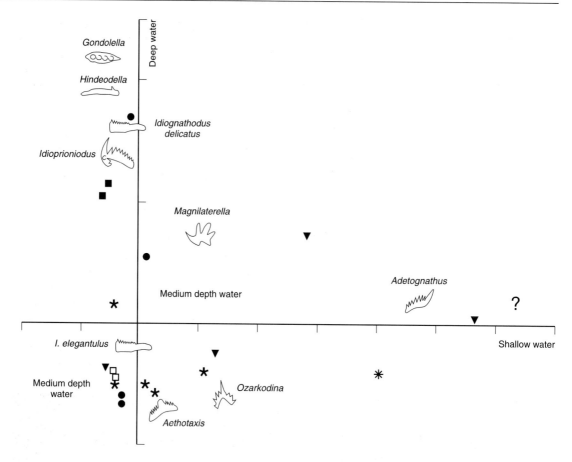

Fig 11.25 Correspondence analysis of conodont assemblages from the Missourian of eastern Kansas. (▼, Outside shale; ★ Shoal limestone; ● Upper limestone; ✳ Middle limestone; □ Phantom black shale; and ■ Black shale).

eigenvector ranges from medium-water to shallow-water depths whereas the second eigenvector ranges from medium-water to deep-water depths. Groups of genera are differentiated along the two eigenvectors and indicate that associations with *Gondolella*, *Idioprioniodus* and *Hindeodella* (possibly synonymous with *Ozarkodina*) occupy deep-water environments at levels of decreasing depths; conversely *Adetognathus*, *Aethotaxis* and *Ozarkodina* are shallow-water taxa.

Evolution

Paraconodonts have been reported from uppermost Proterozoic strata on the Siberian Platform. Simple coniform elements may have been some of the first apatite skeletons. The early and mid Cambrian faunas were exclusively paraconodonts, but during the late Cambrian, the distacodontids evolved, presumably from a simple conical ancestor; their apparatuses had mainly a unimembrate arrangement of coniform elements.

Although many early Ordovician conodonts possessed multimembrate apparatuses with coniforms, some multimembrate assemblages evolved with a variety of chiefly coniform and ramiform element types. Conodont diversity peaked during the mid Ordovician, with the first appearances of over 60 genera. During this interval of experimentation, there was a huge diversity of apparatus patterns never again matched; later Palaeozoic apparatuses are relatively uniform and monotonous, perhaps after the stabilization of feeding modes. Pectiniform elements were common from the early Ordovician onwards, together with a wide diversity of blades and platforms during the mid to late Ordovician. Late Ordovician conodonts had a variety of membrate patterns; but the end-Ordovician extinction of the group removed much of this diversity. Silurian faunas are less variable, with mainly multimembrate apparatuses with ramiform and pectiniform elements.

The conodonts again radiated during the late Devonian, with specialized ramiform and pectiniform elements; over 1000 conodont taxa have been named from the Upper Devonian. Carboniferous conodonts were characterized by a lack of coniform elements, and pectiniform elements lay in the P apparatus position, and ramiform elements in the M and S positions. Most conodonts disappeared during the early Permian and only a few species were represented in the saline waters of the Zechstein Sea across Europe.

Elsewhere, late Permian conodont faunas were more diverse, although most late Permian and Triassic species had small commonly multimembrate apparatuses. The group as a whole disappeared at the very end of the Triassic.

Ichthyoliths

Ichthyoliths are mainly microvertebrate remains such as scales and teeth (Figure 11.26). Although, like many other microsclerites, ichthyoliths are difficult to classify, they are relatively easy to relate to complete articulated specimens; a great deal is known about most fishes, with the exception of the earliest members of the group in the Upper Cambrian and Ordovician. Nevertheless, complete fishes are rare, whereas ichthyoliths are relatively common and are of increasing biostratigraphical and ecological value.

Recently, a number of case studies have served to highlight the potential of ichthyolith studies. For example, Turner (1993) has described an early Carboniferous microvertebrate fauna from the Narrien Range, Queensland, Australia. The ichthyolith assemblage contains teeth, toothplates, scales, lepidotrichia and ornamented bones of both rhipidistian and palaeoniscoid fishes (see also Chapter 9). Moreover acanthodian scales are also common. Less common are the microremains of chrondrichthyans: elasmobranchiform teeth of the *Acanthodes* type, xenacanthoid teeth of the *Diploselache* type, together with rarer brachydont and hybodontiform teeth are represented. The microvertebrate fauna can be used to reconstruct a diverse early Carboniferous fish fauna in a nonmarine to marine-marginal setting commonly associated, elsewhere, with basal tetrapods.

Fig 11.26 Some microvertebrate taxa: (a) thelodont scale (Devonian); (b) thelodont body scale (Devonian); (c) protacrodont shark tooth (late Devonian–early Carboniferous); (d) acanthodian scale (Devonian); (e) shark tooth (Triassic); (f) hybodont shark scale (Triassic). (Scale bars on photographs.)

Plant palynomorphs

Palynomorphs are generally united on the basis of a broadly common chemical composition which makes them resistant to treatment by various acids, including hydrochloric, hydrofluoric and nitric acids. Palynofacies are thus based on spores, pollens and fragments of higher plants, together with the cysts of micro-organisms such as the acritarchs and dinoflagellates; these assemblages also contain residues of indeterminate organic matter, useful in sedimentary geochemistry and in thermal maturation studies.

Spores and pollen

Spores and pollen are part of the reproductive system of plants. Plant material is not usually mineralized, but the polymerized, organic outer coat (exine) of palynomorphs is extremely durable, and can be destroyed by oxidation. Spores and pollen are generally widespread; after dispersal from the parent plant, they are carried by wind and water across both marine and nonmarine environments. The distinctiveness and durability of palynomorphs has ensured their value in the biostratigraphy of the Upper Palaeozoic, particularly the Coal Measures as well as most parts of the Mesozoic and Cenozoic. Geological exploration, particularly by oil companies, relies heavily on palynology, based on spore and pollen analysis. Moreover, pollen analysis is now a routine part of palaeoenviromental reconstructions of the Quaternary.

Most spores show bilateral symmetry; megaspores range in size from several micrometres to about 4 mm, microspores are less than 200 µm in diameter (Figure 11.27). The proximal pole is marked by the germinal aperture; it may be a rectilinear slit (monolete condition) or it may have a triad of branches (trilete condition). The laesurae are the contact scars with neighbouring spores, commonly converging at a point or commissure (see also Chapter 10).

Pollen grains are usually smaller than spores, ranging in size from 2 to 150 µm. Inaperturate spores lack a germinal aperture (Figure 11.27). A single aperture at the distal pole characterizes the gymnosperms, the monocotyledons and primitive dicotyledons. Acolpate or asulcate pollen grains lack an obvious germinal aperture. Many pollen grains, however, such as pine and spruce are saccate with both a body or corpus and vesicles or sacci. The terms colpus and sulcus are often used for similar depres-

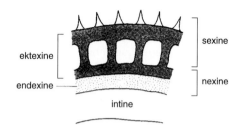

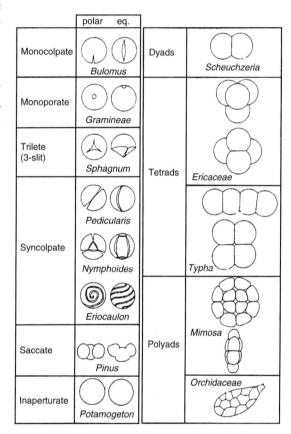

Fig 11.27 Basic morphology and terminology of spores and pollen, shown in polar and equatorial (eq.) views.

sions or furrows; strictly speaking, the sulcus refers to a furrow not crossing the equator of the pollen. Monosulcate pollen with a single distal sulcus developed during a series of meioses is typical of the gymnosperms and the monocotyledon angiosperms. The tricolpate pattern, developed in the dicotyledon angiosperms, has three germinal apertures or colpi arranged with triradiate symmetry.

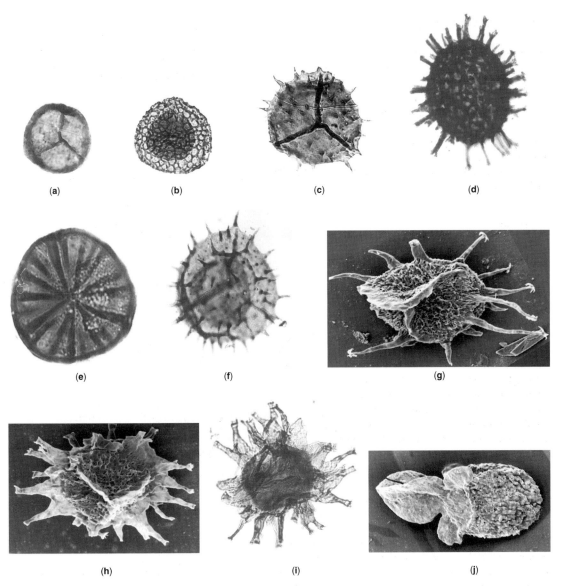

Fig 11.28 Some Devonian and Carboniferous spore taxa: (a) *Retusotriletes*, (b) *Retispora*, (c) *Spinozonotriletes*, (d) *Raistrickia*, (e) *Emphanisporites*, (f) *Grandispora*, (g) *Hystricosporites*, (h), (i), *Ancyrospora*, (j) *Auritolagenicula* (a)–(d) (f), (i) ×400; (g) ×90; (h) ×125; (e) ×750; (j) ×40.

Land plant life cycles

Vascular plants are similar to their algal ancestors in having a two-stage life cycle: the sporophyte phase reproduces asexually through the generation of spores whereas the gametophyte phase reproduces sexually (see also Chapter 10). Land plant reproduction favours the sporophyte mode, i.e. the sporophyte generation is the dominant one. This generation produces sporangia within which spores develop by meiosis. This results in a tetrad of four new cells (spores) which are haploid, i.e. each new cell has half the number of chromosomes compared with the parent plant. When ripe, the sporangium bursts open releasing the spores for dispersal. If conditions are favourable for germination, a gametophyte plant (prothallus) grows, which produces gametes. Fertilization, i.e. the fusion of male and female gametes, leads to the

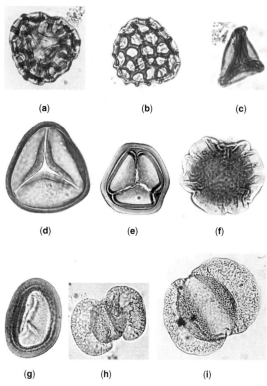

Fig 11.29 Some Jurassic spore and pollen taxa: (a), (b) *Klukisporites*, (c) *Dettmanites*, (d) *Dictyophyllidites*, (e) *Retusotriletes*, (f) *Callialasporites*, (g) *Classopollis*, (h) *Podocarpidites*, (i) *Protopinus* (×400).

new sporophyte generation which is diploid. However, this basic cycle can take a variety of forms, each characteristic of particular plant groups (Figure 11.28).

Early vascular plants, such as the rhyniophytes and primitive clubmosses, are homosporous. The sporophyte produces numerous identical isospores which give rise to the gametophyte generation with both male and female gametes. Heterosporous plants, however, such as the giant clubmosses and giant horsetails, produce both microspores (male) and megaspores (female), often on different plants. On germination, these spores produce separate gametophyte plants, which in turn produce the male and female gametes. Both the homosporous and heterosporous conditions require water to transport the gametes and allow fertilization to take place. Colonization of drier land away from aquatic environments demanded modifications of plant reproductive cycles. Megaspores first develop as an ovule within a megasporangium. The spore, in a protracted gametophyte phase, then produces a micro-prothallus with only a few cells acting as a male microspore; these pollen

grains are then dispersed in search of a female cone for fertilization.

The gymnosperms, such as the conifers, have megasporangia in cones and similar structures (Figure 11.29). Pollen reaching the ovule joins the eggs through a pollen tube and an opening or micropyle. Fertilization in the cone results in a seed, armed with a protective and nutritous coating, prepared for dispersal.

In angiosperms, which are more advanced, the ovule is covered by a tough coat or carpel, attached to the style, projecting upwards; the end of the style carries the stigma, designed to entrain pollen. The anthers or the microsporangia are supported by stamens and liberate pollen for dispersal by water, wind or pollinating insects. On pollination, as in the gymnosperms, a pollen tube grows towards the ovule, in this case through the style before penetrating the micropyle. Clearly this system is heavily dependent on pollinators, attracted by colourful flowers with the promise of nectar (see also Chapter 10).

Classification

Virtually all fossil pollen and spores are identified and classified on the basis of the morphology of the resistant outer wall or exine; as is the case with a number of other palynomorph groups, only a parataxonomy is possible. In one scheme the palynomorphs are grouped together into 'turma' categories; thus spores belong to the Anteturma Sporites and pollen to the Anteturma Pollenites. Other methods favoured by exploration geologists involve merely cataloguing the shapes in special-

---

Box 11.3 A spore and pollen code

A detailed shorthand notation or code has been developed to describe rapidly and communicate pollen and spore morphology. A simplified abstract of the scheme is outlined below.

S (spore) classified on laesura (scar) type:
  c = trilete
  a = monolete
  b = dilete
  0 = lacking laesurae
P (pollen) classified on colpation or sulcation type:
  a = monocolpate
  c = tricolpate
  0 = lacking colpation.
Number of pores:
  1 = one pore
  n = n pores.

Box 11.4 Quaternary pollen analysis

Pollen analysis has formed the framework for the biostratigraphical correlation and palaeoenvironmental reconstructions of Quaternary interglacial deposits and communities. For example, pollen ecostratigraphy has been used to monitor the changing composition and structure of the plant cover and woodlands of Connemara, western Ireland, during the last 10 000 or so years (Figure 11.30). Primeval woodland with elm, hazel and oak had developed by about 4100 BP, but the arrival of Neolithic farmers there led to a marked decline first in elm, and then in hazel and oak, as large-scale clearances of woodland took place around 4000 BP. The woodlands were replaced by dandelion, plantain and grasses as the open ground turned into meadows. Pollen data suggest the woodlands then revived about 3700 BP as intensive farming was abandoned.

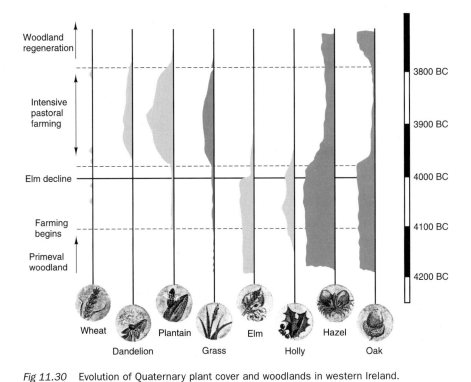

Fig 11.30   Evolution of Quaternary plant cover and woodlands in western Ireland.

ist databases built with reference to the exine structure, germinal aperture, outline, shape, size and ornament of the palynomorph (Box 11.3); their use in biostratigraphy and thermal maturation studies need not rely on a biological classification.

Basal tracheophytes

Tracheophytes are simple free-sporing vascular land plants, and several sporomorphs have been associated with them. Pteridophyte-like spores occur in the Palaeozoic (Upper Ordovician–Devonian); in fact, over 15 sporomorphs have been described from rocks of Silurian age. Cryptospores lack monolete or trilete markings and are commonly found in monads, dyads and tetrads, often with an outer membraneous envelope. Upper Ordovician occurrences, such as the spores from the Caradoc of North Africa, which also occur with cuticle-like cells, are commonly small smooth-walled forms, dominating assemblages until the end of the Llandovery. However, some of these simple spore types with trilete markings may come from other plant groups, such as the bryophytes.

Both monads and dyads occur. Dyads lack membranes, whereas monads with triradiate markings are smooth-walled forms with an equatorial thickening. *Ambitosporites*, for example, occurs in the sporangia of some *Cooksonia*, suggesting an association with the tracheophytes. It first appears during the Llandovery, suggesting the arrival of axially organized tracheophytes.

The Lower Devonian Rhynie Chert has, to date, yielded the most informative early tracheophyte flora; most of the plants were homosporous, producing isospores having sculpture and spines, thus the gametophyte certainly still required a moist environment for germination and growth. Monolete spores from lycopsids appeared during the mid and late Devonian, larger in size and with a variety of sculptures. Some assemblages, however, apparently have both microsporous and megasporous morphs, suggesting the evolution of heterospory in the Devonian floras. For example, the early Devonian *Chaleuria* shed both micro- and megaspores.

Psilophytopsida

The psilophytopsids or seed-bearing plants became well established by the mid Carboniferous, but the seed-bearing mode had already evolved during the late Devonian in the progymnosperms.

Progymnosperms

The progymnosperms (see p. 233) had fern reproductive modes combined with gymnosperm-like architecture. The late Devonian tree-like plant *Archaeopteris* shows the development of megaspores surrounded by woody tissue. Many primitive pollen cannot be easily distinguished from microspores; *Archaeosperma*, for example, is often classified as a prepollen. The prepollen have monolete and trilete structures and are commonly located within the megaspores of seed ferns.

Gymnosperms

The first gymnosperms, such as the seed ferns, evolved during the latest Devonian–mid Carboniferous. Groups such as the conifers and the seed ferns were derived from progymnosperm ancestors and possessed, for the first time, true seeds. Although the seedless plants, (e.g. the lycopsids and the sphenopsids) dominated many of the coal swamps of the Carboniferous, with large genera such as *Calamites*, *Lepidodendron* and *Sigillaria* supreme, the gymnosperm *Cordaites* and *Glossopteris* were equipped to colonize the higher ground. They produced bisaccate pollen, very common components of Upper Carboniferous palynofacies.

Monosulcate pollen, typical of the cycads and ginkgos, was supplemented, during the Carboniferous and Permian, by both polyplicate and saccate grains.

Angiosperms

The flowering plants, or the angiosperms (see pp. 237–241), diversified during the mid Cretaceous and rose to dominate modern floras. Both the pollen and seeds of angiosperms are distinctive; the pollen has a double outer wall, whereas the seeds also have a double protective casing. The first undoubted angiosperm pollen grains are reported from Lower Cretaceous horizons, where morphs such as *Clavatipollenites* are oval and monosulcate. During the Cretaceous the monosulcate condition was supplemented by tricolpate (e.g. *Tricolpites*) and triporate forms.

Both monocotyledon pollen, monosulcate and bilaterally symmetrical, and dicotyledon pollen, with both furrows and pores, can be recognized, although *Magnolia*, a dicotyledon, has monosulcate grains.

Spores, pollens and plants

Any multi-element organism creates taxonomic problems. Plants themselves consist of leaves, branches, stems and roots, all with quite different morphologies. The problem is even more acute with spores and pollens.

Two main methods have been used to try and associate fossil spores and pollens with their respective hosts. Comparisons of fossil assemblages with the living plants and their spores and pollens allow reconstructions of some fossil plants and associated palynomorphs. Pollen and spores may be found in fossil sporangia. Alternatively, numerical comparisons, based on the degree of correlation between spores and pollens with plant macrofossils may give statistically significant results.

Geological history

There is still some controversy surrounding the age and identity of the first true spores. Nevertheless, by the early Silurian a green turf-like vegetation probably covered many marine marginal belts, shedding simple trilete spores. During the Devonian there was a marked evolution and diversification of spore types and seed habit.

Diversification of floras during the Carboniferous established a range of pollen types. Monosaccate pollen from the Cordaitales and the Pteridospermales, together with disaccate pollen from gymnosperms, were important; monosulcate and polyplicate pollen appeared for the first time, as did monolete spores (Figure 11.31).

During the Permian, spores were less common than

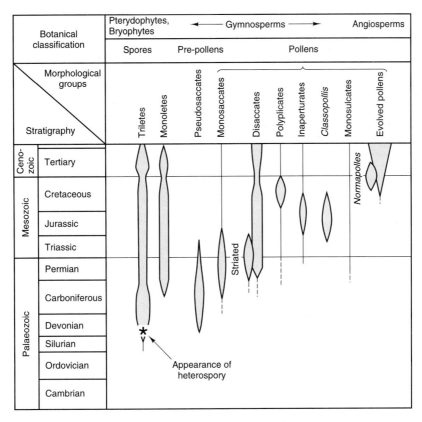

Fig 11.31 Stratigraphical distribution of main pollen and spore types.

saccate pollens, which continued through the Triassic. Some authorities have suggested that global cooling, associated with a major late Palaeozoic glaciation, reduced the abundance of spore-bearing plants.

Gymnosperms continued to dominate the floras of the early Jurassic, including monosaccates from Cordaitales, disaccates from some Coniferales, monosulcates from Bennettitales, Cycadales and Ginkgoales, polyplicates from Gnetales and inaperturates from other Coniferopsida. During the late Cretaceous and the Tertiary, angiosperm pollen, largely similar to that of modern flowering plants, became more abundant.

Further reading

Aldridge, R.J. (1987) *Palaeobiology of conodonts*. British Micropalaeontology Association and Ellis Horwood, London.

Bate, R. and Robinson, E. (eds) (1978) *A stratigraphical index of British Ostracoda*. British Micropalaeontology Association and Ellis Horwood, London.

Bignot, G. (1985) *Elements of micropalaeontology*. Graham and Trotman, London.

Brazier, M.D. (1980) *Microfossils*. George Allen and Unwin, London.

Higgins, A.C. and Austin, R.L. (1985) *A stratigraphical index of conodonts*. British Micropalaeontology Association and Ellis Horwood, London.

Jenkins, D.G. and Murray, J.W. (1989) *Stratigraphical atlas of fossil foraminifera*. 2nd edition. British Micropalaeontology Association and Ellis Horwood, London.

Lipps, J.H. (ed.) (1993) *Fossil prokaryotes and protists*. Blackwell Scientific Publications, Oxford.

Traverse, A. (1988) *Paleopalynology*. Chapman and Hall, London.

References

Aldridge, R.J. (1990) Extraction of microfossils. In D.E.G Briggs and P.R. Crowther (eds) *Palaeobiology – a synthesis*. Blackwell Scientific Publications, Oxford, pp. 502–504.

Aldridge, R.J., Briggs, D.E.G., Smith, M.P., Clarkson, E.N.K. and Clark, N.D.L. (1993a) The anatomy of conodonts. *Philosophical Transactions of the Royal Society of London*, **B340**, 405–421.

Aldridge, R.J., Jeppsson, L. and Dorning, K.J. (1993b) Early Silurian oceanic episodes and events. *Journal of the Geological Society of London*, **150**, 501–513.

Bergström, S.M. (1990) Relations between conodont provincialism and the changing palaeogeography during the early Palaeozoic. In W.S. McKerrow and C.R. Scotese (eds) *Palaeozoic palaeogeography and biogeography*. Geological Society Memoir 12, pp. 105–121.

Dzik, J. (1990) Conodont evolution in high latitudes of the Ordovician. *Courier Forschungs-Institut Senckenberg*, **117**, 1–28.

Olempska, E. (1989) Gradual evolutionary transformations of ontogeny in an Ordovician ostracod lineage. *Lethaia*, **22**, 159–168.

Turner, S. (1993) Early Carboniferous microvertebrates from the Narrien Range, central Queensland. *Memoir of the Association of Australasian Palaeontologists*, **15**, 289–304.

12 Trace fossils

- Trace fossils represent the activities of organisms.
- Trace fossils may be treated as fossilized behaviour, or as biogenic sedimentary structures.
- Trace fossils include tracks and trails, burrows and borings, faecal pellets and coprolites, root penetration structures, and other kinds of pellets.
- Trace fossils are named on the basis of shape and ornamentation, not on the basis of the supposed maker.
- One animal may produce many different kinds of trace fossils, and one trace fossil type can be produced by many different kinds of animals.
- Trace fossils may be produced within a sedimentary layer, or on the surface; trace fossils may be preserved in the round, or as moulds and casts on the bottoms and tops of beds.
- Trace fossils may be classified according to the mode of behaviour represented: movement, feeding, farming, dwelling, escape or resting.
- Certain trace fossil assemblages (ichnofacies) appear to repeat through time, and may indicate broad aspects about the conditions of deposition.
- Trace fossils often occupy particular levels (tiers) in the sediment column, and the depth of tiering has increased through time.
- Trace fossils are of limited use in stratigraphy, except in some special cases, such as in the definition of the Precambrian –Cambrian boundary.

Introduction

Trace fossils are the preserved remains of the activity and behavioural patterns of organisms. Common examples are burrows of bivalves and worms that live in estuaries and shallow seas, complex feeding traces of deep-sea animals on the ocean floor, and the footprints of dinosaurs and other land animals preserved in mud and sand beside rivers and lakes. In all of these cases, the trace fossils can be interpreted in two different ways:

1. as evidence of the behaviour of organisms, and hence interpreted as part of their palaeobiology;
2. as evidence of sedimentary environments, and hence interpreted in the same way as sedimentary structures.

For example, a trackway of dinosaur footprints can be used to give information about the shape of the soft parts of the feet of the dinosaur that made them, the pattern of scales on the skin, the running speed, and the environment in which the animal lived. The dinosaur tracks can equally be used to indicate that the sedimentary environment was continental, perhaps low-lying, and probably situated in a warm climatic belt.

Trace fossils are immensely common in many types of sedimentary rocks, and they have been observed by geologists for centuries. Indeed, many trace fossils were given zoological and botanical names from early in the 19th century, since they were thought to be fossilized seaweeds or worms. The only trace fossils that were correctly interpreted from the start were dinosaur footprints, although many of these had been thought to be the products of flocks of huge birds.

The modern era of trace fossil studies began in the 1950s with the work of the German palaeontologist Adolf Seilacher. He established a classification for trace fossils based on behaviour, and he discovered that certain assemblages of trace fossils indicate particular water depths. In addition, trace fossils were used widely by exploration geologists in the 1960s and 1970s when the study of depositional environments revolutionized our understanding of the sedimentary rock record. These contributions gave a strong scientific basis to the study of trace fossils, often called ichnology (from the Greek *ichnos*, 'a trace').

Trace fossils as fossilized behaviour

Types of trace fossils

There are many kinds of trace fossils, and many of the words used to describe them (tracks, trails, burrows, borings) are in common use. There are also a variety of cryptic fossils and sedimentary structures that might be regarded as trace fossils, but perhaps should not. The main trace fossil types are given in Table 12.1.

Many ichnologists might also include other examples of biological interaction with sediments as trace fossils, such as stromatolites (see p. 67), some kinds of mud mounds, dinosaur nests, heavily bioturbated or reworked sediments and the like. Not included are eggs, which are body fossils, or physical sedimentary structures such as tool marks produced by bouncing and rolling shells and pieces of wood.

Table 12.1   The main types of trace fossils, with definitions of the main terms

A. Traces on bedding planes
  1. *Tracks*: sets of discrete footprints, usually formed by arthropods or vertebrates.
  2. *Trails*: continuous traces, usually formed by the whole body of a worm, mollusc or arthropod, either travelling or resting.

B. Structures within the sediment
  3. *Burrows*: structures formed within soft sediment, either for locomotion, dwelling, protection or feeding, by moving grains out of the way.
  4. *Borings*: structures formed in hard substrates, such as limestone, shells or wood, for the purpose of protection, dwelling or carbonate extraction, by cutting right through the grains. Includes bioerosion feeding traces, such as drill holes in shells produced by gastropods.

C. Excrement
  5. *Faecal pellets and faecal strings*: small pellets, usually less than 10 mm in length, or strings of excrement.
  6. *Coprolites*: discrete faecal masses, usually more than 10 mm in length, and usually the product of vertebrates.

D. Others
  7. *Root penetration structures*: impressions of the activity of growing roots.
  8. *Non-faecal pellets*: regurgitation pellets of birds and reptiles, excavation pellets of crustaceans, and the like.

Principles of naming trace fossils

Trace fossils are given formal names, often based on Latin and Greek, just like living and fossil plants and animals (see p. 4). However, there are some fundamental differences between the nomenclature of trace fossils and that of body fossils and modern organisms. Trace fossil genera are called ichnogenera (singular: ichnogenus), and trace fossil species are called ichnospecies.

The key to understanding the naming of invertebrate trace fossils is to realize that the names usually say nothing about the organism that made the trace. In the early days of ichnology, the common U-shaped burrow *Arenicolites* was named after the burrow of *Arenicola*, the lugworm, and the meandering deep-sea trail *Nereites* was named after another polychaete annelid, *Nereis*. However, most *Arenicolites* burrows and most *Nereites* trails have nothing to do with the modern worms *Arenicola* and *Nereis*. Vertebrate footprints, on the other hand, can often be matched more readily with their producers, and track names frequently indicate the zoological affiliation of the track-maker.

The principle that invertebrate trace fossils should not be named after the supposed maker is based on two observations:

1. one animal can make many different kinds of traces; and
2. one trace fossil may be made by many different kinds of organisms.

The fiddler crab *Uca* is observed today to make four quite distinct kinds of trace fossils (Figure 12.1): a J-shaped living burrow, a running track, a star-shaped feeding pattern, and faecal pellets, each with its own name, as well as excavation pellets and feeding pellets.

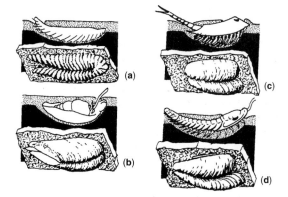

Fig 12.2   One trace fossil may be produced by many different organisms. Here, all the traces are resting impressions, cubichnia, of the ichnogenus *Rusophycus*, produced by the polychaete worm *Aphrodite* (a), a nassid snail (b), a notostracan branchiopod shrimp (c), and a trilobite (d).

An example of one trace fossil made by many different animals is the ichnogenus *Rusophycus*, a bilobed resting impression marked by transverse grooves (Figure 12.2). *Rusophycus* can be made by at least four different animals, belonging to three phyla (an annelid, a mollusc and two arthropods) but the traces are so similar that they must be given the same name.

An additional consideration is that if trace fossils were named after their proposed makers, the name would depend on the validity of that interpretation: trace fossil names could not change at the whim of every palaeobiologist who proposed a different maker for the same trace.

The nature of preservation of a trace fossil may affect its appearance, but the name cannot necessarily take account of this. The appearance of trails and burrows may be altered significantly by the grain size, location with respect to a fine- and coarse-grained horizon, and water content of the sediment in which they are preserved (see p. 279). This can be seen clearly with the example of the . *Nereites–Scalarituba–Neonereites* complex, a series of trace fossil forms produced by a single deep-sea grazing organism (Figure 12.3). The situation is different for many vertebrate traces. For example, it is often possible to follow a single dinosaur trackway for some distance, and the shape of individual foot and hand prints might vary substantially, depending on the sediment type and the animal's behaviour. It would clearly be unhelpful to give each variant print in a single trackway a different name.

In conclusion, *invertebrate trace fossil names should be based only on morphological features including shape and ornamentation, and not on the postulated maker or mode of preservation.*

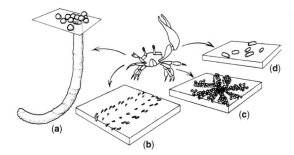

Fig 12.1   One animal may make many different kinds of trace fossils. The modern fiddler crab *Uca* makes a J-shaped living burrow ((a) domichnion, *Psilonichnus*), a walking trail ((b) repichnion, *Diplichnites*), a radiating grazing trace with balls of processed sand ((c) pascichnion), and faecal pellets ((d) coprolites).

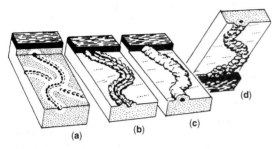

Fig 12.3 Variations in the physical nature of the sediment may create variations in the appearance of a trace fossil. Here, a subsurface patch-feeding burrow, develops different morphologies, and therefore has different names, when preserved in sand (a: *Scalarituba*), at a sand–mud interface in firm sediment (b: *Nereites*), at a sand–mud interface in wetter sediment (c: *Neonereites*), and at a mud/sand interface, seen from below (d: *Neonereites*).

Preservation of trace fossils

Trace fossils may be formed on bedding planes, within sedimentary horizons, or at the contacts of beds. The relationships of the trace fossils to the sediment, and the ways in which they are preserved must be established.

Seilacher's terminology, developed in the early 1960s, is frequently used (Figure 12.4). Burrows preserved in the sediment in three-dimensions are termed 'full relief' structures, while those seen on the surface of a bed are semireliefs. Semirelief burrows and trails may occur on the top of a bed, called epireliefs (*epi* = on), or on the bottom, termed hyporeliefs (*hypo* = under). Hyporelief preservation is very common in sedimentary sequences where sandstones and mudstones are interbedded, a feature of turbidite and storm-bed successions. Here, the traces are best seen on the bottoms of sandstone beds as sole structures, because the mudstones often flake away.

It is important to realize that burrows and surface trails are not always easy to distinguish. Burrows are formed within sediment, and are thus endogenic (*endo* = within; *genic* = made), and they are seen both as full reliefs and as semireliefs along bedding planes. However, if a burrow is formed at a bed boundary, or if subsequent erosion skims top sediment laminae away, the burrow may be seen as a semirelief. Trails are formed on the top of the sediment pile, and are thus exogenic (*exo* = outside) structures, typically seen as semireliefs. Undertracks are impressions formed on sediment layers below the surface on which the animal was moving, and it is important to distinguish these from the true track since the morphology may be different.

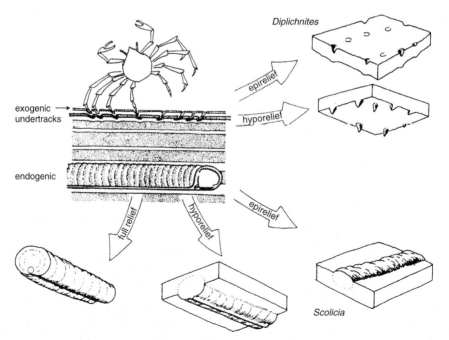

Fig 12.4 Terminology for trace fossil preservation, depending on the relationship of the trace to sediment horizons.

Fig 12.5 Trace fossils of the deep ocean floor. The patch-feeding trace (pascichnia) *Helminthopsis* meanders on one horizon, and the network burrow system (agrichnia) *Paleodictyon* is seen at a different level, in this field photograph from the Lower Silurian Aberystwyth Grits, Wales.

Interpreting ancient behaviour

Studies of the palaeobiology of trace fossils have been of great importance in situations where body fossils are rare, and where the traces were made by soft-bodied organisms that are not otherwise known as fossils. Two examples are in the deep sea and on land. Very little is known from body fossils of the history of life in deep abyssal oceans, and indeed very little is known about life in these zones today because they are inaccessible. Trace fossils, however, are abundant in many deep oceanic settings (Figure 12.5), and they show the diversity of trail-making and burrowing soft-bodied organisms, how many of them built complex shallow burrow systems and efficient patch-feeding trails, and how these assemblages have evolved through the Phanerozoic. On land, some continental sequences preserve very few body fossils, and the only indications of animal life are abundant dinosaur and other vertebrate tracks (Figure 12.6), as well as tracks and burrows made by insects and pond-living animals. These trace fossils can indicate the diversity of animals, where they lived, and what they were doing.

One of the major advances in trace fossil studies was Seilacher's (1967b) classification of behavioural categories. He divided trace fossils into seven behavioural types, depending on the postulated activities represented (Figure 12.7). Tracks and traces representing movement from A to B, such as worm trails or dinosaur trackways, are termed repichnia (*repere* = to creep; *ichnos* =

Fig 12.6 Dinosaur tracks, late Jurassic, Colorado. Here five brontosaur tracks run in parallel, and in the same direction, suggesting that a herd moved across this site.

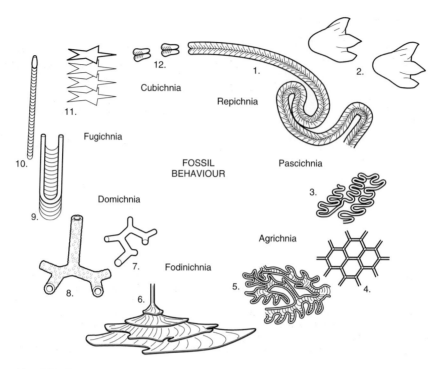

Fig 12.7   The behavioural classification of trace fossils, showing the major categories, and some typical examples of each. Illustrated ichnogenera: 1. *Cruziana*; 2. *Anomoepus*; 3. *Cosmorhaphe*; 4. *Paleodicyton*; 5. *Phycosiphon*; 6. *Zoophycos*; 7. *Thalassinoides*; 8. *Ophiomorpha*; 9. *Diplocraterion*; 10. *Gastrochaenolites*; 11. *Asteriacites*; 12. *Rusophycus*.

trace). Grazing trails that involve movement and feeding at the same time, are called pascichnia (*pascere* = to feed). These are typically coiled or tightly meandering trails found in deep oceanic sediments, where the regular pattern is an adaptation to feeding on restricted patches of food. Some unusual deep-sea horizontal burrow systems appear to have been maintained for trapping food particles, or for growing algae. These are termed agrichnia (a*gricola* = farmer). Feeding burrows, such as those produced by earthworms, as well as many marine examples, are called fodinichnia (*foda* = food). Living burrows and borings are termed domichnia (*domus* = house). Escape structures, or fugichnia (*fugere* = to flee) are traces of upward movement of worms, bivalves or starfish seeking to escape from beneath a layer of sediment that has been dumped suddenly on top of them. Fugichnia are found in cases of rapid sedimentation, in beach, storm bed, and turbidite sediments. Resting traces, or cubichnia (*cubare* = to lie down), may be of many types, and can include impressions of trilobites, starfish and jellyfish.

Trace fossils in sediments

Trace fossils as environmental indicators

One of the most striking discoveries about trace fossils has been that certain forms, and assemblages, can provide an age-independent guide to ancient sedimentary environments. Repeated assemblages of trace fossil types in rocks of different ages seem to correlate with deep abyssal sediments, with deep shelf conditions, with the intertidal zone, and so on. In some cases, members of the trace fossil assemblages are comparable from the Cambrian to the present day, evidence for recurrent behavioural patterns that have lasted for up to 550 Ma. The trace fossils have remained rather constant in appearance, even if their producers might have been quite different.

This palaeoenvironmental scheme of trace fossil, presented by Seilacher (1964, 1967a) has been modified and

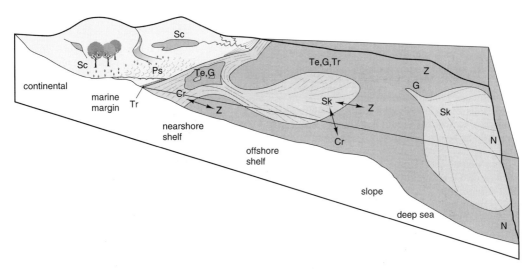

Fig 12.8 The major ichnofacies, and their typical positions in a hypothetical diagram of marine and continental environments. Typical offshore marine soft-sediment ichnofacies, from deep oceanic and basinal locations to the intertidal zone, includes the *Nereites* (N), *Zoophycos* (Z), *Cruziana* (Cr), and *Skolithos* (Sk) ichnofacies, which may occur in various water depths and in different conditions of sedimentation. A storm-sand fan and a turbidite fan are indicated. The *Psilonichnus* (Ps) Ichnofacies occurs in supratidal marshes, and the *Scoyenia* (Sc) Ichnofacies includes all lacustrine, and related continental settings. The *Glossifungites* (G) Ichnofacies is typical of firmgrounds, the *Trypanites* (Tr) Ichnofacies consists of borings in limestone, and the *Teredolites* (Te) Ichnofacies consists of borings in wood.

enlarged since then (Frey *et al.*, 1990), but in principle it divides trace fossil assemblages into a number of ichnofacies (Figure 12.8; see Box 12.1). The ichnofacies are named after a characteristic trace fossil, and they indicate particular sedimentary facies. The ichnofacies is identified on the basis of an assemblage of trace fossils, and it may be recognized even if the name-bearing form is absent.

The classic marine ichnofacies, those named for *Nereites*, *Zoophycos*, *Cruziana* and *Skolithos*, are not simply depth-related, as Seilacher first proposed; they seem to be associated with particular sedimentary regimes, combining aspects of water depth and of rate of deposition. These four ichnofacies include assemblages of trace fossils typical of fairweather normal conditions of deposition, and those characteristic of exceptional storm and turbidite event beds. The *Scoyenia* Ichnofacies is one of several continental trace fossil facies, and depends on the presence of shallow fresh water, while the *Psilonichnus* Ichnofacies is controlled by coastal marine influence on a terrestrial setting. The *Glossifungites*, *Trypanites* and *Teredolites* ichnofacies are controlled by substrate alone, and they could theoretically occur across a range of the depth zones represented by Seilacher's classic bathymetric sequence of ichnofacies. In fact, they are mostly restricted to marginal marine, intertidal and shallow shelf zones, but that is related to the commonest occurrences of the required substrates.

Organisms in sediments

The nature of sediments determines the kinds of organisms that live in or on them, and hence the kinds of trace fossils that may be preserved. The ichnofacies scheme just outlined highlights the important roles of broad sedimentary environment (marine or continental, deep oceanic, or intertidal, lake or terrestrial), salinity and sedimentation rate.

The physical properties of sediments can exert controls on trace fossil distributions, and four factors are particularly important:

1. *The average grain size* of the sediment: this affects sediment-ingesting burrowers, organisms that require particular sediment sizes to line their burrows, and filter-feeders which must avoid fine suspended sediment.

2. *Sediment stability*, particularly in the *Glossifungites* and *Trypanites* ichnofacies, which depend on firm and lithified substrates respectively: some organisms build burrows of different morphology, depending on the stability of the sediment.

Box 12.1 The nine ichnofacies

The *Nereites* Ichnofacies (Figure 12.9a) is recognized by the presence of meandering pascichnia such as *Nereites*, *Neonereites* and *Helminthoida*, spiral pascichnia such as *Spirorhaphe*, and agrichnia such as *Paleodictyon* and *Spirodesmos*. Vertical burrows are almost entirely absent. This ichnofacies is indicative of deep-water environments, and includes ocean floors and deep marine basins. The trace fossils are found in muds deposited from suspension, and in the mudstones and siltstones of distal turbidites.

The *Zoophycos* Ichnofacies (Figure 12.9b) is characterized by complex fodinichnia such as *Zoophycos*, and it may contain other deep traces such as *Thalassinoides* in tiered arrangements. The ichnofacies occurs in a range of water depths between the abyssal zone and the shallow continental shelf. This ichnofacies may occur in normal background conditions of sedimentation, whereas the *Nereites* Ichnofacies may be a matching association found at similar water depths during times of event (turbidite) deposition.

The *Cruziana* Ichnofacies (Figure 12.9c) shows rich trace fossil diversity, with horizontal repichnia (*Cruziana* and *Aulichnites*), cubichnia (*Rusophycus*, *Asteriacites* and *Lockeia*), as well as vertical burrows. This ichnofacies represents mid and distal continental shelf situations which may lie below normal wave base, but may be much affected by storm activity.

The *Skolithos* Ichnofacies (Figure 12.9d) is recognized by the presence of a low diversity of abundant vertical burrows, domichnia such as *Skolithos*, *Diplocraterion* and *Arenicolites*, fodinichnia such as *Ophiomorpha*, and fugichnia. These all typically indicate intertidal situations where sediment is removed and deposited sporadically, and the organisms have to be able to respond rapidly in stressful conditions. The *Skolithos* Ichnofacies was at first seen as occurring only in the intertidal zone, but it is also typical of other shifting-sand environments, such as the tops of storm-sand sheets and the tops of turbidity flows.

The *Psilonichnus* Ichnofacies (Figure 12.9e) is a low-diversity assemblage, consisting of small vertical burrows with basal living chambers (*Macanopsis*), narrow sloping J-shaped and Y-shaped burrows (*Psilonichnus*: a ghost crab burrow), root traces, and, where relevant, vertebrate footprints. It is typical of backshore, dune areas, and supratidal flats on the coast.

The *Scoyenia* Ichnofacies (Figure 12.9f) is typified by a low-diversity trace fossil assemblage, mainly simple horizontal fodinichnia (*Scoyenia* and *Taenidium*), with occasional vertical domichnia (*Skolithos*) and repichnia produced by insects or freshwater shrimps (*Cruziana*/*Isopodichnus*) preserved in fluvial and lacustrine sediments, often in the silts and sands of redbed sequences. Associated subaerial sediments, such as aeolian sands and palaeosols, representing an unnamed ichnofacies, may contain domichnia and repichnia of insects, and dinosaur and other tetrapod footprints. As shown by Lockley *et al.* (1994), these tracks may occur in distinctive recurrent assemblages, allowing the recognition of named vertebrate ichnofacies.

The *Glossifungites* Ichnofacies (Figure 12.9g) is characterized by domichnia such as *Glossifungites* and *Thalassinoides* and sometimes plant root penetration structures, but other behavioural trace fossil types are rare. The sediments are firm, but not lithified, and these may occur in firm compacted muds and silts in marine intertidal and shallow subtidal zones. The firmgrounds may develop in low-energy situations such as salt marshes, mud bars, of high intertidal flats, or in shallow marine environments where erosion has stripped off superficial unconsolidated layers of sediment, exposing firmer beds beneath.

The *Trypanites* Ichnofacies (Figure 12.9h) is characterized by domichnial borings of worms (*Trypanites*), bivalves (*Gastrochaenolites*), barnacles (*Rogerella*) and sponges (*Entobia*) formed in shoreline rocks or in lithified limestone hard-

Fig 12.9   Block diagrams showing typical trace fossils of the major ichnofacies. (a) *Nereites* Ichnofacies, characterized by *Spirorhaphe* (1), *Lorenzinia* (2), *Chondnites* (3), *Paleodictyon* (4), *Nereites* (5), and *Cosmorhaphe* (6), viewed in pelagic carbonate ooze. (b) *Zoophycos* Ichnofacies, characterized by *Phycosiphon* (1), *Zoophycos* (2), and *Spirophyton* (3)(c) *Cruziana* Ichnofacies, characterized by *Asteriacites* (1), *Cruziana* (2), *Rhizocorallium* (3), *Thalassinoides* (4), *Chondrites* (5), *Teichichnus* (6), *Arenicolites* (7). (d) *Skolithos* Ichnofacies, characterized by *Ophiomorpha* (1), *Diplocraterion* (2), *Skolithos* (3), and *Monocraterion* (4). (e) *Psilonichnus* Ichnofacies, characterized by *Psilonichnus* (1), *Macanopsis* (2), and vertebrate footprints (3). (f) *Scoyenia* Ichnofacies, characterized by *Scoyenia* (1), *Ancorichnus* (2), *Cruziana* (3), and *Skolithos* (4). (g) *Glossifungites* Ichnofacies, characterized by *Thalassinoides* (1), bivalve borings (2), polychaete burrows (3), *Rhizocorallium* (4), and *Psilonichnus* (5). (h) *Trypanites* Ichnofacies, characterized by echinoid grooves (1), barnacle borings (2), *Entobia* (sponge borings) (3), *Gastrochaenolites* (bivalve borings) (4), *Trypanites* (polychaete and sipunculid worm borings) (5).(i) *Teredolites* Ichnofacies, characterized by vertical bulbous burrows of bivalves (*Teredolites*), and subhorizontal burrows.

Box 12.1 (Cont.)

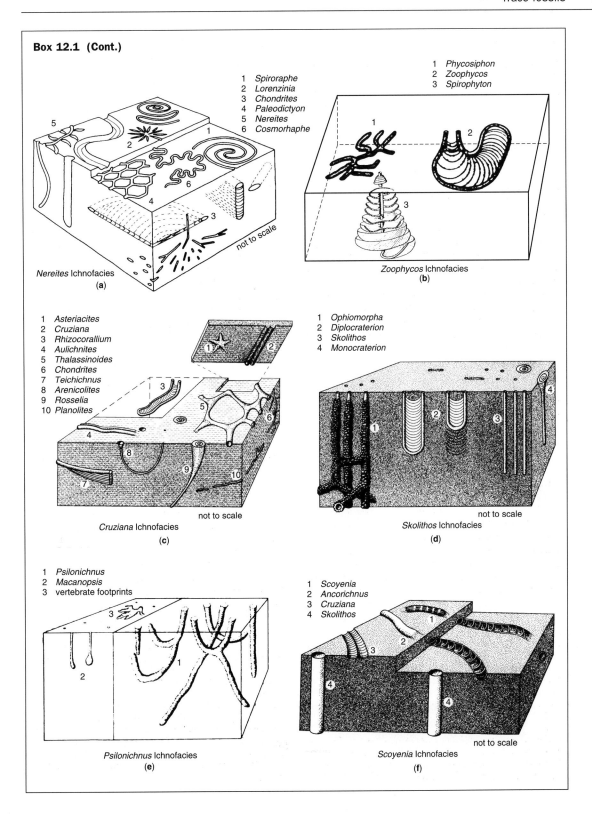

1 Spiroraphe
2 Lorenzinia
3 Chondrites
4 Paleodictyon
5 Nereites
6 Cosmorhaphe

Nereites Ichnofacies
(a)

1 Phycosiphon
2 Zoophycos
3 Spirophyton

Zoophycos Ichnofacies
(b)

1 Asteriacites
2 Cruziana
3 Rhizocorallium
4 Aulichnites
5 Thalassinoides
6 Chondrites
7 Teichichnus
8 Arenicolites
9 Rosselia
10 Planolites

not to scale

Cruziana Ichnofacies
(c)

1 Ophiomorpha
2 Diplocraterion
3 Skolithos
4 Monocraterion

not to scale

Skolithos Ichnofacies
(d)

1 Psilonichnus
2 Macanopsis
3 vertebrate footprints

Psilonichnus Ichnofacies
(e)

1 Scoyenia
2 Ancorichnus
3 Cruziana
4 Skolithos

not to scale

Scoyenia Ichnofacies
(f)

Box 12.1 (Cont.)

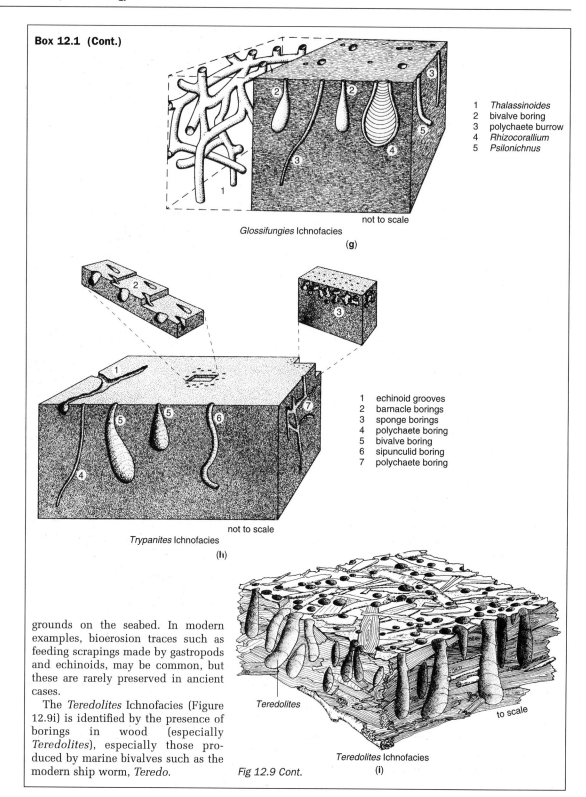

1   *Thalassinoides*
2   bivalve boring
3   polychaete burrow
4   *Rhizocorallium*
5   *Psilonichnus*

not to scale

Glossifungies Ichnofacies

(g)

1   echinoid grooves
2   barnacle borings
3   sponge borings
4   polychaete boring
5   bivalve boring
6   sipunculid boring
7   polychaete boring

not to scale

Trypanites Ichnofacies

(h)

Teredolites

to scale

grounds on the seabed. In modern examples, bioerosion traces such as feeding scrapings made by gastropods and echinoids, may be common, but these are rarely preserved in ancient cases.

The *Teredolites* Ichnofacies (Figure 12.9i) is identified by the presence of borings in wood (especially *Teredolites*), especially those produced by marine bivalves such as the modern ship worm, *Teredo*.

Fig 12.9 Cont.

Teredolites Ichnofacies

(i)

3. *Water content*: depending on the water content, sediments may range from soupy in consistency to totally lithified sediments with zero porosity, whose sole trace fossils are borings (*Trypanites* Ichnofacies). Firmgrounds contain relatively little water, and again they are characterized by particular burrows of the *Glossifungites* Ichnofacies.

4. *Chemical conditions* in sediments, particularly oxygen levels: trace fossils are absent in completely anoxic situations, but a surprising array of animals can survive in dysoxic (very low oxygen) conditions.

Burrowing organisms divide up the different strata of unconsolidated sediment in rather precise ways. Each burrower is restricted to particular depths of burrowing, some exploiting the near-surface oxygenated zone, and others extending ever deeper into the sediment. Most burrowers are restricted to the top 20 mm of sediment, since this minimizes the physical effort for animals that are simply moving from A to B. The surface layers are also favoured by organisms that are feeding on organic matter. Deeper burrowers are mainly those forming domichnia, where the body of the organism is large, or where it possesses long siphons, in order to keep contact with oxygenated waters. Deeper layers are also safer from predators, whether those operating from the surface, or other burrowers. There are also feeding opportunities at depth, at the redox layer, where the oxygenated surface sediments meet the deeper anoxic sediments, a horizon that is characterized by unusual shelly faunas and sulphur-oxidizing bacteria.

Such tiering patterns increased in complexity through time, as shown by Bromley and Ekdale (1986) and Ekdale and Bromley (1991). In a Middle Ordovician example (Figure 12.10a), the subsurface layers are filled with simple horizontal burrows, *Planolites*. These are cut by branching fodinichnia, *Chondrites*, exploiting an organic-rich layer at a depth of 20–30 mm below the surface of the sediment. The deepest burrows may be *Teichichnus*, showing spreiten structure, and extending down to 100 mm at the deepest. An early Jurassic example (Figure 12.10b) shows a substantial increase in depth burrowed, to perhaps 0.5 m, with small *Chondrites* in upper layers, a new large *Chondrites* extending to lower layers, and domichnia, *Thalassinoides*, at the deepest levels. Finally, in a late Cretaceous example (Figure 12.10c), there are at least nine tiers, three horizons of shallow burrows near the surface, *Planolites*, *Thalassinoides*, *Taenidium*, *Zoophycos*, and large and small *Chondrites* going deepest, perhaps to a maximum depth of 1 m.

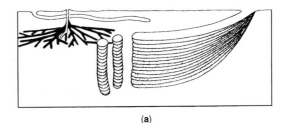

(a)

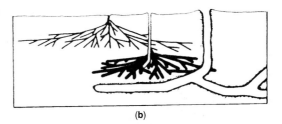

(b)

(c)

Fig 12.10   Examples of trace fossil tiering, in which burrowers choose specific depth horizons below the sediment–water interface. In the middle Ordovician limestones of Öland, Sweden (a), there are three tiers. In the early Jurassic Posidonienschiefer of Germany (b), there are also three tiers. In the late Cretaceous Chalk of Denmark (c), there are at least nine tiers.

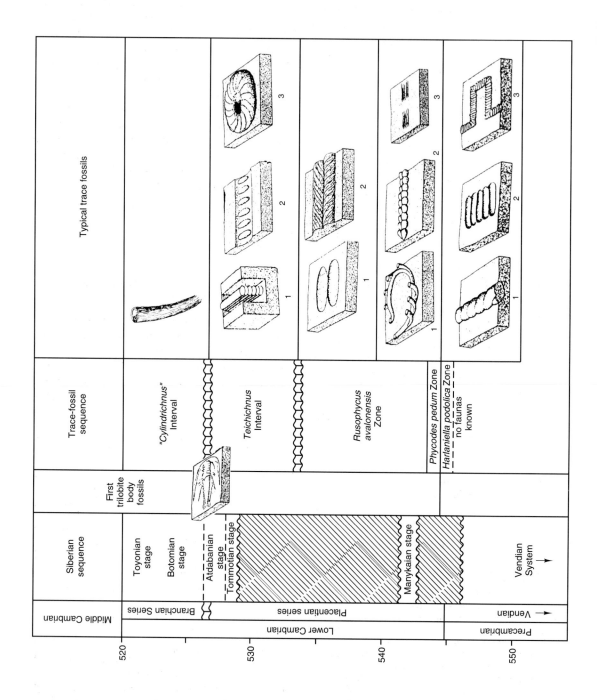

Trace fossils and time

Trace fossils are generally not good stratigraphic indicators since they are largely facies-controlled. In other words, the very properties that make trace fossils good indicators of depositional environments, make them generally of limited value in biostratigraphy. There are one or two exceptions, and one of these is the Precambrian–Cambrian boundary interval.

Metazoan body fossils are known sporadically in the late Precambrian; for example, the Ediacara faunas and the Vendian small shelly fossil assemblages (see pp. 76–82). The base of the Cambrian System, and therefore the Precambrian–Phanerozoic boundary has generally been marked at the first occurrence of trilobite body fossils (Figure 12.11). However, the oldest known trilobites are reported from the Atdabanian of various parts of the world, whereas trilobite trace fossils, such as *Monomorphichnus*, *Rusophycus*, *Cruziana* and *Diplichnites*, occur at lower levels, and these have now been used as reasonably sure indicators of Cambrian strata (Crimes, 1987), although some workers still demand body fossil evidence. The Precambrian–Cambrian boundary was determined in 1993 on the basis of the first appearance of complex feeding burrows, coelenterate impressions, and arthropod traces (Landing, 1993). This boundary, marking one of the key stratigraphic boundaries in the history of the Earth, falls between the *Harlaniella podolica* Trace Fossil Zone, a unit containing the unique ichnogenera *Harlaniella*, *Palaeopascichnus* and *Nenoxites*, and the *Phycodes pedum* Zone, characterized by *Phycodes*, *Nereites*, *Monomorphichnus*, *Skolithos*, *Arenicolites*, *Protopaleodictyon* and *Conichnus*.

There is also some evidence for evolution of trace fossils through time. For example, pascichnia such as

Nereites and agrichnia such as *Paleodictyon* seem to have become smaller and more regular through time – perhaps evidence for improvements in feeding efficiency. Trace fossil assemblages as a whole also seem to have evolved through time. Body fossil communities became more diverse during the course of the Phanerozoic, as new groups arose and explored new feeding modes and occupied new habitats. The same seems to be broadly true of trace fossils (Figure 12.12). Typical shallow-marine trace fossil assemblages of the Cambrian consisted of 15–20 ichnospecies, and diversities have remained constant ever since. However, in deep oceanic settings, the *Nereites* Ichnofacies, there was an increase in diversity from about five ichnospecies in the Cambrian, to about ten from the Silurian to Jurassic, and 10–50 thereafter.

Case studies

Aberystwyth Grits, Lower Silurian, Wales

The Lower Silurian clastic rocks of central Wales and the Welsh coast have long been known as a source of trace fossils. Indeed, the first specimens of *Nereites* were named from the Welsh Basin turbidite succession in Murchison's 'Silurian System' of 1839. Since then, a variety of trace fossil associations have been identified,

Fig 12.11 The base of the Cambrian is defined by the first appearance of the feeding burrow *Phycodes pedum*, and associated forms. Arthropod trace fossils appear in the next trace fossil zone, pre-dating the first occurrence of trilobite body fossils by 15 Ma. This is a rare, but important, example of a stratigraphic role for trace fossils. The standard stratigraphy across the Precambrian–Cambrian boundary may be dated by some body fossil faunas, but trace fossil assemblages offer a more complete succession. Typical trace fossils shown are: *Harlaniella podolica* Zone (1, *Harlaniella*; 2, *Palaeopascichnus*; 3, *Nenoxites*), *Phycodes pedum* Zone (1, *Phycodes*; 2, *Nereites*; 3, *Monomorphichnus*), *Rusophycus avalonensis* Zone (1, *Rusophycus*; 2, *Cruziana*), *Teichichnus* Interval (1, *Teichichnus*; 2, *Plagiogmus*; 3, *Astropolichnus*), '*Cylindrichnus*' Interval (*Cylindrichnus*).

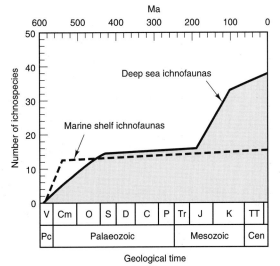

Fig 12.12 Trace fossil diversity has increased through time, especially in the deep sea, while marine shelf trace fossil diversity has remained more constant since the Cambrian.

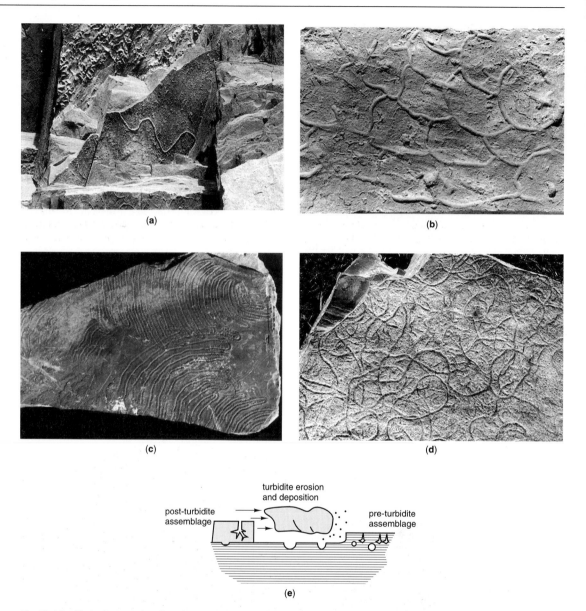

Fig 12.13 Typical trace fossils of the Lower Silurian sediments of the Welsh Basin (*Nereites* Ichnofacies). (a) *Helminthopsis*, (b) *Paleodictyon*, (c) *Nereites*, (d) *Gordia*, (e) the pre- and post-turbidite trace fossil assemblages.

each associated with a different sub-environment of the deep basin, and all belonging to the broad *Nereites* Ichnofacies. Crimes and Crossley (1991) identified 25 ichnogenera from the sandstone turbidites of the Aberystwyth Grits Formation, the commonest forms being *Helminthopsis, Paleodictyon* and *Squamodictyon* (Figure 12.13a, b). The finer-grained pelagic sediments of similar age yielded different ichnofaunas, consisting mainly of *Nereites, Dictyodora, Gordia* and

Helminthoida (Figure 12.13c, d).

One clear distinction in the Welsh Basin ichnofaunas was probably the result of minor turbidite activity at the toe of spreading fans. Pre-turbidite and post-turbidite assemblages have been identified, representing the trace fossils that are formed in normal background times, and those that were formed after a turbidity flow event. Before the flow, Orr (1995) identified an assemblage of surface trails and shallow burrows. After passage of a low-energy turbidite flow, the top layers of the existing sediment were stripped off, removing surface and subsurface traces, and casting the deeper preturbidite burrows as convex hyporeliefs on the sole of the turbidite sand. After the flow had waned, a post-turbidite trace fossil assemblage of burrows was developed within the turbidite sand (Figure 12.13e).

Cardium Formation, Upper Cretaceous, Alberta

Many sedimentary sequences show a mix of ichnofacies, as would be expected, since none of the ichnofacies is exclusive to a single location or water depth. The Cardium Formation of Alberta, Canada, has produced abundant trace fossils from a sequence of muds and sandstones (Pemberton and Frey, 1984). The normal quiet-water sedimentation produced mud, silt and fine sand layers, and diverse trace fossils of the *Cruziana* Ichnofacies (Figure 12.14), mainly representing the activities of mobile carnivores and deposit feeders exploiting relatively nutrient-rich fine-grained sediments. These sediments were interrupted sporadically by storm beds, thick units of coarse sand washed back from the shore region into deeper water by storm surge ebb

currents. The trace fossils of these units are *Skolithos*, *Ophiomorpha*, *Diplocraterion* and various fugichnia (Figure 12.14), all of typical elements of the *Skolithos* Ichnofacies.

One view of this alternation between trace fossils of the *Cruziana* and *Skolithos* ichnofacies might be that there had been repeated changes in sea level from deep to shallow offshore conditions. However, the control is more probably the dramatic changes in energy of deposition. The opportunistic members of the *Skolithos* Ichnofacies colonized the storm beds, probably having been washed down from the intertidal zone, and they were able to cope with the rapid fluctuations in unconsolidated sediment depth. The storm events doubtless killed off most of the members of the *Cruziana* Ichnofacies, or displaced them to the margins of the affected area. After the storm surge ebb currents waned, and slow sedimentation resumed, the surface-feeding organisms re-colonized the whole area.

Morrison Formation, Upper Jurassic, Colorado

The final trace fossil case study is one of the most spectacular. Continental sediments of the Permian and the Mesozoic have often produced rich horizons bearing foot-

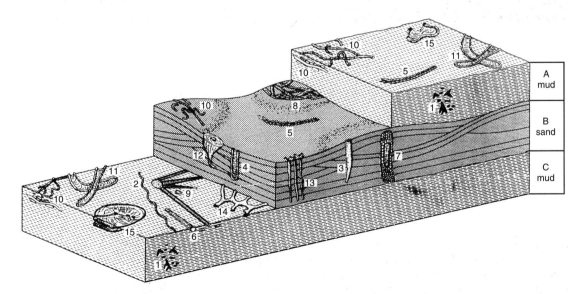

Fig 12.14 Sediments and trace fossils in the late Cretaceous Cardium Formation of Alberta. Normal fine-grained sediments (A, C) are associated with *Cruziana* Ichnofacies trace fossils, while intermittent coarse sandstone storm beds (B) show trace fossils of the *Skolithos* Ichnofacies. 1. *Chondrites*; 2. *Cochlichnus*; 3. *Cylindrichnus*; 4. *Diplocraterion*; 5. *Gyrochorte*; 6. *Palaeophycus*; 7. *Ophiomorpha*; 8. *?Phoebichnus*; 9. *Taenidium*; 10. *Planolites*; 11. *Rhizocorallium*; 12. *Rosselia*; 13. *Skolithos*; 14. *Thalassinoides*; 15. *Zoophycos*.

prints of dinosaurs, and other fossil reptiles, and some of the richest date from the late Jurassic of North America. This was the time of such giant plant-eating sauropod dinosaurs as *Apatosaurus*, *Camarasaurus*, *Diplodocus* and *Brachiosaurus*, smaller plant-eaters such as the ornithopod *Camptosaurus*, and the stegosaur *Stegosaurus*, and the meat-eating theropods *Allosaurus* and *Ceratosaurus*.

The world's largest mapped dinosaur trackway assemblage comes from the Purgatoire River, Colorado, where Lockley *et al.* (1986) recorded 1300 footprints, belonging to more than 100 trackways, along a 340 m stretch of a single bedding plane (Figure 12.15a). The footprints are preserved in a succession of lacustrine clastic sediments: shallow-water shales, shoreface limestones with ripple marks and mudcracks, and quartz-rich sandstones. The succession represents changing lake levels, and footprints are most abundant at times of low lake-level stand when the shoreface sediments developed (Figure 12.15b). The tracks themselves are good indicators of the shoreline of the lake in terms of their depths in the sedi-

ment (shallow imprints on dry sediment; deep imprints on submerged surfaces). Associated fossils include bivalves, gastropods, fishes, and plant remains, but invertebrate trace fossils appear to be rare.

The tracks represent sauropods and theropods, perhaps five species in all. Most of the tracks were heading west-north-west, and there is evidence that the sauropod tracks indicate movement of herds of animals along the side of the Morrison Lake. Such huge dinosaur track sites have now been located in many parts of North America, and elsewhere in the world, mainly in rocks of Jurassic and Cretaceous age. These can indicate such aspects of dinosaur behaviour as running speed, herding and hunting strategies. Many of them may indicate segments of long-distance migratory routes, documenting movements of herds of large dinosaurs over hundreds or thousands of kilometres each year, analogous to the movements of modern ungulates today, as they search for fresh plant food.

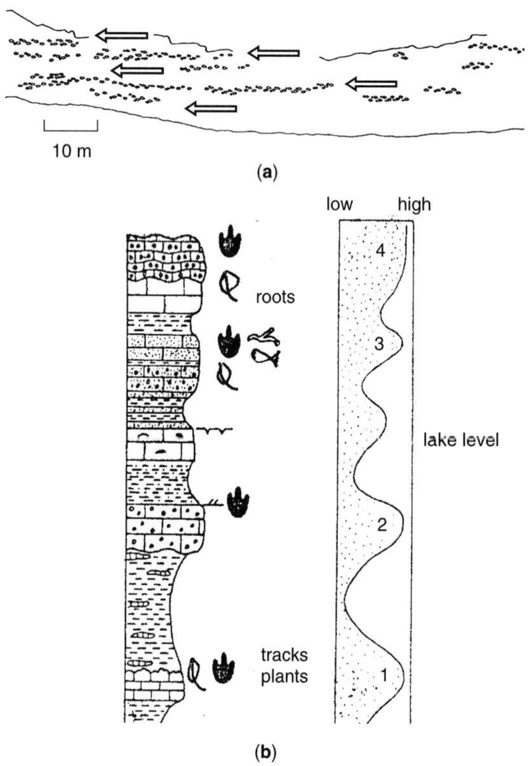

Fig 12.15   Map of a dinosaur megatrack site, on a bedding plane in the late Jurassic Morrison Formation of the Purgatoire River, Colorado (a). Directions of locomotion are indicated. (b) Dinosaur tracks occur at times when the lake shoreline migrated over the locality.

Further reading

Bromley, R. G. (1996) *Trace fossils: biology, taphonomy and applications*, 2nd edition. Chapman & Hall, London.

Donovan S. K. (ed.) (1994) *The palaeobiology of trace fossils*. Wiley, New York.

Ekdale, A. A. and Bromley, R. G. and Pemberton, S. G. (1984) *Ichnology; the use of trace fossils in sedimentology and stratigraphy*. Society of Economic Paleontologists and Mineralogists, Tulsa, Oklahoma.

Lockley, M. G. (1991) *Tracking dinosaurs*. Cambridge University Press, Cambridge.

Maples, C. G. and West, R. R. (eds) (1992) *Trace fossils. Short courses in paleontology, No. 5*. Paleontological Society, Tulsa, Oklahoma.

References

Bromley, R. G. and Ekdale, A. A. (1986) Composite ichnofabrics and tiering of burrows. *Geological Magazine*, **123**, 59–65.

Crimes, T. P. (1987) Trace fossils and correlation of late Precambrian and early Cambrian strata. *Geological Magazine*, **124**, 97–119.

Crimes, T. P. and Crossley, J. D. (1991) A diverse ichnofauna from Silurian flysch of the Aberystwyth Grits Formation, Wales. *Geological Magazine*, **26**, 27–64.

Ekdale, A. A. and Bromley, R. G. (1991) Analysis of composite ichnofabrics: an example in the uppermost Cretaceous Chalk of Denmark. *Palaios*, **6**, 232–249.

Frey, R. W., Pemberton, S. G. and Saunders, T. D. A. (1990) Ichnofacies and bathymetry: a passive relationship. *Journal of Paleontology*, **64**, 155–158.

Landing, E. (1993) Precambrian–Cambrian boundary global stratotype ratified and a new perspective of Cambrian time. *Geology*, **22**, 179–182.

Lockley, M. G., Houck, K J., and Prince, N. K. (1986) North America's largest dinosaur trackway site: implications for Morrison Formation paleoecology. *Geological Society of America Bulletin*, **97**, 1163–1176.

Lockley, M. G., Hunt, A. P. and Meyer, C. A. (1994) Vertebrate tracks and the ichnofacies concept: implications for palaeoecology and palichnostratigraphy. In S. K. Donovan (ed.) *The palaeobiology of trace fossils*. Wiley, New York, pp 241–268.

Orr, P. J. (1995) A deep-marine ichnofaunal assemblage from Llandovery strata of the Welsh Basin, west Wales, U.K. *Geological Magazine*, **132**, 267–285.

Pemberton, S. G. and Frey, R. W. (1984) Ichnology of storm-influenced shallow marine sequence: Cardium Formation (Upper Cretaceous) at Soebe, Alberta. *Canadian Society of Petroleum Geologists, Memoir*, **9**, 281–304.

Seilacher, A. (1964) Biogenic sedimentary structures. In J. Imbrie. and N. Newell (eds) *Approaches to paleoecology*. Wiley, New York, pp. 296–316.

Seilacher, A. (1967a) Bathymetry of trace fossils. *Marine Geology*, **5**, 413–428.

Seilacher, A. (1967b) Fossil behavior. *Scientific American*, **217**, 72–80.

13   Major events

Key Points

- The fossil record gives accurate information about diversifications and extinctions.
- There may be 5–50 million species on Earth today, almost certainly higher than at any time in the past.
- Many examples of evolutionary trends, one-way changes in a feature or features, are in reality more complex.
- The idea of progress in evolution, change with improvement in competitive ability, is hard to demonstrate.
- The 'big five' mass extinctions occurred in the late Ordovician, the late Devonian, the end of the Permian, the end of the Triassic, and the end of the Cretaceous. There were many smaller mass extinction events.
- During mass extinctions, 20–90% of species were wiped out; these include a broad range of organisms, and the events appear to have happened rapidly.
- The history of life on Earth is marked by a number of major expansions that followed the acquisition of new characters, and the mastery of new environments.

Introduction

The record of fossils gives a rich and spectacular picture of the history of life. Palaeontologists have been as successful as archaeologists and historians in piecing together a detailed picture of the events of the past, even though palaeontologists have a very much longer timescale to deal with. It is likely that the last 200 years of palaeontological research have given a broadly correct picture of the order of appearance of major groups of plants and animals through geological time, their distributions over the continents and oceans of the past, their life strategies and adaptations, and their patterns of evolution (see Chapters 2 and 3).

The diversification of life

Onward and upward

An accepted principle of evolution is that all modern and ancient life on Earth is part of a single great phylogenetic tree (synapomorphies of the clade 'Life on Earth' are the DNA–RNA protein synthesis system, and possibly the homeobox, a gene sequence that codes early developmental stages). If that is the case, there must have been a time in the Precambrian when there was only a single species. Today, there are between 5 and 50 million species (see Box 13.1), so a plot of species numbers through time must show a pattern of phenomenal increase over the past 3500 Ma. But just what sort of pattern?

It is impossible to plot an accurate diagram of species numbers through time, because so many species were never fossilized, and others are yet to be found and identified. Palaeontologists have focused on those parts of the fossil record where the results might be believable. Valentine (1969) presented the first serious effort, when he plotted the numbers of families of skeletonized shallow marine invertebrates through the Phanerozoic (Figure 13.1a). The pattern showed a jerky increase, with several declines, and a particularly dramatic rate of increase from the Cretaceous to the present. Valentine argued that this pattern might be representative of the pattern of diversification of all of life.

However, Raup (1972) argued that the graph showed more about the sources of error in the fossil record than it did about the true pattern of the diversification of life. He suggested that the low diversity values in the early Palaeozoic reflected the fact that such ancient rocks were rare, the fossils in them were often metamorphosed or eroded away, and palaeontologists devoted too little attention to them. Raup suggested, then, that the true pattern of diversification of marine invertebrates had been a rapid rise to modern diversity levels during the Cambrian and Ordovician, and a steady equilibrium level since then (Figure 13.1b).

Various tests have confirmed that although Raup identified important sources of error, the fossil record is good enough to uphold the general outline of Valentine's graph. Other data sets have produced the same pattern of diversification: slow rates at first, many set-backs, and a rapid rate of increase over the past 100 Myr, with no sign of a levelling-off. This is true of vertebrates and plants, and the latest plots for marine animals are comparable with Valentine's original effort (Figure 13.2).

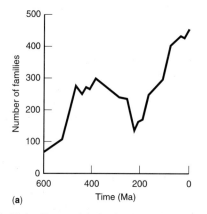

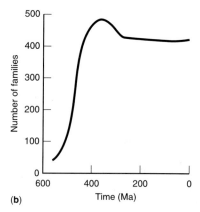

Fig 13.1   Two models for the diversification of marine invertebrate life over the past 600 Ma of good-quality fossil record: (a) the empirical model, in which the data from the fossil record are plotted directly, and (b) the bias simulation model, in which corrections are made for the supposedly poor fossil record of ancient rocks.

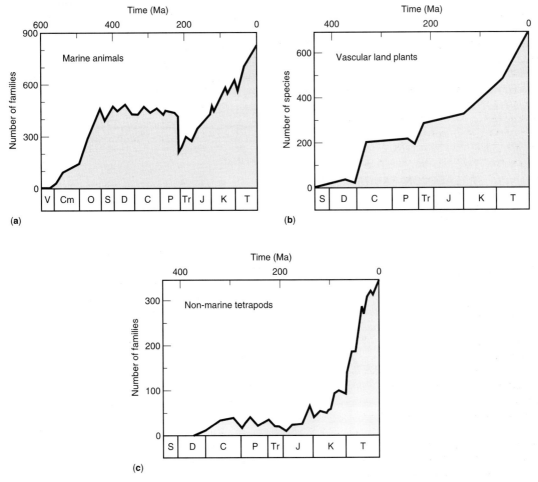

Fig 13.2   The diversification of three groups of multicellular organisms during the Phanerozoic: (a) marine animals; (b) vascular land plants; (c) nonmarine tetrapods. All graphs show similar shapes, with a long initial period of low diversity, and then rapid increase since the Cretaceous. Geological period abbreviations are standard, running from Vendian (V) to Tertiary (T).

Is the history of life progressive?

Sepkoski (1984) suggested a progressive interpretation for the diversification of animal life in the sea. He argued that the pattern of increase in family numbers divided into three phases, each of which was characterized by a plateau in diversity (Figure 13.3), one in the Cambrian, one during the rest of the Palaeozoic, and one during the Mesozoic and Cenozoic, the 'modern' fauna.

Sepkoski (1984) suggested that the apparent sequence of three global marine animal faunas through Phanerozoic time was an example of large-scale progress. Each fauna represents a set of animal groups which possessed particular arrays of adaptations, and each of which had different competitive abilities. The large-scale replacements, which happened during the Ordovician and during the Triassic, were the result of new forms entering adaptive zones that were already occupied, taking those over, and then radiating out into new modes of life. The greater adaptability of the Palaeozoic fauna allowed it to reach a higher global equilibrium of family diversity than the Cambrian fauna could achieve. The Modern fauna has yet to reach its global equilibrium diversity level.

This progressive interpretation of the diversification of life has been challenged (Benton, 1987, 1995). The first criticism is that Sepkoski's three faunas are human con-

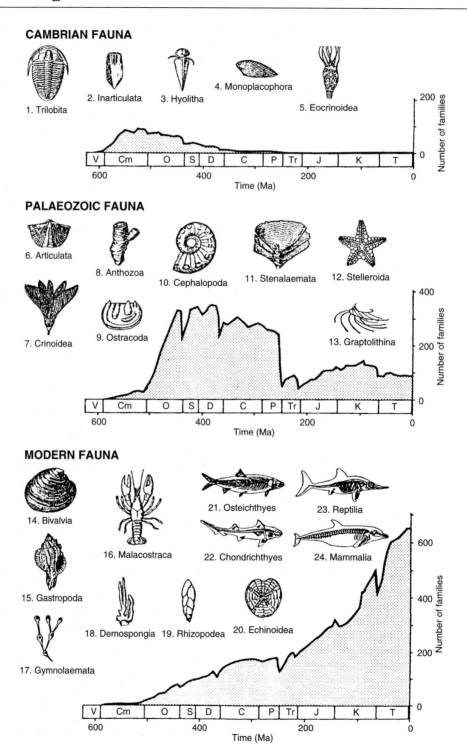

CAMBRIAN FAUNA

1. Trilobita
2. Inarticulata
3. Hyolitha
4. Monoplacophora
5. Eocrinoidea

PALAEOZOIC FAUNA

6. Articulata
8. Anthozoa
10. Cephalopoda
11. Stenalaemata
12. Stelleroida
7. Crinoidea
9. Ostracoda
13. Graptolithina

MODERN FAUNA

14. Bivalvia
16. Malacostraca
21. Osteichthyes
23. Reptilia
22. Chondrichthyes
24. Mammalia
15. Gastropoda
18. Demospongia
19. Rhizopodea
20. Echinoidea
17. Gymnolaemata

Fig 13.3 The history of family diversity of the three great 'faunas' of marine animals, showing a Cambrian phase, a Palaeozoic phase, and a 'modern' phase. The three phases add together to produce the overall pattern of diversification in Figure *13.2(a)*.

1. The environment is forever changing, and it is unlikely that selection pressure for change in a particular feature would be maintained for millions of years.
2. Palaeontological evidence rarely supports the simple explanations of evolutionary trends (see Box 13.2).
3. The occurrence of changes through time does not mean progress; progress involves demonstrated improvement of adaptation.

Biotic replacements: testing for progress

An obvious feature of the history of life is the way in which great groups of plants and animals have come and gone. The replacement of brachiopods by bivalves is a famous example. This had always been seen as a pro-

structs, and there is no evidence for any cohesion within them. Secondly, there is no evidence that equilibrium diversities exist, whether at a local or a global scale. It is not necessary to assume that the origin of a new species always drives another one out of existence. It is more likely that the patterns of increase in diversity have no top limit, and the changes in diversity level, both drops and rises, are the results of chance events, such as mass extinctions and radiations into new habitats.

The idea of progress

The idea of progress in the history of life has a long and chequered history. Evolution is progressive only in that advantageous adaptations may be inherited by the offspring of successful parents. Hence, generation by generation, some feature may seem to show a trend of change, such as the elongation of the neck of giraffes, or the increase in size of horses. The problems in transferring such ideas to the fossil record are threefold:

Box 13.2 (Cont.)

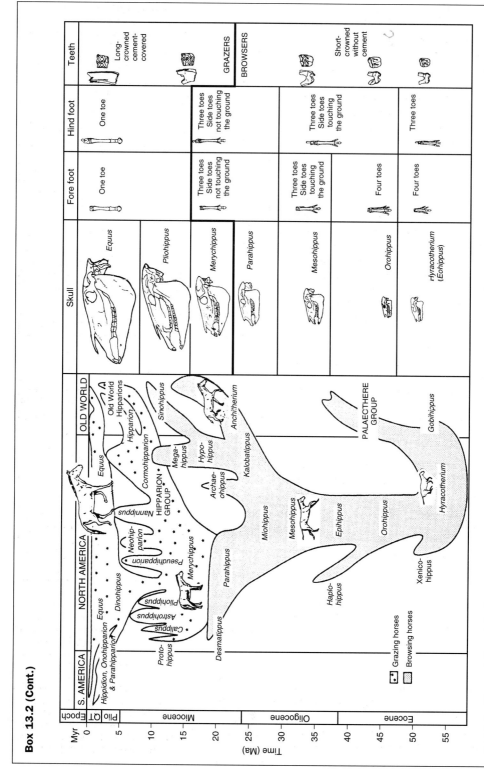

Fig 13.4 The evolution of the horses has been interpreted as a simple one-way trend towards large size, single toes and deep teeth. The reality is more complex: horse evolution has followed a branching pattern, and the line to the modern horses, *Equus*, was not pre-ordained: notice the diversity of North American horses in the late Miocene and Pliocene. The evolutionary steps did not all occur in parallel: *Merychippus* was a grazing horse, with deep-rooted teeth, but retained a three-toed foot.

gressive process: brachiopods are less adaptable than bivalves, and they clearly succumbed to long-term competition, perhaps lasting for tens or hundreds of millions of years. A re-study of this question by Gould and Calloway (1980) suggested that the take-over was more complex. Brachiopods and bivalves had maintained fairly constant diversities through the Palaeozoic, with brachiopods being more diverse (Figure 13.5). The end-Permian mass extinction event, around 250 Ma, drove their diversities right down. The bivalves recovered, and began to radiate rapidly during the Triassic and Jurassic, while the brachiopods have remained at the same low post-extinction diversity level ever since.

Another major biotic replacement attributed to competition was the two-step relay among Triassic vertebrates: the mammal-like reptiles were supposedly ousted by thecodontians as carnivores and rhynchosaurs as herbivores, and these were then out-competed by the dinosaurs (Figure 13.6a). The whole process supposedly took 30–40 Myr, and the dinosaurs succeeded in the end because they ran faster and had bigger teeth than the groups they replaced. A re-study of the evidence (Benton, 1983) suggests that the replacement happened more rapidly (Figure 13.6b) than was expected in the competitive model. The thecodontians never ousted the mammal-like reptiles as carnivores, although the rhynchosaurs were highly successful as herbivores. The dinosaurs then rose rapidly to dominance about 225 Ma, after a mass extinction event during which most of the thecodontians, mammal-like reptiles, and the rhynchosaurs died out.

Perhaps the majority of major biotic replacements were mediated in this way by mass extinctions which removed the old players from the field, and left the way clear for new groups to radiate. If this is the case, then it is hard to sustain a view that each new radiation of plants or animals marks a step upwards and forwards in the great relay race of evolutionary progress.

Mass extinctions

Definition

There have been many mass extinctions during the history of life, some large, and some very large but the meaning of the phrase 'mass extinction' is not entirely clear. All mass extinctions share certain features in common, but differ in others. The common features are as follows:

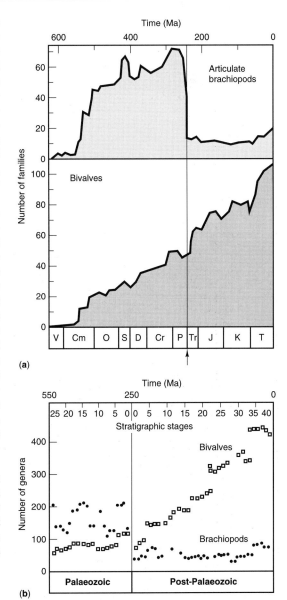

(a)

(b)

Fig 13.5 A classic example of competitive replacement? Articulate brachiopods were the dominant seabed shelled animals in the Palaeozoic, whereas bivalves take that role today. It was assumed that the bivalves competed long-term with the brachiopods during the Palaeozoic, even in the Permian, and eventually prevailed. A plot of the long-term fates of both groups (a) shows a steady rise in bivalve diversity, and a drop in brachiopod diversity. However, brachiopods were also diversifying during the Palaeozoic, and they were hard hit by the Permo-Triassic extinction event (a, b). The bivalves managed to recover afterwards, while the brachiopods did not (b).

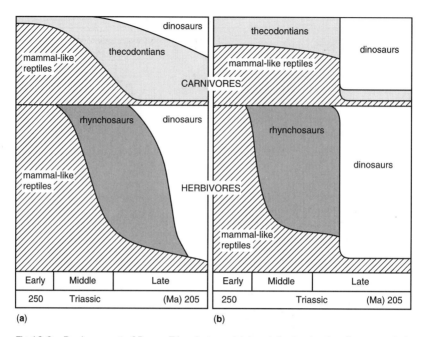

Fig 13.6 Replacement of Permo-Triassic terrestrial vertebrates by the dinosaurs during the late Triassic. (a) Classically, the rise of the dinosaurs was viewed as part of a long-term competitive relay of replacements, in which the mammal-like reptiles gave way to the thecodontians and rhynchosaurs, and these in turn to the dinosaurs. (b) New evidence suggests that the mammal-like reptiles, thecodontians, and rhynchosaurs were largely wiped out by a mass extinction event around 225 Ma, and that the dinosaurs radiated opportunistically afterwards.

1. many species became extinct, perhaps more than 30% of the extant biota;
2. the extinct forms span a broad range of ecologies, and typically include marine and nonmarine forms, plants and animals, microscopic and large forms;
3. the extinctions all happened within a short time, and hence relate to a single cause, or cluster of interlinked causes.

It has proved impossible so far to define these terms more precisely for many substantial biological and geological reasons, which will be explored below.

The confusion about definitions has meant that there is no accepted list of all mass extinctions, although palaeontologists agree on the 'big five': late Ordovician, late Devonian, Permian–Triassic (PTr), late Triassic, and Cretaceous–Tertiary (KT). These five, and the smaller mass extinctions may be sorted into major, intermediate, and minor events, based on their magnitudes (Figure 13.7; see Box 13.3).

The PTr mass extinction is in a class of its own, since it is known that 50% of families disappeared at that time, and this scales to a loss of 80–95% of species. The assumption that a higher proportion of species than families are wiped out is based on the observation that families contain many species, all of which must die for the family to be deemed extinct. Hence, the loss of a family implies the loss of all its constituent species, but many families will survive even if most of their contained species disappear. The 'intermediate' mass extinctions (Figure 13.7) are associated with losses of 20–30% of families, and perhaps 50% of species, while the 'minor' mass extinctions experienced perhaps 10% family loss and 20–30% species loss.

Patterns of mass extinctions

Good-quality fossil records indicate a variety of patterns of extinction. Detailed collecting of planktonic microfossils based on centimetre-by-centimetre sampling up to, and across, crucial mass extinction boundaries offers the best evidence of the patterns of mass extinctions. In detail, some of the range charts (Figure 13.8) reveal a stepped pattern of decline over a time interval of 0.5 – 1.0 Ma, during which 53% of the foraminifer species

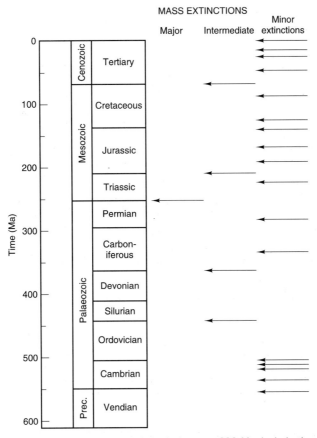

Fig 13.7 Mass extinctions through the past 600 Myr include the enormous Permo-Triassic (PTr) event around 250 Ma which killed twice or three times as many families, genera and species (50% of families and up to 96% of species) as the 'intermediate' events. These were global in extent, and involved losses of 20% of families, and 50% of species. Some of the minor mass extinctions were perhaps global in extent, causing losses of 10% of families and up to 20–30% of species, but many may have been regional in extent, or limited taxonomically or ecologically.

died out. However, should a palaeontologist describe this as an example of catastrophic or gradualistic extinction? A gradualist would argue that the extinction lasts for more than 0.5 Myr – too long to be the result of an instant event. A catastrophist would say that the killing lasted for 1–1000 years, and would argue that the stepped pattern in Figure 13.8 is the result of incomplete preservation, incomplete collecting, or reworking of sediment by burrowers.

This kind of detailed sampling is not possible for organisms such as dinosaurs. They are preserved in continental sediments, which are deposited sporadically, and specimens are large and rare. Nevertheless, two teams attempted large-scale field sampling in Montana to establish once and for all whether the dinosaurs had drifted to extinction over 5–10 Myr, the view of the gradualists, or whether they had survived in full vigour to the last minute of the Cretaceous Period, when they were catastrophically wiped out. Needless to say, one team found evidence for a long-term die-off, and the other team demonstrated sudden extinction.

Selectivity of mass extinctions

The second defining character of mass extinctions (see above) was that they should be ecologically catholic, i.e. that there should be no selectivity. This seems to be the

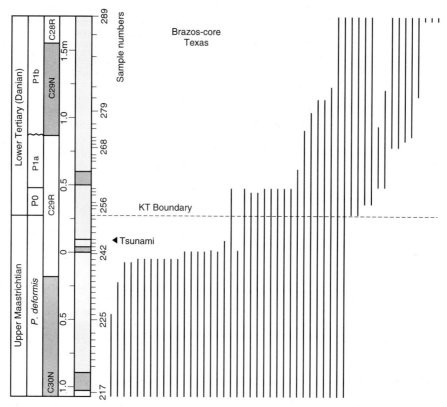

Fig 13.8 Patterns of extinction of foraminifera in a classic KT section, spanning about 1.5 Myr. A species loss of 53% occurred in two steps close to the KT boundary, and iridium (Ir) anomaly. Dating is based on magnetostratigraphy, and the KT boundary falls in the C29R (reversed) zone. Planktonic zones (P0, P1a, P1b) are indicated.

case, even though it might seem that, for example, large reptiles were specially selected for extinction during the KT event. The dinosaurs and some other large reptiles certainly died out then, but a larger number of microscopic planktonic species also died out.

The only evidence of selectivity during mass extinctions has been against genera with limited geographic ranges. Jablonski and Raup (1995) could find no evidence for selectivity during the KT event for ecological characters of bivalves, such as mode of life, body size, or habitat preference. They did find that the probability of extinction for bivalve genera declined predictably, depending upon the number of major biogeographic realms they occupied.

Timing of mass extinctions

The third defining character of mass extinctions (see above) concerns their timing. It is clearly important to

know whether a particular extinction event lasted for 5 Myr or one year. Biostratigraphic techniques are not ideal for determining the timing of mass extinction events since there is the risk of a circular argument. Exact age dating using radiometric and other techniques may give precise dates, but the error may still be too large to distinguish biologically slow and rapid processes.

Fossil sampling is also a key issue. Even if a palaeontologist can prove that dinosaur fossils suddenly disappear from the rock record at a particular horizon, it cannot simply be assumed that the disappearance is the result of extinction; there might have been an environmental change at that point, and the animals moved elsewhere, or there might have been a substantial hiatus in deposition, or depositional processes may have changed in such a way that bones are no longer buried and preserved. The intensity of collecting effort may also affect the apparent timing of species extinctions.

Box 13.3 The major mass extinction events

The *late Precambrian* event is ill-defined in terms of timing, but this may mark the end of metazoans of the Ediacara type, although some are known from the Cambrian, and the beginning of late Vendian and Cambrian faunas.

An extinction at the end of the *early Cambrian* marked the disappearance of previously widespread archaeocyathan reefs.

A series of mass extinctions during the *late Cambrian*, perhaps as many as five, are marked by major changes in trilobite faunas in North America and other parts of the world. Inarticulate brachiopods were also affected. During these events, and just after, animals in the sea became more diverse, and articulate brachiopods, corals, fishes, gastropods and cephalopods diversified dramatically.

In the *late Ordovician*, substantial turnovers occurred in marine faunas, with the extinction of up to 70% of species. Reefs were disrupted, and many families of brachiopods, echinoderms, ostracodes and trilobites died out. These extinctions are associated with a major glaciation over the South Pole, which locked water into the ice and lowered sea levels globally. Polar faunas moved towards the tropics, and warm-water faunas died out as the whole tropical belt disappeared. Oceanic overturn killed off many deep-water taxa.

The *late Devonian* mass extinction was a succession of extinction pulses lasting over 10 Myr in all. The abundant free-swimming cephalopods were decimated, as were the armoured fishes. Substantial losses occurred also among rugose and tabulate corals, articulate brachiopods, crinoids, stromatoporoids, ostracodes and trilobites. Causes could be a major cooling phase associated with anoxia on the seabed, or massive impacts of extraterrestrial objects.

The largest of all extinction events, the *Permo-Triassic* (*PTr*) event, is one of the least-known (Erwin, 1993). The dramatic changeover in faunas and floras at this time marks the boundary between the Palaeozoic and Mesozoic eras. Dominant groups in the sea disappeared, or were much reduced: rugose and tabulate corals, articulate brachiopods, stenolaemate bryozoans, stalked echinoderms, trilobites and goniatite ammonoids. There were also dramatic changes on land, with widespread extinctions among plants, insects and tetrapods. Causes seem to have been earthbound, perhaps related to massive basaltic volcanic eruptions in Siberia, with associated release of gases, global warming, and possible oceanic anoxia.

The *late Triassic* events were major, but not so extensive. A marine mass extinction event at the Triassic–Jurassic boundary is marked by the loss of most ceratite ammonoids, many families of brachiopods, bivalves, gastropods and marine reptiles, as well as the final demise of the conodonts. An earlier event, near the beginning of the late Triassic, also had effects in the sea, with major turnovers among reef faunas, ammonoids and echinoderms, but it was particularly important on land. There were large-scale changeovers in floras, and many amphibian and reptile groups disappeared. Causes of these events may have been climatic changes associated with the onset of rifting of Pangaea and the opening of the Atlantic, together with drift of continents away from the tropical belt.

Mass extinctions during the *Jurassic* Period (Figure 13.7) were minor in extent. The early Jurassic and end-Jurassic events at least seem to have been largely restricted to Europe, and to have involved losses of benthic bivalves, gastropods, brachiopods and free-swimming ammonites as a result of major phases of anoxia. Many animals were unaffected, and the events are undetectable on land. The mid Jurassic mass extinction is poorly documented, but may involve losses of cephalopods.

The *early Cretaceous* mass extinction event is similarly a minor blip in the overall pattern of extinction. The *Cenomanian–Turonian* mass extinction is more substantial, with extinctions among dinoflagellates and foraminifera, as well as cephalopods, echinoids, sponges, bony fishes and ichthyosaurs. These disappearances may relate to a major rise in sea-level and cooling phase, or to impacts.

The *Cretaceous–Tertiary* (*KT*) mass extinction is by far the best-known, both to the public, because of the loss of the dinosaurs, but also to researchers, because of the wealth of excellent geological sections available for study. As well as the dinosaurs, the pterosaurs, plesiosaurs, mosasaurs, ammonites, belemnites, rudist, trigoniid and inoceramid bivalves, and most foraminifera disappeared. The postulated causes range from long-term climatic change to instantaneous wipeout following a major extraterrestrial impact (see below).

The *Eocene–Oligocene* event is marked by substantial extinctions among plankton and open-water bony fishes in the sea, and by a major turnover among mammals in Europe at least. The event was rapid, and there is evidence for impact, as well as for temperature change.

Later *Tertiary* events are poorly defined. There was a dramatic extinction among mammals in North America in the mid Oligocene, and minor losses of plankton in the mid Miocene, but neither

Box 13.3 (Cont.)

event was large. Planktonic extinctions occurred during the Pliocene, and these may be linked to disappearances of bivalves and gastropods in tropical seas.

The latest extinction event, at the end of the *Pleistocene*, while dramatic in human terms, barely qualifies for inclusion. As the ice sheets withdrew from Europe and North America, large mammals such as mammoths, mastodons, woolly rhinos and giant ground sloths, died out. Some of the extinctions were related to major climatic changes, and others may have been exacerbated by human hunting activity. The loss of large mammal

species was, however, minor in global terms, amounting to a total loss of less than 1% of species.

One mass extinction event often ignored, occurring at the *present day*, may pass the numerical test when it is assessed in the future. Rates of species loss are hard to calculate: for birds, it is known that about 1% of all 9000 species have gone extinct since 1600, but some 20% of bird species are endangered, and could disappear in the next century. If these rates of species loss are scaled to millions of years, they equal the extinction rates found in ancient mass extinctions.

Periodicity of mass extinctions

There are many viewpoints on the causes of mass extinctions, but a primary divergence in opinions is between those who seek a single explanation for all mass extinctions, and those who believe that each event had its own unique causes. If there was a single cause, it might be sporadic changes in temperature (usually cooling) or in sea level, or periodic impacts on the Earth by asteroids (giant rocks) or comets (balls of ice).

The search for a common cause gained credence with the discovery by Raup and Sepkoski (1984) of a regular spacing of 26 Myr between extinction peaks during the last 250 Ma (Figure 13.9). It was argued that regular periodicity in mass extinctions implies an astronomical cause, and three suggestions were made:

1. the eccentric orbit of a sister star of the Sun, dubbed Nemesis (but not yet seen);

2. tilting of the galactic plane; or
3. the effects of a mysterious planet X which lies beyond Pluto on the edges of the Solar system.

These hypotheses involve a regularly repeating cycle which disturbs the Oört comet cloud and sends showers of comets hurtling through the solar system every 26 Myr.

The jury is still out on periodicity. Further statistical analyses have tended to confirm Raup and Sepkoski's original finding, but the search for Nemesis and Planet X has not been successful. Critics argue that each mass extinction was a one-off, and that there is no linking principle. The 26 Myr cycle discovered by Raup and Sepkoski is, they argue, a statistical artefact or the result of limited data analysis. Indicators of impact have been found for only two or three of the 10 or 11 mass extinction peaks that are elements of the periodic cycle, and the evidence is only really strong for the KT event (see below). On the other hand, study of craters has shown

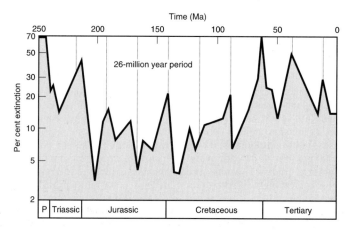

Fig 13.9   Periodic extinctions of marine animal families over the past 250 Ma. Extinction rate is plotted as per cent extinction per million years.

that the rates of asteroid and comet strikes through geological time are similar to the rates of extinctions of different magnitudes.

Causes of mass extinctions

The causes of mass extinctions are still unclear, despite a great deal of study: 500–1000 publications are produced each year about aspects of the KT event. This intensity of research has been maintained since 1980, when Alvarez *et al.* (1980) published their view that the extinctions had been caused by the impact of a 10 km-diameter asteroid on the Earth. The impact caused massive extinctions by throwing up a vast dust cloud which blocked out the sun and prevented photosynthesis, and hence plants died off, followed by herbivores, and then carnivores.

The two key pieces of evidence for the impact hypothesis are an iridium anomaly world-wide at the KT boundary, and associated shocked quartz. Iridium is a platinum-group element that is rare on the Earth's crust, and

reaches the Earth from space in meteorites, at a low average rate of accretion. At the KT boundary, that rate increased dramatically, giving an iridium spike (Figure 13.10). Shocked quartz has also been found, grains of quartz bearing criss-crossing lines produced by the pressure of an impact.

A catastrophic extinction is indicated by sudden plankton and other marine extinctions in certain sections (Figure 13.8), and by abrupt shifts in pollen ratios. The shifts in pollen ratios show a sudden loss of angiosperm taxa and their replacement by ferns, and then a progressive return to normal floras. This fern spike (Figure 13.10), found at many terrestrial KT boundary sections, is interpreted as indicating the aftermath of a catastrophic ash fall: ferns recover first and colonize the new surface, followed eventually by the angiosperms after soils begin to develop. This interpretation has been made by analogy with observed floral changes after major volcanic eruptions.

The main alternative to the extraterrestrial catastrophist explanation for the KT mass extinction is

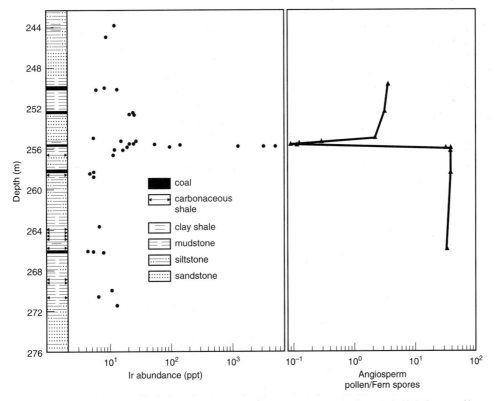

Fig 13.10 The iridium spike and fern spike, as recorded in continental sediments in York Canyon, New Mexico. The iridium spike, measured in parts per trillion (ppt), an enhancement of 10 000 times normal background levels, is generally interpreted as evidence for a massive extraterrestrial impact. The fern spike indicates sudden loss of the angiosperm flora, and replacement by ferns.

the gradualist model, in which extinctions occurred over long intervals of time as a result of climatic changes. On land, subtropical lush habitats with dinosaurs gave way to strongly seasonal temperate conifer-dominated habitats with mammals. Further evidence for the gradualist scenario is that many groups of marine organisms declined gradually through the late Cretaceous. Climatic changes on land are linked to changes in sea level and in the area of warm shallow-water seas.

Recent evidence of the site of impact has strengthened the catastrophist model for the KT event. A putative crater, the Chicxulub Crater, has been identified deep in late Cretaceous sediments on the Yucatán peninsula, Central America (Figure 13.11), and it seems to have produced a range of physical effects in the proximity. A ring of coeval coastline deposits show evidence for tsunami (massive tidal wave) activity, presumably set off by a vast impact into the proto-Caribbean. Further, the KT boundary clays ringing the site also yield abundant shocked quartz and glassy spherules (Figure 13.12) that supposedly match geochemically the bedrock under the crater site. Further afield, the boundary layer is thinner, there are no tsunami deposits, spherules are smaller or absent, and shocked quartz is less abundant.

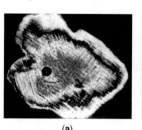

(a)   (b)

Fig 13.12   Evidence for KT impact in the Caribbean. (a) Shocked quartz from a KT boundary clay. (b) A glassy spherule from the KT boundary section at Mimbral, northeast Mexico, evidence of fall-out of volcanic melts from the Chicxulub crater. The spherule is about 1.5 mm in diameter.

A third school of thought has gained some ground in recent years, the view that most of the KT phenomena may be explained by volcanic activity. The Deccan Traps in India represent a vast outpouring of lava which occurred over the 2–3 Myr spanning the KT boundary. Supporters of the volcanic model seek to explain all the physical indicators of catastrophe (iridium, shocked quartz, spherules and the like) and the biological consequences as the result of the eruption of the Deccan Traps. In some interpretations, the volcanic model explains instantaneous catastrophic extinction, while in others it allows a span of 3 Myr or so, for a more gradualistic pattern of dying off caused by successive eruption episodes.

Thus, the geochemical and petrological data such as the iridium anomaly, shocked quartz and glassy spherules, as well as the Chicxulub Crater give strong evidence for an impact on Earth 65 Myr ago, although some aspects might also support a volcanic model. Some palaeontological data support the view of instantaneous extinction, but the majority still indicates longer-term extinction over 1–2 Myr. Key research questions are whether the long-term dying-off is a genuine pattern, or whether it is partly an artefact of incomplete fossil collecting, and, if the impact occurred, how it actually caused the patterns of extinction that occurred. Available killing models are either biologically unlikely, or too catastrophic: recall that a killing scenario must take account of the fact that 75% of families survived the KT event, many of them seemingly entirely unaffected. Whether the two models can be combined so that the long-term declines are explained by gradual changes in sea-level and climate and the final disappearances at the KT boundary were the result of impact-induced stresses is hard to tell.

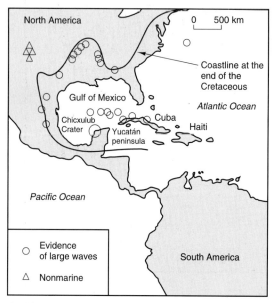

Fig 13.11   The KT impact site identified? Location of the Chicxulub Crater on the Yucatán Peninsula, Central America, and sites of tempestite deposits around the coastline of the proto-Caribbean (squares). Continental KT deposits are indicated by triangles.

After the mass extinction: radiation

After mass extinctions, the recovery time is proportional to the magnitude of the event. Biotic diversity took some 10 Ma to recover after major extinction events such as the late Devonian, the late Triassic, and the KT. Recovery time after the massive PTr event was much longer: it took some 100 Myr for total global marine familial diversity to recover to pre-extinction levels (Figure 13.2).

It is possible to examine the recovery phase after mass

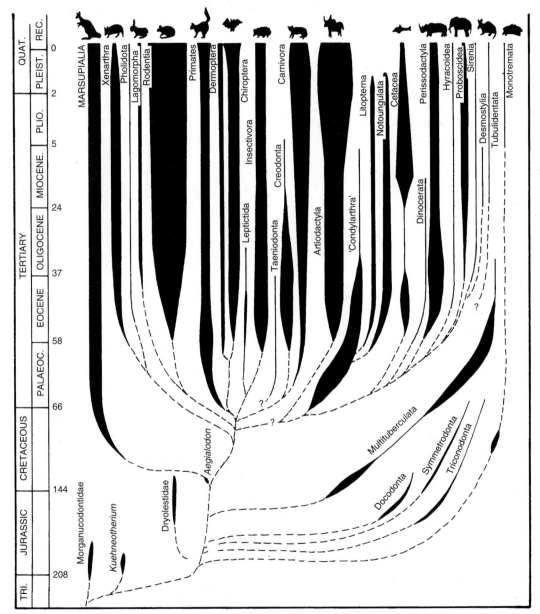

Fig 13.13   Radiation of the placental mammals after the KT extinction event. Within the first 10 Myr. after the event (Palaeocene, early Eocene), numerous orders of mammals had arisen, including all modern orders, and some that since went extinct. This shows that it took about 10 Myr for ecosystems to recover after the devastating loss of the dinosaurs, and other taxa, during the KT event.

extinctions in more detail. After the KT event, the dinosaurs and other land animals were gone, and vertebrate communities were much impoverished. The placental mammals, which had diversified a little during late Cretaceous times, and which ranged in size up to that of a cat, underwent a dramatic radiation (Figure 13.13). Within the 10 Myr of the Palaeocene and early Eocene, 20 major clades evolved, and these include the ancestors of all modern orders, ranging from bats to horses, and rodents to whales. During this initial period, overall ordinal diversity was much greater than it is now: it seems that during the ecological rebound from a mass extinction, surviving clades may radiate rapidly, and many body forms and ecological types arise. Half of the dominant placental groups of the Palaeocene became extinct soon after, during a phase of ecospace filling and competition, until a more stable community pattern became established 10 Myr after the mass extinction.

Ten major steps

The study of palaeontology reveals a great deal about how life has achieved its present astonishing diversity. Some major events in the history of life may be identified. The ten examples chosen here are not the only selection that could be made, but they represent some major adaptations that enabled substantial increases in diversity to occur (Figure 13.14).

1. The origin of life

The most widely accepted view about the origin of life, the biochemical model (see Chapter 4), is that complex organic molecules were synthesized naturally in the Precambrian oceans. The first living organisms were probably bacteria recognized in rocks about 3.5 billion years old. These were prokaryotic cells, small and lacking nuclei. The initial life forms operated in the absence of oxygen, and they caused one of the most significant changes in the history of the planet, i.e. the raising of oxygen levels in the atmosphere.

2. Eukaryotes and the origin of sex

In marked contrast to the prokaryotes, eukaryote cells are usually large, with organelles and membrane-bounded nuclei containing the chromosomes (see Chapter 4). Eukaryotes differ from prokaryotes in another funda-

mental respect: their cells reproduce sexually. The oldest fossil eukaryotes are hard to identify (see p. 67), but the clade was well established by 1000 Ma. Asexual reproduction tends to propagate only clones, but the mixing of genetic material during sexual reproduction opened the door to the exchange of genetic material, mutation and recombination, and the development of variation in populations, the basic material for evolution.

3. Multicellularity

Multicellular organisms are clusters of eukaryote cells organized into different tissue types and organs, where different parts of the organism are responsible for particular functions and tasks. Some of the oldest multicellular eukaryotes have been reported from rocks 1260–950 Ma in Canada; these red algae (see p. 70) were one of nearly 20 multicellular eukaryote lineages. Molecular evidence suggests all these groups had a common origin about 1200 Ma. Some 400 Ma later, true metazoans had arrived.

Four lines of study have aided the search for the first metazoans. Firstly, body fossils of possible worms, *Sinosabellidites* and *Protoarenicola*, carbonaceous tubes with growth lines, have been reported from Precambrian rocks predating the appearance of the Ediacara fauna (see Chapter 5), dated at more than 1000 Ma. Secondly, trace fossils imply the development of a grade of organization and structures capable of locomotion. Ichnofossils from the Medicine Peak Quartzite of Wyoming may be 2400 Ma, hinting at the presence of a mobile metazoan around the Archaean–Proterozoic boundary. This seems unlikely, and other interpretations are possible. Thirdly, the marked decline of stromatolites about 1000 Ma suggests the presence of grazing metazoans. Finally, molecular phylogenies (see p. 69) suggest an initial divergence of metazoans around 800–1000 Ma.

4. Skeletons

During an interval of a few million years in the earliest Cambrian, a wide variety of mineralized skeletons appeared (see Chapter 5), and these are seen spectacularly in the Mid Cambrian Burgess Shale fauna (see Chapter 8). Mineralized hard parts conferred distinct advantages, by providing protection, support, and areas for the attachment of muscles. Predator pressure may have been the main driving force behind the acquisition of hard parts. Marked changes in ocean chemistry during the early Cambrian marine transgression may have

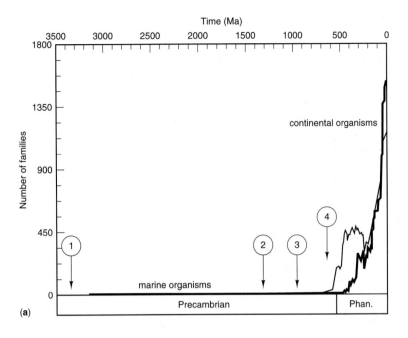

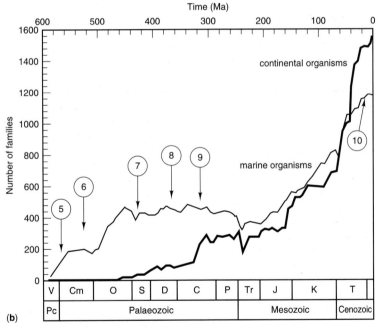

Fig 13.14 The diversification of life, with the timing of the ten major biological advances indicated: 1, origin of life; 2, eukaryotes and the origin of sex, 3, multi-cellularity; 4, skeletons; 5, predation; 6, biological reefs; 7, terrestrialization; 8, trees and forests; 9, flight; 10, consciousness. The diversification of life is plotted for (a) the whole of the past 4000 Myr, and (b) for the Phanerozoic.

forced organisms to ingest and excrete large quantities of minerals, and enhanced oxygen levels in the world's oceans made precipitation of minerals easier.

Diverse mineralized tubes characterize many Early Cambrian fossil assemblages. Apart from these tubes, of uncertain taxonomic affinities, better known organisms also drew on a variety of substances to construct skeletal structures: calcite (e.g. brachiopods, bryozoans, trilobites, ostracodes and corals), aragonite (e.g. hyoliths and molluscs), apatite (e.g. conulates, lingulate brachiopods, conodonts and vertebrates), opal (e.g. hexactinellid sponges and radiolarians), and agglutinated material (e.g. foraminiferans).

5. Predation

The late Precambrian Ediacara fauna was soft-bodied, and it appears likely that predators and scavengers were absent. This changed during the early Cambrian, when evidence from borings in shells suggests that this form of predation was an important part of the ecosystem. The rapid diversification of armoured and protective strategies was a feature of the Cambrian radiation. It is probable that predation had an important influence on evolutionary processes by the coevolution of predator–prey systems, where prey and predator organisms evolve ever-better defensive and offensive strategies respectively. Such 'arms races' became intensified several times during the Phanerozoic, most notably in the Cretaceous when new efficient predators such as crabs, hole-boring gastropods, and teleost fishes caused dramatic changes in the lifestyles of their prey (see Chapters 8 and 9).

6. Biological reefs

Biological reefs are the marine equivalents of tropical rain forests. Modern reefs are highly diverse colourful frameworks of brain, horn and staghorn corals, together with organ pipes, sea fans and sea whips, providing food and accommodation for thousands of species from the mantis shrimp to the carpet shark. These large carbonate structures form the basis of many types of tropical island, from barrier reefs to fringing atolls. The origin of reefs was a major event.

Throughout the Phanerozoic, the main reef-builders have changed, with major changes punctuated by mass extinctions. The first reefal frameworks appeared during the early Cambrian, constructed first by solitary clustered polychaete worm tubes, and then by archaeocy-

athans (see Chapter 5). Ordovician reefs were dominated by algae, bryozoans, stromatoporoids, and rugose and tabulate corals (see Chapters 5 and 6). Tabulate corals and stromatoporoids, together with algae and bryozoans, dominated reefs from the Silurian until near the end of the Devonian. Carboniferous and Permian reefs were made from bryozoans, algae and calcareous sphinctozoan sponges. During the mid Triassic, frameworks of algae, scleractinian corals, bryozoans and sphinctozoan sponges developed. Jurassic and Cretaceous reefs were constructed by scleractinian corals, lithothamnian algae, and siliceous and sphinctozoan sponges, to which were possibly added rudist bivalves in the late Cretaceous. During the Tertiary, scleractinian corals expanded to their present diversity, where they now dominate biological frameworks.

7. Terrestrialization

The colonization of the land added major new environments to those previously occupied by life. It is hard to date the first move of life on to land. Soils have been reported from mid Precambrian sequences, but colonization then by microbes was unlikely in the hot, anaerobic and irradiated landscape. Probably by the mid Ordovician, soils were coated by a green scum of cyanobacteria. By Silurian times, small vascular plants, such as *Cooksonia*, with a well-developed vascular system, stomata, a waxy covering and trilete spores was well established (see Chapter 10). The new land plants relied upon the soil for some of their nutrients, but they also generated modern-style soils over landscapes that had previously been bare rock. This stabilization of the land by plant growth slowed down the rate of erosion, and it was one of the most dramatic effects that life has had on the physical nature of the Earth.

Colonizing invertebrates were faced with problems of dehydration, respiration, and to a lesser extent, support. These problems were overcome by the development of waterproof skins, lungs and skeletal support. Although hydrostatic skeletons, such as those of slugs, have been successful, the toughened exoskeleton of arthropods was an ideal protective covering, providing support and attachment for the soft parts. By the early Devonian, the low green vegetation was inhabited by myriapods, insects and possibly arachnids. During the Carboniferous, these faunas were supplemented by oligochaete worms and scorpions, together with both prosobranch and pulmonate gastropods (see Chapter 7). Vertebrates moved on to the land during the Devonian, presumably to exploit the new sources of plant and invertebrate food, and full terrestrialization occurred with the reptiles in the Mid Carboniferous, when the amniotic egg evolved (see Chapter 9).

8. Trees and forests

The next major expansion of living space on land took place during the Carboniferous, with the development of forests. The first tree-sized plants arose in the late Devonian, forms such as the progymnosperm *Archaeopteris*, which reached a height of 8.4 m and had secondary woody tissue and growth rings. Trees became abundant and diverse in the Carboniferous, with large lycopods such as *Lepidodendron* (50 m tall) and *Sigillaria* (30 m tall), equisetaleans such as *Calamites* (20 m tall), ferns such as *Psaronius* (3 m), progymnosperms, seed ferns such as *Medullosa* (10 m), and early conifers (see Chapter 10).

The significance of the tree habit was that it created a vertically tiered range of new habitats. Trees, with their long roots, gained access to nutrients that were not available to smaller plants, and the addition of new storeys of vegetation allowed more plants to pack into an available space than before. Insects and other invertebrates sheltered and fed in the bark and among the leaves and fruits, as well as in the new leaf litter habitats in the soil. The evolution of trees and forests led to a dramatic burst in the rate of diversification of vascular land plants, as well as a radiation of insects, and their predators, such as spiders, amphibians and reptiles.

9. Flight

The next major expansion of land life was into the air. Perhaps the evolution of trees led directly to this further leap in adaptations: having tempted various insect groups to move off the ground in search of edible leaves and fruits, gliding and true flight became inevitable. Insects arose in the early Devonian, but the first true fliers are Carboniferous in age (see Chapter 7). There was a dramatic diversification of insect groups in the late Carboniferous. Flight has doubtless been the clue to the vast success of insects: today there are millions of species, too many to identify and count accurately, representing perhaps 70% of all living animals.

Flight has arisen more than 30 times among vertebrates (see Chapter 9). Gliding diapsid reptiles are known from the late Permian and late Triassic. The first true flapping vertebrates, the pterosaurs, arose in the late Triassic, dominating Mesozoic skies, and reaching wingspans of 11–15 m, the largest flying animals of all time. Modern vertebrate fliers include birds and bats, both successful groups, and both largely feeding on insects, the birds by day and the bats by night. The other aerial vertebrates today are gliders, animals that swoop through the air supported passively on expanded membranes of some sort: flying fishes, frogs, lizards and snakes, as well as various gliding marsupial and placental mammals.

Plants too exploited flight and flying animals. Numerous plants use the wind for dispersal of pollen and seeds (see Chapter 10). Certain plant growth habits could also be argued to be analogous to flight; for example, the lianas of tropical forests which descend from tall trees, and sweep from plant to plant, exploiting small patches of sunlight.

10. Consciousness

Human beings probably have to feature somewhere in a list of major biotic advances, although how to view the role of humans in evolution has been a question that has dogged philosophers for centuries. What is the nature of the dramatic changes wrought by humanity? Are they to be seen in a negative light, in terms of destruction of the Earth and its inhabitants, or are they to be regarded positively, in terms of the attainment of a level of intelligence and creativity never before achieved?

Because of their brains, and the ability to adapt the environment, humans are a hugely successful species, living in vast abundance in every part of the Earth, from the poles to the equator. Humans live at all altitudes, and could theoretically create appropriate enclosed environments to allow humans to survive indefinitely underwater, at high altitudes, and on other planets. *Homo sapiens* is the first species to leave the Earth and come back alive (numerous insects and microbes have doubtless been swept up into the upper reaches of the atmosphere · and lost into space, but they have not benefited from the experience).

The human ability to make things is not unique, in that many species can make constructions, nests and the like, or even use tools (birds and apes use twigs to reach otherwise unavailable food). Language is also not unique, in that most animals have some form of communication, often rather complex. The extent to which humans make things and communicate is, however, unique. Perhaps also unique to *Homo sapiens*, but perhaps we shall never know, is consciousness, the ability to think about one's existence, about the past, and about the future.

A direct result of consciousness is the ability to plan. One person, or a team, may invent something after years of work, and that appears to be uniquely human, but the creative process may continue through many generations. No other species has invented writing, the means whereby each human brain has access to virtually infinite stores of knowledge accumulated by other people and recorded in books and in electronic form. Humans have also extended their physical capabilities in ways not achieved by any other species: it is possible to travel from New York to London in a few hours in an aeroplane, a useful adjunct to our rather feeble legs; it is possible to see the surface of the moon through a telescope, an extension of our poor eyesight; it is possible to speak to someone on the other side of the world without shouting; it is possible to estimate the phenetic resemblances of many specimens of brachiopods in a few minutes using a computer, a task that would take many days using the unaided brain.

Value judgements of human activities are not relevant in a purely evolutionary view of the history of the Earth. Success in evolutionary terms is measured by the abundance of a species, its dominance of the physical and biotic environment, and its longevity. Humans are outrageously successful on the first two counts, but the duration of the species *Homo sapiens* cannot yet be judged. Consciousness seems to have permitted *Homo sapiens* to achieve results that no other species has remotely approached. After all, was it possible until the geological Recent, for a palaeontological textbook to be written, purchased (one hopes) and read? The pursuit of a knowledge of the history of life on Earth is a part of human consciousness.

Further reading

Allen, K. C. and Briggs, D. E. G. (1989) *Evolution and the fossil record*. Belhaven, London.

Alvarez, W. and Asaro, F. (1990) An extraterrestrial impact. *Scientific American*, Oct. 1990, 44–52.

Briggs, D.E.G. and Crowther, P. R. (eds) (1990) *Palaeobiology, a synthesis*. Blackwell Scientific Publications, Oxford.

Courtillot, V. E. (1990) A volcanic eruption. *Scientific American*, Oct. 1990, 53–60.

Donovan, S. K. (ed.) (1989) *Mass extinctions*. Belhaven Press, London.

Gould, S. J. (ed.) (1993) *The book of life*. Ebury, London.

Hildebrand, A. R. and Boynton, W. V. (1991) Cretaceous ground zero. *Natural History*, **6/91**, 46–53.

Raup, D. M. (1991) *Extinction: bad genes or bad luck*. Norton, New York.

Sharpton, V. L. and Ward, P. D. (eds) (1990) Global catastrophes in earth history. *Geological Society of America Special Paper*, **247**, 1–631.

Skelton, P. (ed.) (1993) *Evolution; a biological and palaeontological perspective*. Addison Wesley, Wokingham.

References

Alvarez, L. W., Alvarez, W., Asaro, F. and Michel, H. V. (1980) Extraterrestrial cause for the Cretaceous–Tertiary extinction. *Science*, **208**, 1095–1108.

Benton, M. J. (1983) Dinosaur success in the Triassic: a non-competitive ecological model. *Quarterly Review of Biology*, **58**, 29–55.

Benton, M. J. (1987) Progress and competition in macroevolution. *Biological Reviews*, **62**, 305–38.

Benton, M. J. (1995) Diversification and extinction in the history of life. *Science*, **268**, 52–58.

Erwin, D. H. (1993) *The Great Paleozoic Crisis: life and death in the Permian*. Columbia University Press, New York.

Gould, S. J. and Calloway, C. B. (1980) Clams and brachiopods – ships that pass in the night. *Paleobiology*, **6**, 383–396.

Jablonski, D. and Raup, D. M. (1995) Selectivity of end-Cretaceous marine bivalve extinctions. *Science*, **268**, 389–391.

Raup, D. M. (1972) Taxonomic diversity during the Phanerozoic. *Science*, **177**, 1065–1071.

Raup P, D. M. and Sepkoski, J. J., jr (1984) Periodicities of extinctions in the geologic past. *Proceedings of the National Academy of Sciences*, USA, **81**, 801–805.

Sepkoski, J. J., jr (1984) A kinetic model of Phanerozoic taxonomic diversity. III. Post-Paleozoic families and mass extinctions. *Paleobiology*, **10**, 246–267.

Valentine, J. M. (1969) Patterns of taxonomic and ecological structure of the shelf benthos during Phanerozoic time. *Palaeontology*, **12**, 684–709.

Geological time scale

Eono-them	Era-them	Sub-erathem, System, Sub-system			Series	
Phanerozoic	Cenozoic	Quaternary			Holocene	
				Q	Pleistocene	
		Tertiary	Neogene		Pliocene	PLI
					Miocene	UMI
						MMI
				Ng		LMI
			Paleogene		Oligocene	OLI
					Eocene	EOC
		T		Pg	Palaeocene	PAL
	Mesozoic	Cretaceous			Senonian	SEN
					Gallic	GAL
				K	Neocomian	NEO
		Jurassic			Malm	MLM
					Dogger	DOG
				J	Lias	LIA
		Triassic			Upper	u
					Middle	m
				Tr	Scythian	SCY
	Palae-ozoic	Permian			Zechstein	ZEC
				P	Rotliegendes	ROT
		Carboniferous	Pennsylvanian		Gzelian	GZE
					Kasimovian	KAS
					Moscovian	MOS
				C(u)	Bashkirian	BSK
			Mississippi		Serpukhovian	SPK
					Viséan	VIS
		C		C(l)	Tournaisian	TOU

Eono-them	Era-them	Sub-erathem, System, Sub-system		Series	
Phanerozoic	Palaeozoic	Devonian		Upper	u
				Middle	m
			D	Lower	l
		Silurian		Pridoli	PRD
				Ludlow	LUD
				Wenlock	WEN
			S	Llandovery	LLY
		Ordovician		Ashgill	ASH
				Caradoc	CRD
				Llandeilo	LLO
				Llanvirn	LLN
				Arenig	ARG
			O	Tremadoc	TRE
		Cambrian		Merioneth	MER
				St David's	STD
			Є	Caerfai/Comley	CRF
Precambrian	Proterozoic	Sinian	Vendian	Ediacara	ED
			V	Varanger	VAR
			Sturtian		STU
			Riphean	Karatau	KAR
				Yurmatin	YUR
			RIF	Burzyan	BUZ
			Animikean		ANI
			Huronian		HUR
	Archaean	Randian			RAN
		Swazian			SWZ
		Isuan			ISU
PЄ		Hadean			HDE

Glossary

Technical terms used in describing fossils and palaeontological phenomena are listed here. Use the glossary in conjunction with the index to find fuller explanations in the text. Group names of organisms are not listed: see the index. Where necessary, the plural form (pl.), and the adjective (adj.) are given. When there are several related terms in the glossary, these are cross referenced by the term cf. (= compare, *confere*). Many technical terms used in palaeontology, as in science generally, are pure Latin or Greek words: their original meanings are indicated by inverted commas.

abdomen 'belly'; the posterior body segment of an arthropod (cf. thorax).

aboral 'opposite the mouth'.

abrasion 'wearing down'; removal of edges and processes by tumbling in a current.

abundance numbers of individuals of a species in a population or a sample.

abyssal 'bottomless'; of the deepest oceanic zones.

accretion (in sedimentology and skeletal growth) 'growth' by the addition of new material.

acoelomate lacking a coelom.

adapical 'opposite the apex' or top.

adaptation 'fitting'; any feature of an organism that has a function; also, the processes associated with the acquisition of an adaptation.

adductor muscle/ adductor 'pull towards'; a muscle that closes the valves of a brachiopod (cf. diductor muscle).

aerobic 'air-containing'; oxygen-rich.

agglutinated 'stuck together'.

agrichnion (pl. **agrichnia**) 'farming trace'.

ahermatypic (of corals) living below the photic zone, generally in cold waters (cf. hermatypic).

alar 'wing-like'.

algal bloom a huge growth of algae, often in the surface waters of a lake.

alimentary canal the gut.

allochthonous 'other soil'; derived from elsewhere (cf. autochthonous).

allometry 'other measure'; change in proportions during growth (cf. isometry).

allopatric speciation 'other homeland'; formation of new species by splitting of the geographic range occupied by the parent species.

alveolus (pl. **alveoli**) 'little hollow'; a tooth socket; a deep depression, for example in the guard of a belemnite.

ambulacral area (**amb**) 'walking'; one of five zones of narrower plates in the echinoid skeleton (cf. interambulacral area).

amerous 'no parts'; with an undivided coelom (cf. metamerous, oligomerous, pseudometamerous).

amino acid basic building block of proteins; there are 21 amino acids, which may be strung together in a huge range of different sequences to produce all the multitudes of protein types.

amoeba 'change'; an irregularly-shaped single-celled organism.

anaerobic oxygen-poor (cf. aerobic).

analogue a feature of two or more organisms that is superficially similar in morphology or function, but which arose from different ancestries (adj. **analogous**).

anapsid 'no arches'; a tetrapod skull with no temporal openings (cf. diapsid, synapsid)

anoxic oxygen-poor, or with no oxygen.

antenna (pl. **antennae**) a 'feeler', or anterior long sensory appendage, in an arthropod.

anterior front (cf. posterior).

anther 'flower'; the part of a flower that produces pollen.

apatite calcium phosphate ($CaPO_4$), typical mineralized constituent of skeletons of vertebrates, conodonts, some brachiopods, and some worms.

aperture 'opening'; specifically the opening of a gastropod shell.

apex 'tip' or top (adj. **apical**).

apomorphy (pl. **apomorphies**) 'from shape'; a

derived character, a feature that arose once only in evolution.

appendage a limb, or limb-like, projection from the side of the body, used mainly for locomotion and feeding.

appendiculate possessing appendages.

aptychus (pl. **aptychi**) a paired covering plate over the aperture of a cephalopod.

aragonite a form of calcium carbonate ($CaCO_3$) that occurs commonly as needles in shelly skeletons of various organisms and in lime muds.

arborescent 'tree-like-growing'.

archegonium (pl. **archegonia**) 'founder of a race'; the fertile female structures within the ovule of a plant.

Aristotle's lantern the jaw system of echinoids, consisting of hinged plates and muscles.

articular the tiny bone at the back of a reptile lower jaw which articulates with the quadrate in the skull.

articulation 'connection'; refers typically to parts of skeletons that remain in natural contact in fossils.

asexual reproduction reproduction in the absence of sex.

assemblage a collection of fossil specimens or species found together, which may partly represent a naturally-occurring community or population, but which has also been overwritten by processes of sedimentary accumulation.

astogeny growth of the compound structure of a colonial coral.

astragalus a major ankle bone.

astrorhiza (pl. **astrorhizae**) 'star root'; a star-like pattern of radiating canals on the outer surface of a stromatoporoid, the exhalent canal system.

atavism a 'throw-back', an ancestral feature that reappears in an organism, often by an error in development.

atoll a reef that entirely surrounds a volcanic island which may, or may not, still be visible above the waves (cf. fringing reef, barrier reef).

authigenic preservation casting of the outer shape of a fossil.

autochthonous 'self soil'; *in situ*, not moved from elsewhere (cf. allochthonous).

autotheca 'self case'; one of the types of thecae in a dendroid graptolite, larger than a bitheca.

autotroph 'self-feeder'; an organism that converts inorganic matter into food (cf. heterotroph).

axis the line of symmetry running through an organism; the middle stalk of a frond (adj. **axial**).

barrier reef a reef that lies offshore on the shelf, separated from the land by a lagoon (cf. atoll, fringing reef).

basal at the bottom; plates lying low in the calyx of a crinoid.

bathymetric 'depth measure'; relating to depth of water.

benthos 'depth'; the organisms that live in and on the seabed (adj. **benthic** or **benthonic**).

biconvex 'two-convex'; a shape that is convex in both directions.

bilateral 'two-sided'.

bilobed 'two-lobed'.

binomen 'two name'; the standard generic and specific name of an organism.

bioerosion 'life removal'; the removal of skeletal materials by boring organisms using chemical and physical means.

biogeochemical cycle the movement of carbon, and other organic chemicals, through organisms and sediments.

biogeographical province a large-scale geographic region that is inhabited by specific plants and animals.

biological species concept the definition of a species as a group of organisms that are capable of interbreeding and of producing viable offspring.

biomass the mass of biological material, plant and animal, represented in a specific place at a specific time.

biostratigraphy 'life stratigraphy'; the dating of rocks by means of fossils.

bioturbation 'life disturbed'; disturbance of sediments by the activity of animals or plants.

bipedal 'two-footed'; walking on the hindlimbs only (cf. quadrupedal).

biramous 'two-branched' (cf. uniramous).

biserial 'in a double row' (cf. uniserial, triserial).

bitheca one of the types of thecae in a dendroid graptolite, smaller than an autotheca.

blastopore 'sprout opening'; a deep depression on the side of a blastula-stage embryo that eventually becomes the mouth, in protostomes, or the anus, in deuterostomes.

blastula 'small sprout'; early embryonic phase, shaped like a hollow ball with a deep depression on one side, the blastopore.

body chamber the last-formed chamber of a cephalopod shell in which the animal lives.

body fossil the remains of an organism, usually termed simply 'fossil' (cf. trace fossil).

bone the characteristic skeletal tissue of vertebrates,

formed mainly from apatite and collagen.

brachial 'of the arm'; plates lying at the base of the arms in the calyx of a crinoid.

brachial valve the valve of a brachiopod that does not contain the pedicle foramen (cf. pedicle valve).

brachiole a pore perforating the plate of a blastozoan.

byssus a bundle of fibres used for attachment to hard substrates in some bivalves.

caecum (pl. **caeca**) 'blind'; a blind sac or side projection of the gut or mantle.

calcified 'lime-made'; invested with carbonate, either calcium carbonate or calcium phosphate.

calcite a variety of forms of calcium carbonate ($CaCO_3$) that occur commonly in the shelly skeletons of various organisms and in limestones.

calice 'cup'; the upper part of the skeleton of a coral, the corallum, in which the polyp sits.

calyx 'covering'; the outer covering of a flower, formed from the sepals; the cup-like skeleton containing the body of a crinoid.

canine teeth 'dog-like'; pointed teeth in mammals, used for piercing food items (cf. incisor teeth, cheek teeth, premolar teeth, molar teeth).

carbonate made from calcium carbonate.

cardinal major or key.

carnassial teeth 'flesh-eating'; specialized cheek teeth in carnivorous mammals, used for tearing flesh.

carnivore 'flesh-eater'.

carpel 'fruit'; the specialized structure that encloses the ovule in an angiosperm flower.

cartilage flexible supporting tissue in chordates, formed from collagen.

catastrophe a sudden event (adj. **catastrophic**; cf. gradualistic).

cateniform 'chain-like'.

cellulose a carbohydrate that is the main component of the cell walls of plants.

cephalic 'of the head'.

cephalis 'head'; the upper whorl of a radiolarian test.

cephalon 'head'; the anterior segment of a trilobite (cf. thorax, pygidium).

chamber a walled compartment of the cephalopod or foraminiferan shell.

cheek teeth the teeth in the back of the jaw of mammals, divided into premolars and molars, used for chewing the food (cf. incisor teeth, canine teeth).

chelicera (pl. **chelicerae**) 'crab's claw-horn'; the pincer of a chelicerate arthropod.

chitin a protein that forms most of the hard parts of arthropods.

chitinophosphate a hard tissue composed of chitin and phosphate.

chloroplast 'pale green-moulded'; organelle in a eukaryote plant cell that carries out photosynthesis.

choanocyte a collar cell in a sponge that moves water by beating its flagellum.

chromosome 'colour body'; a long strand of DNA, paired with a matching sequence, and typically forming an elongate X shape, composed of sequences of genes.

chronospecies 'time species'; a species that is part of a lineage, whose origin and extinction are defined somewhat arbitrarily by gaps in the fossil record, or by major morphological changes.

chronostratigraphy 'time stratigraphy'; established international standard divisions of geological time.

cilium (pl. **cilia**) 'eyelash'; a hair-like lash borne by a cell (adj., **ciliate/ ciliated**).

cirrus (pl. **cirri**) 'curl'; a small flexible projection.

clade a monophyletic group.

cladistic analysis, cladistics the classification of taxa, or of biogeographic regions, in terms of shared derived characters (synapomorphies).

cladogram a dichotomously (two-way) branching diagram indicating the closeness of relationship of a number of taxa.

class a division in classification; contains one or more orders, and is contained in a phylum.

classification the process of naming organisms and arranging them in a meaningful pattern; also, the end-result of such a procedure, a sequential list of organism names arranged in a way that reflects their postulated relationships.

clastic 'broken'; sedimentary rocks formed from eroded and transported material.

clavate 'club-shaped'.

cleidoic 'closed'.

climax community the final stable community established after a certain time of adjustment.

cluster analysis multivariate mathematical techniques for finding clusters of most-similar organisms or communities.

cnidoblast 'sea anemone sprout'; a poisonous stinging cell of a cnidarian.

coaptive 'fitting together'.

coelom 'cavity'; the body cavity in animals found between the gut and the outer body surface (adj. **coelomate**).

coenosteum the skeleton of a stromatoporoid.

coevolution 'evolution together' of two or more organisms that interact in some way.

collagen a flexible protein that makes up cartilage, and which forms the flexible framework of bone, upon which apatite crystals precipitate.

colonial (of corals, graptolites, bryozoans, etc.) living in fixed association with other individuals, and forming a unified 'super-organism' (cf. solitary).

colpus 'depression'.

columella (pl. **columellae**) a pillar-like part of the skeleton of a coral that runs up the middle of the corallum.

columnal one segment of the stalk of a crinoid, an ossicle.

columnar column-like.

commensalism 'feeding together'; a biological interaction where smaller species live on larger ones, and feed on debris from their eating activities, but the small organisms do not damage the larger ones (adj. **commensal**).

commissure 'joining'; the line of contact of the two valves of a brachiopod or bivalve, excluding the hinge; point of contact of laesurae of a spore.

common ancestor the individual organism, or species, that gave rise to a particular monophyletic group, or clade.

common descent the shared ancestry of all organisms, tracing back through common ancestors to the origin of life.

community a group of plants and animals that live together and interact, in a specified area.

competition the interaction between two individuals or two species where both require the same limiting resource (food, space, shelter), and success by one implies a disadvantage to the other.

compound eye an eye composed of many separate units, typical of arthropods (cf. ocellus).

concavoconvex concave on one side and convex on the other; hence, rather C-shaped in cross-section.

concentration deposit a rich accumulation of fossils produced by physical sedimentary processes that bring the specimens together (cf. conservation deposit).

concentric repeated circular pattern, running parallel to an outer circular margin (cf. radial).

conch 'shell'.

concretion an irregular concentration, commonly of calcite or siderite, formed by chemical precipitation within the sediment.

congulum (pl. **congula**) sealing ring of the diatom skeleton.

conservation deposit a rich accumulation of fossils produced on the spot by processes that prevent decay and scavenging (cf. concentration deposit).

conservation trap a specific location where organisms are trapped and preserved instantly.

conterminant 'near the end'.

continental drift relative movements of continental and oceanic plates through time.

coprolite 'excrement stone'; fossilized dung.

corallite 'coral stone'; the skeletal part of a coral.

corallum (pl. **coralla**) the skeletal part of a coral.

corpus 'body'.

correlation matching of rocks of equivalent age.

corrosion chemical destruction of hard tissues.

cosmopolitan 'citizen of the world'; living world-wide, or nearly world-wide (cf. endemic).

costa (pl. **costae**) 'rib'.

costation pattern of ribbing, which gives a zig-zag pattern to the commissure line in brachiopods and bivalves.

cotyledon 'cup'; the food storage area of a seed.

cubichnion (pl. **cubichnia**) 'resting trace'.

cusp pointed projection of top of a mammalian tooth or a conodont element.

cuticle horny protein outer covering in many plants and animals.

cyst the fertile 'resting stage' of an alga.

cytology the study of cells.

decay break-down of tissues by chemical means or by microbial attack.

deciduous 'fall from'; shedding leaves each winter.

declined 'sloping down'.

deformation distortion of rocks and fossils by physical stretching and compression.

delthyrium 'delta door'; a small triangular-shaped zone in the hinge region of the pedicle valve of a brachiopod, lying between the pedicle foramen and the margin of the valve (cf. notothyrium).

deltidial plates plates that cover the delthyrium in some brachiopods.

dendrogram a 'tree diagram'; a branching diagram that shows relationships.

dendroid 'tree-like'.

dentary 'tooth-bearing'; the tooth-bearing bone in the lower jaw of a vertebrate (cf. maxilla, premaxilla).

denticle 'toothlet'; a small tooth-like structure.

dentine the main constituent of teeth, a form of apatite.

dermal bone 'skin'; bone formed initially within the endoderm.

deuterostome 'posterior mouth'; animal in which the embryonic blastopore often develops into the anus (cf. protostome).

dextral 'right-handed' (cf. sinistral).

diagenesis the physical and chemical processes that affect rocks and fossils after burial.

diagnosis a brief outline of the distinguishing (= diagnostic) features of an organism or a group.

diapsid 'two arches'; a tetrapod skull with two temporal openings (cf. anapsid, synapsid).

dibranchiate 'two-armed'; possessing one pair of gills, as in coleoids.

dichotomous 'two cut'; branching in two directions.

dicyclic 'double cycle' (cf. monocyclic).

diductor muscle/ diductor 'pull away'; a muscle that opens the valves of a brachiopod (cf. adductor muscle).

digit finger or toe.

diploblastic 'two-layered'; the two-layered body plan seen in cnidarians, in which the endoderm and ectoderm are separated by the mesogloea, but there is no coelom (cf. triploblastic).

diploid (in cell biology) the normal double complement of chromosomes (cf. haploid).

disarticulation breaking apart and losing natural connections; typically of parts of a skeleton.

dissepiment 'partition'; a horizontal, or nearly horizontal, plate of tissue supporting a tabula in an archaeocyathan or coral skeleton; a connecting structure in the rhabdosome of a dendroid graptolite.

dissepimentarium the area occupied by dissepiments in a coral.

dissoconch 'apart shell'; the initial shell of a rostroconch mollusc.

dissolution chemical breakdown of a solid element or compound by solution in a liquid.

distal 'far' from the source (cf. proximal).

diversification increase in numbers of species of a group, or of life as a whole.

diversity the number of species, genera, or families in a defined geographic area, or in the world.

diverticulum 'byway'; a blind-ending side branch (usually of the gut).

DNA deoxyribose nucleic acid, the organic chemical that makes up the genes and chromosomes, and which stores genetic material.

dolomite a form of limestone containing magnesium, and commonly some iron.

domichnion (pl. **domichnia**) 'dwelling trace'.

dominance relative abundance of species within a community; certain species are dominant if they are much commoner than others (cf. evenness).

dorsum 'back'; upper side (adj. **dorsal**; cf. ventral).

doublure the thickened outer margin of the trilobite cephalon or pygidium.

dyad 'double unit'; a double-unit spore (cf. monad, tetrad).

dysoxic 'badly oxic'; containing low levels of oxygen (cf. aerobic, anaerobic, anoxic).

ecdysis moulting.

ecophenotypic change change in the phenotype during an organism's life time, induced by local environmental changes, but not coded in the genotype.

ecospace range of habitats occupied by certain organisms.

ecosystem the combination of habitats and organisms in a particular place at a particular time.

ectoderm 'outside skin'; the outer skin (cf. endoderm).

ectoplasm 'outside mould'; the outer layer of proteinaceous material in a cell (cf. endoplasm).

ectotherm 'outside heat'; an animal that controls its temperature solely from external sources (cf. endotherm).

embryology the study of embryos.

encruster an organism that grows over the substrate, or other organisms, by creating a hard attached skeleton.

encystment turning into a cyst.

endemic 'in district'; restricted to a particular geographic area (cf. cosmopolitan).

endoderm 'inside skin'; the lining of the gut (cf. ectoderm).

endogastric 'within the stomach' (cf. exogastric).

endogenic 'formed within' the sediment, such as a burrow (cf. exogenic).

endoplasm 'inside mould'; the inner portion of proteinaceous material in a cell (cf. ectoplasm).

endoskeleton 'inside skeleton' (cf. exoskeleton).

endostyle 'inside pen'; mucus organ in the gut of chordates.

endosymbiont 'inside together life'; an organism that lives in symbiotic relationship with another, and is entirely enclosed within its structure.

endotherm 'inside heat'; an animal that controls its body temperature by internal means (cf. ectotherm).

enrollment 'rolling up'.

enteron 'gut'; the gut and respiratory cavity of a cnidarian.

epifauna 'top fauna'; animals that live on the seabed, not within the sediment (cf. infauna).

epirelief 'top relief'; a trace fossil on the top of a bed (cf. hyporelief).

epitheca 'top case'; the upper half of a diatom theca (cf. hypotheca).

epithelium (pl. **epithelia**) cell layer forming outer tissues.

epoch a division of geological time, such as Eocene, Oligocene, or Miocene, a subdivision of a period, and composed of several stages.

equilibrium 'equal balance'; a fixed level.

eukaryote 'well kernel'; single- and multi-celled life form with a nucleus, including algae, fungi, plants, and animals (cf. prokaryote).

euryotopic 'wide place'; of wide ecological preferences (cf. stenotopic).

eustatic 'well standing'; relating to simultaneous worldwide changes in sea level.

eutrophication 'healthy nutrition'; oxygen starvation, usually in a lake, caused by decaying algae after an algal bloom.

evenness the approach to equal abundance of species within a community (cf. dominance).

evolute 'rolling out'; coils of a gastropod or cephalopod shell that are all at least partially exposed (cf. involute).

evolution 'unrolling'; change in organisms through time.

exceptional preservation preservation of soft parts and of soft-bodied organisms.

exine the tough outer wall of pollen and spores.

exogastric 'outside the stomach' (cf. endogastric).

exogenic 'formed outside', or on the surface of the sediment, such as a trail (cf. endogenic).

exoskeleton 'external skeleton' (cf. endoskeleton).

extinction the disappearance of a species, genus, or family.

exuviae 'thrown off'; cast-off moulted skins.

facial suture the dividing line on the cephalon of a trilobite along which the exoskeleton splits during moulting (cf. free cheek).

facies a characteristic association of sedimentary features that may indicate a particular environment of deposition.

facies fossil a fossil that is characteristic of a particular sedimentary facies.

faecal (fecal) referring to excrement or dung.

family a division in classification; contains one or more genera, and is contained in an order.

fasciculate 'bundle-like'.

fauna the characteristic animals of a particular place and time.

femur the thigh bone (cf. fibula, tibia).

fibula one of the shin bones (cf. tibia, femur).

firmground a sea or lake floor composed of semi-consolidated calcareous sediment (cf. hardground).

flagellum (pl. **flagella**) 'whip'; hair-like organelle in a eukaryote cell that is used for swimming.

flattening compression of a fossil by pressure from above.

flora the characteristic plants of a particular place and time.

fluvial/ fluviatile referring to rivers.

flysch an accumulation of alternating sandstones and mudstones, generally formed in a deep basin from turbidity flows.

fodinichnion (pl. **fodinichnia**) 'feeding trace'.

foliated 'leaf-like'; consisting of thin layers.

food chain the unidirectional links between food and consumers within a community.

food pyramid the pattern of biomass distribution within a community, typically with large biomass of primary producers, and smaller and smaller biomasses of sucessive consumers.

food web the complex feeding interactions among members of a community.

foramen (pl. **foramina**) 'pierce'; a small opening.

formation (in stratigraphy) a rock unit that may be identified and mapped in a regional context; subdivided into members, and combined with other formations into a group.

fossil 'dug up'; the remains of a plant or animal that died in the distant past.

fragmentation breaking of a shell or skeleton into small pieces.

framework reef a reef whose basic structure is formed entirely from organic skeletons (corals, archaeocyathans, bryozoans, crinoids, etc.).

free cheek the lateral portion of the cephalon of a trilobite which is divided from the central portion by a facial suture, and which separates during moulting.

fringing reef a reef that lies on the margins of a landmass, with no intervening lagoon (cf. atoll, barrier reef).

frustule 'a bit'; the skeleton of a diatom.

fugichnion (pl. **fugichnia**) 'escape trace'.

full relief (of a trace fossil) seen in three dimensions (cf. semirelief).

furca (pl. **furcae**) 'fork'; a backwards-pointing flexible spine in an arthropod.

fusellar tissue the bandage-like tissues composing the periderm of graptolites.

fusiform 'spindle-shaped'.

gamete 'marriage'; a sex cell, such as an egg or sperm.

gametophyte 'marriage plant'; in plants that show alternation of generations, the stage that produces gametes and which engages in sexual reproduction (cf. sporophyte).

genal spine 'cheek'; the pointed spine at the posterior lateral margin of the trilobite cephalon.

gene an identifiable sequence within a chromosome that codes for a particular feature of an organism.

gene flow the movement of genes through a population by interbreeding.

gene pool the sum total of the genotypes of all individual organisms in a defined population.

geniculation 'little knee'; bent at right angles.

genotype the sum of the features of an organism, or population, contained in the genes.

genus (pl. **genera**) the category in classification above the species.

geographic range the complete area within which a species, or other taxon, lives.

germinal aperture opening for the passage of gametes in a spore or pollen grain.

gill bars bars of cartilage or bone that support the gill slits.

gill slits gill openings behind the head, found in chordates.

glabella (pl. **glabellae**) 'bald'; the raised middle portion of a trilobite cephalon.

golden spike (in stratigraphy) a point in a rock section, equivalent to an instant in geological time, that marks the internationally accepted base of a stratigraphic division (e.g. member, formation, system/ period, or series/ epoch).

gonad 'generation'; the organ that produces sex cells; the ovary or testis.

Gondwana ancient supercontinent composed of South America, Africa, India, Australia, and Antarctica.

gradualism see phyletic gradualism.

gradualistic steady change (cf. catastrophic).

group (in stratigraphy) a number of formations occurring in sequence that share some broad-scale features.

guard the bullet-shaped solid terminal part of the belemnite shell (cf. phragmocone, pro-ostracum).

habitat the environmental setting within which a species, or a community, lives.

haploid (in cell biology) a half complement of chromosomes, as found in the sex cells (cf. diploid).

haptonema 'fasten-thread'; a flagellum-like structure in coccolithophores.

hardground a sea or lake floor composed of consolidated calcareous sediment (cf. firmground).

herbaceous low-growing, bushy.

herbivore 'plant-eater'.

hermatypic (of corals) restricted to the photic zone, generally in tropical waters (cf. ahermatypic).

heterocercal tail 'different tail'; tail fin, as in sharks, which is asymmetrical and has a large upper lobe.

heterochrony 'different time'; changes in the timing and rate of development that affect evolution.

heteromorph 'different form' (adj. **heteromorphic**); supposed female ostracod (cf. tecnomorph).

heteropygous 'different-rumped'; trilobite with a pygidium slightly smaller than the cephalon (cf. macropygous, micropygous).

heterosporous 'different spores'; producing microspores and megaspores (cf. homosporous).

heterotroph 'different feeder'; an organism that feeds on a variety of materials (cf. autotroph).

hierarchy a system consisting of smaller and smaller categories which, in the case of biological classifications, are inclusive, that is, the smaller units fit within larger ones.

hinge the zone of attachment along which the two valves of a brachiopod or bivalve shell open and shut.

holaspid 'true shield'; the final larval stage in trilobites (cf. nauplius larva, protaspid, meraspid).

holdfast the rooting structure that fixes an archaeocyathan, a crinoid, a dendroid graptolite, or a seaweed to a rock.

holochroal eye an eye with many small closely-packed lenses, seen in trilobites and many other arthropods (cf. schizochroal eye).

homeobox 'same box'; a gene found in all organisms that controls orientation in early stages of embryonic development.

homeomorphic 'same form'.

homology a feature that arose once only; an apomorphy (adj. **homologous**).

homosporous 'producing the same spores' (cf. heterosporous).

humerus the upper arm bone in vertebrates (cf. radius, ulna).

hyaline 'glassy'; composed of tiny aligned calcite crystals.

hydrostatic 'water-standing'; water-supported.

hypha (pl. **hyphae**) 'web'; a branching tissue strand that forms part of a fungus.

hyponome wide tube lying beneath the head of a cephalopod through which water is squirted to achieve propulsion.

hyporelief 'under relief'; a trace fossil on the bottom of a bed (cf. epirelief).

hypostoma 'under hole'; a plate underneath the

trilobite cephalon which may have supported the mouth region.

hypotheca 'under-case'; the lower half of a diatom theca (cf. epitheca).

ichnofacies a facies based on characteristic trace fossils.

ichnofossil 'trace fossil'.

ichnogenus (pl. **ichnogenera**) a genus of trace fossil.

ichnology 'trace study'; the study of trace fossils

ichnospecies a species of trace fossil.

ilium upper bone of the typical tetrapod pelvis (cf. ischium, pubis).

impendent 'hanging down'.

imperforate 'lacking holes' (cf. perforate).

incisor teeth 'cutting'; the front teeth in mammals, used for snipping food off (cf. canine teeth, cheek teeth, premolar teeth, molar teeth).

incongruence (in cladistics) lack of matching of character sets.

infaunal an animal that lives within the sediment (cf. epifaunal).

infrabasal 'below basal'; plates lying below the basals in the calyx of a crinoid.

ingroup the organisms of interest in a cladistic study, as opposed to the outgroup, which is everything else.

integument 'covering'; skin.

interambulacral area (**interamb**) one of five zones of broader plates in the echinoid skeleton (cf. ambulacral area).

interarea flattened parts of the brachiopod hinge region that are exposed externally.

intertidal 'between tides'; between normal high and low water marks (cf. supratidal).

intervallum 'between walls'; the space between the inner and outer walls in archaeocyathans.

involute 'rolling in'; coils of a gastropod or cephalopod shell that are all concealed by the outermost coil (cf. evolute).

ischium lower posterior bone of the typical tetrapod pelvis (cf. ilium, pubis).

isometry 'same measure'; maintenance of identical proportions during growth (cf. allometry).

kingdom the highest division in classification; contains one or more phyla.

laesura (pl. **laesurae**) 'scar'; contact scars between neighbouring spores.

Lagerstatte (pl. **Lagerstätten**) 'lying place'; a deposit containing large numbers of exceptionally preserved fossils.

lamella (pl. **lamellae**) 'small thin plate'; a thin plate or layer.

lamina (pl. **laminae**) 'thin plate'; a thin plate or layer.

lancet 'sharp knife'; the pointed area of small plates in the calyx of a blastozoan.

lappet 'small lobe'; a side flap seen in some graptolites and ammonoids.

larva (pl., **larvae**) juvenile which has a different form to the adult.

lateral side.

lateral line canal a canal that runs along the side of the body in fishes, and which bears sensory cells that can detect movements in the water.

lepidotrichium (pl. **lepidotrichia**) 'scale-hair'; a thin bony rod in the fin of a fish.

ligament 'bind'; a bundle of fibrous tissues linking skeletal elements (in molluscs and vertebrates).

lignification deposition of lignin.

lignin 'wood'; woody tissue.

lineage an evolving line, consisting of one or more species that have direct genetic links through time.

lithostratigraphy 'rock stratigraphy', the sequence and correlation of rocks.

littoral 'shore'; coastal.

locus (pl. **loci**) 'place'.

lophophore 'crest-bearing'; a specialized feeding and respiratory organ found in brachiopods and bryozoans (adj. **lophophorate**).

lorica 'leather corslet'; the outer covering of a tintinnid.

lumbar 'loin'; of the lower back.

lung book the air-breathing lung of a spider, arranged in many layers like the pages of a book.

macroconch 'large shell'; the larger of two morphs of a cephalopod species, probably the female (cf. microconch).

macroevolution 'large evolution'; evolution at species level and above, including those evolutionary topics (speciation, lineage evolution, trends, diversification, extinction events) that may be studied by palaeontologists; those parts of evolution excluded from microevolution.

macrofossil a 'large fossil', one that can be seen with the naked eye, in comparison to a microfossil.

macropygous 'large-rumped'; trilobite with a pygidium larger than the cephalon (cf. heteropygous, micropygous).

madreporite 'mother stone'; a plate in the echinoid skeleton near the anus that connects the water-vascular system to the external environment.

magnetostratigraphy stratigraphy based on magnetic reversals.

mamelon a small 'breast-like' projection on a stromatoporoid.

mantle portion of the body tissues of a brachiopod or mollusc that are involved in secretion of shell material.

mass extinction a major extinction event, typically marked by the loss of 10% or more of families, and 40% or more of species, in a short time.

massive solid.

maxilla 'jawbone'; the main tooth-bearing bone in the upper jaw of a vertebrate (cf. premaxilla, dentary).

median 'middle'.

medusa (pl. **medusae**) 'gorgon'; a free-swimming jellyfish-like stage in cnidarian development (cf. polyp).

megasporangium the structure that contains megaspores or ovules.

megaspore 'big spore'; the larger spore of early seed-bearing plants (cf. microspore).

meiosis 'diminution'; the process of cell division that involves reduction of chromosome numbers from the diploid to the haploid condition, prior to production of eggs or sperm (cf. mitosis).

member (in stratigraphy) a localized rock unit that may be mapped within a limited area; forms part of a formation.

meraspid 'middle shield'; the third larval stage in trilobites (cf. nauplius larva, protaspid, holaspid).

mesentery 'middle gut'; fleshy projection of the endoderm into the gut cavity of a cnidarian.

mesoderm 'middle skin'; the tissue type that forms a variety of organs between the endoderm and ectoderm of many animals.

mesogloea 'middle glue'; a gelatinous substance that separates the ectoderm and endoderm in diploblastic animals.

metacoel 'change cavity'; the body cavity of a bryozoan.

metamerous 'change part'; with a segmented coelom (cf. amerous, oligomerous, pseudometamerous).

metamorphism 'change form'; geological processes involving high temperature and/ or high pressure, usually associated with tectonic activity within the crust.

metamorphosis 'change form'; change, during development, from the larval to the adult form.

metaphyte 'later plant'; multicelled plant.

metasicula 'later sicula'; the main part of the sicula of a graptolite (cf. prosicula).

metazoan 'later animal'; multicelled animal.

micrite 'microscopic calcite'; calcite ($CaCO_3$) that occurs as small crystals (cf. micrite).

microconch 'small shell'; the smaller of two morphs of a cephalopod species, probably the male (cf. macroconch).

microevolution 'small evolution'; processes of evolution below the species level, generally studied in the laboratory and in the field; those parts of evolution excluded from macroevolution.

microfossil 'small fossil'; a fossil that can be seen only with a microscope.

micropygous 'small-rumped'; trilobite with pygidium much smaller than the cephalon (cf. heteropygous, macropygous).

micropyle 'small gate'; the opening through which the pollen tube approaches the ovule, in a flower.

microspore 'small spore'; the smaller spore of early seed plants (cf. megaspore).

microsporophyll the male fertile structures of a flower.

mineralization process of formation of a mineral; in palaeontology, refers typically to the formation of the hard constituent of a skeleton, or to replacement of tissues by mineral material during fossilization.

missing link popular term for an organism, usually fossil, that lies midway between two groups, such as *Archaeopteryx* which shows a mix of 'reptilian' and bird-like features.

mitochondrion (pl. **mitochondria**) 'thread granule'; organelle in a eukaryote cell that assists in energy transfer.

mitosis 'thread'; simple cell division involved in normal growth (cf. meiosis).

Modern Synthesis the current view of evolution, based on a combination of Darwin's insights into geographic variation and natural selection, palaeobiology, and genetics, as established in the 1930s and 1940s.

molar teeth 'grinder'; one of the back teeth of a mammal, used for chewing food (cf. incisor teeth, canine teeth, cheek teeth, premolar teeth).

Molecular Clock Hypothesis the assumption that each protein molecule has a constant rate of amino acid substitution: the amount of difference between two homologous molecules indicates distance of common ancestry, and hence closeness of relationship.

monad 'single unit'; a single-unit spore (cf. dyad, tetrad).

monocyclic 'single cycle' (cf. dicyclic).

monolete of a spore with a single slit (cf. trilete).

monophyletic group 'single origin'; a group that includes all the descendants of a common ancestor (cf. paraphyletic and polyphyletic groups).

morphology 'shape study'; shape and form of an organism.

morphospace the theoretical maximum range of shapes of an organism.

mosaic evolution variable rates of evolution of different parts of an organism.

motile capable of movement.

multicellular composed of more than one cell.

multilocular 'many-chambered'.

multimembrate 'many-member' (cf. unimembrate).

mural in or of a wall.

mycelium (pl. **mycelia**) 'mushroom'; the mat-like structure composed from fungal hyphae.

myophore 'muscle-bearer'; an internal plate in a bivalve to which muscles attach.

myotome 'muscle slice'; discrete muscle block along the trunk of a chordate.

nacreous like mother-of-pearl.

naris nostril.

natant 'swimming'.

natural selection the 'survival of the fittest', a process that causes evolution, first proposed by Charles Darwin in 1859; in highly variable populations, organisms with the best adaptations survive best and pass on their winning attributes to their offspring; a cumulative process, but a process that is subject to minor vicissitudes of environmental change.

nauplius larva 'ship sail'; an early larval stage in many arthropods, including trilobites (cf. protaspid, meraspid, holaspid).

nekton 'swimming'; organisms that swim in the open water (cf. plankton).

nema 'thread'; thread-like structure at the top of the sicula of a graptolite.

nematocyst 'thread bladder'; the sting within the cnidoblast of a cnidarian.

neoteny 'new stretch'; paedomorphosis by retention of juvenile morphological characters in the adult.

nephridium 'kidney'; a kidney-like structure for processing waste materials.

neural spine the spine on the upper surface of a vertebra.

niche the lifestyle and ecological interactions of an organism.

node branching point in a cladogram.

notochord 'back string'; the flexible rod-like structure that supports the body of basal chordates, and is a precursor of the backbone in vertebrates.

notothyrium 'back door'; a small triangular-shaped zone in the hinge region of the brachial valve of a brachiopod, lying opposite the delthyrium.

obrution deposit a rich accumulation of fossils produced by very rapid rates of sedimentation that bury the organisms almost instantaneously.

occipital of the back of the head.

ocellus (pl. **ocelli**) 'small eye'; a single eye in an arthropod (cf. compound eye).

oesophagus the part of the gut between the mouth and the stomach.

oligomerous 'few parts'; with a coelom divided longitudinally into two or three zones (cf. amerous, metamerous, pseudometamerous).

omnivore 'eats all'; an animal that feeds on plant and animal food.

ontogeny development from egg to adult.

opal non-crystalline silica.

operculum (pl. **opercula**) 'cover'; a cover or lid that closes an opening (adj. **opercular**).

ophiolite 'snake stone'; a complex of oceanic crustal rocks associated with a subduction zone.

opisthosoma 'behind body'; the abdomen of certain arthropods.

orbit eye socket.

oral 'of the mouth'.

order a division in classification; contains one or more families, and is contained in a class.

orogeny mountain-building (adj. **orogenic**).

osculum 'litttle mouth'; the opening into the central cavity of a sponge.

ossicle 'little bone'; one segment of the stalk or arms of a crinoid, a columnal; or, one segment of the arms of an ophiuroid, a vertebra.

ossify 'turn into bone'.

ostium (pl. **ostia**) 'mouth'; a small perforation in the wall of a sponge.

outgroup all the organisms that lie outside the clade of interest, the ingroup.

ovary 'egg'; egg-producing organ in female animals; structure that contains the ovules, in plants.

ovule an undeveloped (unfertilized) seed.

paedomorphocline a paedomorphic trend in evolution.

paedomorphosis 'juvenile formation'; achievement of sexual maturity in a juvenile body (cf. peramorphosis).

palaeoautecology the study of the ecology of single fossil organisms (cf. palaeosynecology).

palaeobotany the study of fossil plants.

palaeoecology 'ancient ecology'; the life and times of fossil organisms; also, the study thereof.

palaeogeography 'ancient geography', the layout of continents and oceans in the geological past.

palaeontology 'ancient life study'; the study of the life of the past.

palaeosol 'ancient soil'.

palaeosynecology the study of communities of fossil organisms (cf. palaeoautecology).

pallial line 'mantle'; the line that marks the outer margins of attachment of the mantle to the shell in molluscs.

pallial sinus the infolding of the pallial line in molluscs to accommodate the siphons.

palynology the study of fossil pollen and spores.

Pangaea (Pangea) 'all world'; ancient supercontinent composed of all the modern continents.

paragaster 'beside stomach'; the central cavity of a sponge.

paraphyletic group 'parallel origins'; a group that includes some, but not all, the descendants of a common ancestor (cf. monophyletic and polyphyletic groups).

parasitism 'beside food'; a biological interaction where one species lives in or on another, and does it harm.

parataxonomy 'parallel taxonomy'; a non-evolutionary taxonomic system.

parazoan 'beside animal'; the simple body plan found in sponges in which there is no coelom, and cells are not differentiated into tissue types.

parietal 'of a wall'.

parsimony simplicity; in cladistics, the requirement that a cladogram represents the shortest possible tree linking all taxa (adj. **parsimonious**).

pascichnion (pl. **pascichnia**) 'feeding trace'.

pectiniform 'comb-like'.

pectocaulus the linking tubes between zooid housings in pterobranchs.

pectoral of the shoulder girdle.

pedicle 'footlet'; a fleshy stalk that attaches a brachiopod to the substrate.

pedicle valve the valve in a brachiopod shell that contains the pedicle foramen (cf. brachial valve).

pedipalp 'foot stroking'; a second paired appendage, or 'feeler', in arthropods.

pelagic of the open sea; refers to habitats and organisms that are not on the sea bed (cf. benthic).

pelvic of the hip girdle.

pendent 'hanging'.

pentameral 'five-part'.

peramorphocline a peramorphic trend in evolution.

peramorphosis 'over-development'; the achievement of sexual maturity relatively late (cf. paedomorphosis).

perforate 'possessing holes' (cf. imperforate).

periderm 'surrounding skin'; the outer tissue layer of graptolites.

perignathic girdle 'around the jaws'; region of the echinoid skeleton around the Aristotle's lantern.

period (in stratigraphy) the major divisions of geological time, such as Cambrian, Ordovician, and Silurian, which are composed of epochs; equivalent to the system as a division of the rock column.

periostracum 'around shell'; the horny outer layer of a brachiopod or mollusc shell.

periproct 'around anus'; the anal opening of echinoids.

peristome 'around mouth'; the mouth opening of echinoids.

permineralization near-complete replacement of the tissues of an organism by mineral material.

petrifaction 'turning to rock'; fossilization by complete mineralization.

petrology 'study of rocks'.

phaceloid composed of numerous roughly parallel tubes.

pharynx 'mouth space'; the cavity into which water is pumped in chordates; functions in feeding and in respiration.

phenetic referring to characters; analytical techniques that seek to summarize all aspects of variation in all characters of organisms or communities.

phenotype 'show type'; the sum of the externally expressed features of an organism, or population.

photic zone 'light'; the upper parts of a water body that are penetrated by daylight; typically down to 100 m depth.

photosymbiosis 'light together life'; a mutually beneficial interaction between a photosynthesising plant or alga and some other organism.

photosynthesis 'light manufacture'; the breakdown of carbon dioxide and water in the presence of sunlight to produce sugars and oxygen.

phragmocone the main part of an ammonoid shell, except the protoconch and the body chamber; the conical part of a belemnite shell between the guard and the pro-ostracum.

phyletic gradualism the view that evolution is continuous and gradual, and that speciation occurs as part of the gradual change within lineages (cf. punctuated equilibrium).

phylogeny 'race origin'; the pattern of evolution; an

evolutionary tree of all life or of some clades.

phylum (pl. **phyla**) a division in classification; contains one or more classes, and is contained in a kingdom.

phytoplankton 'plant plankton'; the plant components of the plankton.

pinnule 'small feather'; feather-like side branches of the arms of a crinoid.

placenta 'a flat cake'; the tissue structure in female mammals that transfers food and oxygen to the developing embryo.

planispiral 'spiral in a plane' (cf. trochospiral).

plankton 'wandering'; floating organisms that live in the top few metres of oceans and lakes (cf. nekton).

plate tectonics the theory that the Earth's crust consists of moving and interacting plates; a mechanism for continental drift.

pollen mobile fine-grained material produced in the anthers of flowers, and carrying the sperm.

polymeric 'many-segmented'.

polymorphic 'in many forms'.

polyp an attached sea anemone-like stage in cnidarian development (cf. medusa).

polyphyletic group 'many origins'; a group that contains members that arose from more than one ancestor (cf. monophyletic and paraphyletic groups).

porcellaneous composed of minute randomly-oriented calcite crystals.

posterior back (cf. anterior).

predation a biological interaction where one species feeds on another.

prehensile 'seize'; flexible and grasping.

premaxilla the anterior small tooth-bearing bone in the upper jaw of a vertebrate (cf. maxilla, dentary).

premolar teeth cheek teeth of mammals, lying in front of the molars, and used for chewing food (cf. incisor teeth, canine teeth, cheek teeth, molar teeth).

proboscis 'trunk'; elongate nose or snout-like projection.

process (in descriptions of morphology) projection.

progress change with improvement.

progression the sequence from simple to complex organisms through time.

prokaryote 'before kernel'; basal single-celled life form with no nucleus, including bacteria and cyanobacteria (cf. eukaryote).

pro-ostracum 'in front of shell'; the spatulate thin-shelled component of a belemnite shell that is attached to the phragmocone, and which supported the main part of the body.

prosicula the upper first-formed part of the sicula of a graptolite (cf. metasicula).

prosoma 'before body'; the fused head and thorax found in some arthropods.

prosome 'before body'; the upper part of a chitinozoan.

protaspid 'first shield'; the second larval stage in trilobites (cf. nauplius larva, meraspid, holaspid).

protein a complex organic chemical composed of amino acids, the basic building block of organisms.

prothallus 'before shoot'; the first stage of development of a gametophyte plant.

protoconch 'first shell'; the larval portion of a shell.

protostome 'first mouth'; animal in which the embryonic blastopore develops into the mouth (cf. deuterostome).

protractor muscle/ protractor 'pull forwards'; a muscle that pulls forwards.

provinciality the development of specific biogeographic provinces throughout the world.

proximal/ proximate 'near' to the source (cf. distal).

pseudocoelomate possessing a 'false coelom', common in many embryonic animals, and found in adult nematodes.

pseudometamerous with an undivided coelom, but irregularly duplicated organs (cf. amerous, metamerous, oligomerous).

pseudopodium (pl. **pseudopodia**) 'false foot'; a tissue extension.

pseudopuncta (pl. **pseudopunctae**) 'false hole'.

pseudostome 'false hole'.

pubis lower anterior bone of the typical tetrapod pelvis (cf. ilium, ischium).

puncta (pl. **punctae**) 'hole' (adj. **punctate**).

punctuated equilibrium the view that evolution occurs in two styles, long periods of little change (equilibrium; stasis), punctuated by short bursts of rapid change, often associated with speciation (cf. phyletic gradualism).

pygidium 'small rump'; the posterior segment of a trilobite (cf. cephalon, thorax).

pylome 'gate'; opening in an acritarch wall.

pyriform 'pear-shaped'.

pyrite a form of iron sulphide (FeS), occurring as small gold-coloured crystals, often associated with black mudstones and fossils deposited in anaerobic conditions.

quadrate 'square'; the bone in the posterior lateral corner of a reptile skull which articulates with the articular in the lower jaw.

quadrupedal 'four-footed'; walking on all fours (cf. bipedal).

radial 'ray-like'; branching outwards from a central point, like the spokes in a bicycle wheel (cf. concentric); plates lying above the basals in the calyx of a crinoid.

radialian animal with a radial pattern of cells at early phases of division (cf. spiralian).

radiation (in evolution) diversification or branching of a clade.

radiometric dating dating rocks by measurement of the amount of natural radioactive decay of pairs of elements, the parent (starting element) and daughter (resultant element).

radius one of the forearm bones in vertebrates (cf. ulna, humerus).

radula 'scraper'; the rasping feeding organ of molluscs.

ramified 'branched'.

raphe 'seam'; median gash in the diatom skeleton.

receptacle the structure that contains the ovule in a flower.

reclined 'sloping back'.

recumbent 'lying back'.

redox 'reduction–oxidation'; the junction between reducing and oxidizing conditions.

reef a wholly, or partially, organic carbonate construction (cf. framework reef).

refractory a form of carbon that does not break down readily (cf. volatile).

regression 'passage back'; withdrawal of the sea from the land; may be local or global (cf. transgression).

repichnion (pl. **repichnia**) 'creeping trace'.

replication copying or duplication.

resupination 'bent backward'; lying on the back.

reticulate 'net-like'.

retractor muscle/ retractor 'pulls back'; a muscle which pulls backwards, or pulls a structure into its protective skeleton.

rhabdosome 'rod body'; the whole colony of a graptolite.

rhizome 'root'; an underground stem.

rostrum 'beak'; the snout or anteriormost part of the head (adj. **rostral**).

ruga (pl. **rugae**) 'roughness'; irregular small projections (adj. **rugose**).

ruminant a mammal that digests its food in several stages (e.g. a camel or a cow).

saccus (pl. **sacci**) 'bag'; empty structure on the side of some pollen grains (adj. **saccate**).

Scala naturae the 'chain of being', a sequence of organisms, from simple to complex, once interpreted as evidence for unidirectional evolution.

scandent 'climbing'.

scavenging feeding on organisms that are already dead.

schizochroal eye an eye with reduced numbers of large spaced lenses, seen in trilobites (cf. holochroal eye).

sclerite a 'hard' skeletal plate.

scleroprotein the tough proteinaceous material that makes up the periderm of graptolites.

sclerotized 'hardened'.

selenizone 'moon zone'; infilled track of the apertural slit of a gastropod.

semirelief (of a trace fossil) seen on the surface of a bed (cf. full relief).

sepal one of the outermost parts of a flower, lying outside the petals.

septum (pl. **septa**) 'fence'; a dividing wall within the skeleton of various animals (adj. **septate**).

sequence stratigraphy the organization of sedimentary sequences into major packets corresponding to times of transgression, regression, and nondeposition.

sere a plant or epifaunal community that is one of a succession of unstable assemblages on the way to the establishment of a climax community.

sessile 'sitting'; organisms that live on the seabed, and which do not move.

seta (pl. **setae**) 'bristle'; a stiff hair.

sexual dimorphism 'two forms'; differences in the morphology of males and females of a species.

sicula the small cone that is the first part of a graptolite rhabdosome to form.

siderite a form of iron carbonate ($FeCO_3$) that occurs commonly in concretions around fossils.

sinistral 'left-handed' (cf. dextral).

siphon an extendable tube in a mollusc, used for sucking in water with food particles and for expelling filtered water.

siphuncle connecting strand of soft tissue that extends through the chambers of a cephalopod shell.

skeleton supporting structure in an organism, usually involving some mineralized tissues; may be internal (endoskeleton) or external (exoskeleton).

solitary an organism, usually a coral, that lives in isolation (cf. colonial).

somite 'body'; a body segment in an arthropod.

sparry calcite calcite ($CaCO_3$) that occurs as large crystals (cf. micrite).

speciation the process of formation of a new species,

either by the splitting (branching) of a lineage, or by evolution along a lineage, from a pre-existing species.

species a group of organisms, or populations, that includes all the individuals that normally interbreed, and which can produce viable offspring; typically the smallest unit in the hierarchy of a classification of organisms.

species selection selection at the level of species.

sphincter an opening that may be closed by muscular activity.

spicule 'small ear of corn'; a tiny needle-like calcareous or siliceous structure that forms part of the skeleton of a sponge.

spiracle 'to breathe'; an opening near the mouth in a blastozoan; the breathing hole behind the head in a shark.

spiralian animal with an initial sequence of cell division that follows a spiral track (cf. radialian).

spongin a horny organic material that forms around the skeleton of many sponges.

sporangium (pl. **sporangia**) the spore-bearing structure in a land plant.

sporophyte 'spore plant'; in plants that show alternation of generations, the stage that produces spores and which engages in asexual reproduction (cf. gametophyte).

spreite (pl. **spreiten**) 'trace'; indications of former stages of a single burrow.

squamosal 'scale'; major bone in the side of a tetrapod skull which, in mammals, articulates directly with the dentary (lower jaw).

stage (in stratigraphy) a rock-time unit in stratigraphy, a subdivision of a series, and generally composed of several zones.

stagnation deposit a deposit of fossils preserved in anoxic conditions.

stamen 'stand'; the pollen-producing structure of a flower.

stasis 'standing still'; the long periods of little net evolutionary change within a lineage.

stenotopic 'narrow place'; of narrow ecological preferences (cf. eurytopic).

stereom 'solid'; the internal structure of echinoderm skeletal elements.

sternite 'chest'; armoured body covering over the underside of a segment of a eurypterid (cf. tergite).

stigma 'point'; the part of the carpel in a flower that receives pollen.

stipe 'post'; a branch of a graptolite rhabdosome.

stolon 'sucker'; the linking tubes between the thecae in graptolites.

stolotheca 'sucker case'; one of the types of thecae in a dendroid graptolite (cf. autotheca, bitheca).

stoma (pl. **stomata**) 'mouth'; an opening on the underside of a leaf through which water vapour may pass.

stone canal part of the water-vascular system of an echinoid.

stratification (in sedimentary geology) the layering seen within typical sediments; (in community ecology) the layering of different organisms within, typically, a forest or a reef.

stratigraphical range the time from apparent origin to apparent extinction of a fossil taxon.

stratigraphy 'bedding writing'; the sequence of rocks and of events in geological time.

stratophenetics the use of stratigraphic age of specimens and their overall morphological features to draw up an evolutionary tree.

stratotype the reference section for a member or a formation, identified in a specific location.

stroma 'bed'; a supporting framework of connective tissue.

stromatolite 'bed/ mattress rock'; a layered structure generally formed by alternating thin layers of cyanobacteria and lime mud, typically in shallow warm sea waters.

style 'pen'; in a flower, the slender part above the carpels bearing the stigma.

subaerial 'beneath the air'; formed on land.

subduction 'pulling down'; the process whereby one tectonic plate is forced down beneath another.

substrate the underlying surface.

sulcus 'furrow' (adj. **sulcate**).

superposition 'positioning on top'; the observation that younger rocks lie on top of older rocks (unless they have been inverted subsequently by tectonic activity).

supratidal 'above tides'; above normal high water mark (cf. intertidal).

suspension feeder an animal that feeds on small food particles suspended in the water.

suture 'stitched seam'; the firm junction between two bones; the irregular line that marks the junction between two chambers of a cephalopod shell.

symbiont a participant in a symbiotic relationship.

symbiosis 'living together'; the phenomenon of species living together in close interdependence, where one or both species obtains some benefit, and neither is harmed by the relationship (adj. **symbiotic**; cf. parasitism, competition).

synapomorphy shared derived character.

synrhabdosome a group of graptolite rhabdosomes

living in a linked cluster.

synapsid 'joined arch'; a tetrapod skull with one (lower) temporal opening (cf. anapsid, diapsid)

synonym 'same name'; a redundant name given to an organism that has already been named (adj. **synonymous**).

synonymy equivalence of two names applied to a single species, genera, or families.

system (in stratigraphy) the major divisions of the rock column, such as Cambrian, Ordovician, and Silurian; equivalent to the period as a division of geological time.

systematics the study of relationships of organisms and of evolutionary processes.

tabula (pl. **tabulae**) 'table'; a horizontal division within the skeleton of an archaeocyathan or a coral.

tabular flattened.

taphonomy 'death study'; the study of biological and geological processes that occur between the death of an organism and its final state in the rock.

taxon a group of organisms, such as a species, a genus, a family, an order, a class, or a phylum.

taxonomy 'arrangement'; the study of the morphology and relationships of organisms.

tecnomorph supposed male ostracod (cf. heteromorph).

tectonic activity 'building'; physical movements within the Earth's crust, often associated with mountain building, such as faulting, and folding.

tegmen 'covering'; the roof of the calyx of a crinoid.

telson the pointed tail-portion of various arthropods.

temporal opening opening in the skull of a tetrapod behind the orbit.

tendon a sheet of fibrous tissue that attaches a muscle to a bone.

teratological 'monstrous'; relating to abnormalities in development.

tergite 'back plate'; armoured body covering over the back of a segment of a eurypterid (cf. sternite).

terrane a tectonic plate that had a specific geological history.

test 'pot'; the skeleton of an echinoid, foraminifer, or radiolarian.

tetrad 'four unit'; a four-unit spore (cf. monad, dyad).

thallus (pl. **thalli**) 'young shoot'; the skeleton of a calcareous alga.

theca 'case'; the skeletal wall of a coral, dinoflagellate, or diatom; the calyx of a crinoid; the individual living chamber of a graptolite zooid.

thermocline 'temperature slope'; the level at which

the water temperature in the sea, or in a lake, changes rapidly.

thorax the middle 'body' portion of an arthropod (cf. abdomen).

tibia one of the shin bones (cf. fibula, femur).

tiering a special form of stratification seen among benthos and trace fossils, where different ichnotaxa, or taxa, occupy different height or depth zones above and in the sediment.

time-averaging the accumulation of fossils from a variety of time horizons into a single horizon.

tissue cast an impression of the tissues of an ancient organism.

tool mark impression on a sediment surface made by a transported object.

torsion 'twisting'.

trabecula (pl. **trabeculae**) 'little beam'; rod-like structure that crosses a space.

trace fossil remains of the activity of an ancient organism, such as a burrow or track (cf. body fossil).

trachea (pl. **tracheae**) 'artery'; small tube through the cuticle of an arthropod, used in respiration and water-control.

tracheid 'artery'; a water-conducting strand in a land plant.

transgression 'passage across'; advance of the sea on to land; may be local or global (cf. regression).

transpiration 'breath across'; the process whereby fluid is drawn up through a plant by the suction effect of evaporation of water from the leaves.

trend (in evolution) sustained change in a feature through time.

trilete of a spore with a three-branched slit (cf. monolete).

triploblastic 'three-layered'; the body arrangement found in most animals where the ectoderm and endoderm are separated by a third tissue class, the mesoderm (cf. diploblastic).

triserial 'in three rows' (cf. uniserial, biserial).

trochospiral 'wheel spiral'; spiral and pyramidal (cf. planispiral).

trophic of food or feeding.

tsunami a large tidal wave.

tube foot a small fleshy muscular structure that projects through the skeleton of an echinoderm and functions in cleaning, feeding, and locomotion.

tubercle 'small root'; a small projection from a skeleton.

turbidite a rock formed from turbidity flows, mass movements of sand and mud down a slope and into deep water.

turma (pl. **turmae**) 'a troop'; the category term used in classifications of spores.

type specimen the specimen that is selected as the name-bearer, to represent all the characteristic features of a species.

ulna one of the forearm bones in vertebrates (cf. radius, humerus).

umbilicus 'navel'; a cavity in the centre of a gastropod shell.

umbo (pl. **umbones**) the 'shoulder' region of the pedicle valve of a brachiopod; the 'beak' of a bivalve.

unconformity a gap in a sequence of rocks that apparently corresponds to the passage of a considerable amount of time.

undertrack the impression of a track preserved below the surface on which the animal was moving.

uniformitarianism 'the present is the key to the past'; the basic assumption in geology and palaeontology that ancient phenomena may be interpreted in the light of observations of the modern world.

unimembrate 'single-member' (cf. multimembrate).

uniramous 'single-branched' (cf. biramous).

uniserial 'in a single row' (cf. biserial, triserial).

valve one half of a brachiopod or bivalve, each of which consists of two valves.

variation the differences between individuals that normally occur in a population, assessed either at genotypic or phenotypic level.

venter 'belly'; the underside (adj. **ventral**; cf. dorsum).

vertebra (pl. **vertebrae**) an element of the backbone of a vertebrate; or, an element in the arms of an ophiuroid, an ossicle.

vesicle 'bladder'; a fluid-filled sac.

vestigial structure a feature that is incomplete or has no clear function, but which appears to be homologous with something that once functioned in the ancestors.

virgella 'twig'; pointed structure at the base of the sicula of a graptolite.

viscera the internal organs (adj. **visceral**).

volatile 'fleeing'; a form of carbon that breaks down readily (cf. refractory).

volcaniclastic deposits sedimentary deposits derived directly from volcanic eruptions.

whorl a single turn in a spiral shell; a circular array of leaves around a stem.

xenomorphic 'foreign form'; of different form in different regions.

xylem the woody tissue in vascular plants in which tracheids conduct fluids and which also acts as a support.

zone a small unit of geological time, generally identified on the basis of one or more zone fossils, and a subdivision of a stratigraphic stage.

zone fossil a fossil species that indicates a particular unit of time.

zooarium (pl. **zooaria**) the stick-like skeleton of certain kinds of bryozoan colony.

zooecium (pl. **zooecia**) a box-like living chamber within a bryozoan colony.

zooid 'small animal'; an individual animal that lives in part of a colony.

zooplankton 'animal plankton'; the animal components of the plankton.

zooxanthella (pl. **zooxanthellae**) 'animal yellow'; photosynthesizing alga that lives in intimate association with a coral or bivalve.

zygote 'yoke'; the first stage of embryonic development, the product of the fusion of two gametes.

Index

Numbers in **bold** refer to explanations in boxes, although the word may also occur elsewhere on the page. Numbers in *italics* refer to figures or tables; the word may also occur in the text on that page.